TEXT BOOK
OF
MATRIX

DPH MATHEMATICS SERIES

TEXT BOOK OF MATRIX

(For B.A., B.Sc., I.A.S., P.C.S.)

By

A.K. Sharma

DISCOVERY PUBLISHING HOUSE
NEW DELHI-110002

First Published - 2004

Reprinted - 2018

ISBN: 978-81-7141-894-7

Text Book of Matrix

Published by:

DISCOVERY PUBLISHING HOUSE PVT. LTD.

4383/4B, Ansari Road, Darya Ganj
New Delhi-110 002 (India)
Phone: +91-11-23279245, 43596064-65
Fax: +91-11-23253475
E-mail: discoverypublishinghouse@gmail.com
sales@discoverypublishinggroup.com
web: www.discoverypublishinggroup.com

Printed at:
Infinity Imaging Systems
Delhi

Preface

This book "Text Book of Matrix" is written for B.A./B.Sc. students of Indian Universities. The subject matter of the book is prescribed in a very straight forward manner so that the students are able to prepare this paper on their own and score well in examination. Each chapter of this book contains complete theory and fairly large number of solved examples sufficient problems have also been given in the exercises.

A.K. Sharma

Contents

1

Some Basic Concept of Matrix

1.1 INTRODUCTION

Let us consider the system of equations:

$$x + y + z = 1$$

$$x + 2y\ 3 + z = 4$$

$$x + 3y + 5z = 7.$$

Here x, y, z are all unknown and their co-efficient are numbers. Arranging the coefficients in the order in which they occur in the equations and enclosing them in square brackets. We get a rectangular array of the form,

$$\begin{vmatrix} 1 & 1 & 1 \\ 1 & 2 & 3 \\ 1 & 3 & 5 \end{vmatrix}$$

The rectangular array is an example of a matrix. The horizontal lines (→) are called rows or row vectors, and the vertice lines are called column. There are three rows and three column in this matrix. Hence, it is a matrix of order 3 × 3.

We shall use the capital letters to denote matrix,

Thus $$A = \begin{bmatrix} 5 & 1 & 2 \\ 2 & 1 & 3 \end{bmatrix}_{2\times 3}$$

and $$B = \begin{bmatrix} 1 & 2 & 3 \\ 3 & 2 & 1 \\ 2 & 1 & 3 \end{bmatrix}_{3\times 3}$$

are both matrix. They are of the type 2 × 3 and 3 × 3 respectively.

1.2 DEFINITION OF A MATRIX

A system of any mn numbers arranged in a rectangular array of m rows and n columns is called a matrix of order m × n or an m × n matrix (which is read as m by n matrix).

Or

A set of mn elements of a set S arranged in a rectangular array of m rows and n columns is called an m × n matrix over S.

A m × n matrix is usually written as

$$\begin{bmatrix} a_{11} & a_{12} & a_{13} & \cdots & a_{1n} \\ a_{21} & a_{22} & a_{22} & \cdots & a_{2n} \\ \cdots & \cdots & \cdots & \cdots & \cdots \\ a_{m1} & a_{m2} & a_{m3} & \cdots & a_{mn} \end{bmatrix} \text{ is an } m \times n \text{ matrix.}$$

where the symbols a_{ij} represent any numbers (a_{ij} lies in the ith row and jth column).

Note 1: A compact form of the above matrix may be represented by the symbols $[a_{ij}]$, (a_{ij}) $\| a_{ij} \|$ or by a single capital letter A, say.

The element a_{ij} belong to ith row and th column and sometime called the (;)th element of the matrix.

Note 2: Each of the mn numbers constituting an m × n matrix is known as an *element of the matrix.*

The elements of matrix may be scalar or vector quantities.

Note 3: The plural of 'matrix' is 'matrices'.

Example 1:

The results of a music competition are given in the following matrix:

$$\begin{bmatrix} 3 & 2 & 1 & 0 \\ 0 & 3 & 2 & 4 \\ 5 & 0 & 3 & 0 \\ 2 & 1 & 4 & 3 \end{bmatrix}$$

Here the rows represent the teams A, B, C, D in heat order and the columns represent the number of wins, first place, second plane third place and fourth place scored by the teams.

From the above matrix find (a) How many events did the team A win? (b) How may first places did the team B win? (c) How many third places did the team C win? (d) what does 0 represent in second row?

Solution:

(a) $\because$ The first row represents the team A. So the required number = sum of the elements of first row = 3 + 2 + 1 = 6. **Ans.**

(b) As first element of second row (which represents the team B) is zero, os the team B did not win any first place. **Ans.**

(c) The third row represents the team C and third column represents the third place scored by the teams, so the number of third places won by the team C is 3. **Ans.**

(d) The second row represents the team B and the first column represents the first place scored by teams. So 0 in the second row represents that the team B did not score any first place. **Ans.**

Example 2:

Write down the orders of the matrices:

(a) $\begin{bmatrix} 2 & 3 & 5 \\ 1 & 0 & 3 \end{bmatrix}$; *(b)* $\begin{bmatrix} 2 \\ 3 \end{bmatrix}$,

(c) $[3, 4, 5]$; *(d)* $[1]$.

Solution:

The order of the given matrix are:

(a) 2×3; (b) 2×1; (c) 1×3; (d) 1×1. **Ans.**

1.3 TYPES OF MATRICES

(a) Horizontal Matrix: If in a matrix the number of columns is more than the number of rows then it is called a horizontal matrix.

For example $\begin{bmatrix} 1 & 3 & 2 & 3 \\ 2 & 5 & 7 & 9 \end{bmatrix}$ is a horizontal matrix.

(b) Vertical Matrix: If in a matrix the number of rows is more than the number of columns it is called a vertical matrix.

For example $\begin{bmatrix} 2 & 3 \\ 3 & 5 \\ 4 & 6 \\ 5 & 7 \end{bmatrix}$ is a vertical matrix.

(c) Column Matrix: If there if only one column in a matrix, it is called a column matrix.

For example $\begin{bmatrix} 2 \\ 3 \\ 4 \end{bmatrix}$ This is also called column vector.

(d) Square Matrix: If m = n *i.e.*, the number of rows and columns of a matrix are equal, then the matrix is of order $n \times n$ and is called a square matrix of order n.

For example $\begin{bmatrix} 2 & 3 & 1 \\ 1 & 5 & 2 \\ 7 & 6 & 9 \end{bmatrix}$ is a square matrix and $\begin{bmatrix} 1 & 3 & 2 & 3 \\ 2 & 5 & 7 & 9 \end{bmatrix}$

is a rectangular matrix.

(e) Rectangular Matrices: When the number of rows and columns of the array are not equal, then the matrix is known as a rectangular matrix.

(f) Row Matrix: If in a matrix, there is only one row it is called a row matrix. For example [1, 2, 3]. This is also called a row vector.

(g) Null (or zero) Matrix: If all the elements of an $m \times n$ matrix are zero, then it is called a null or zero matrix and denoted by $O_{m \times n}$ or simply O.

For example $\begin{bmatrix} 0 & 0 & 0 \\ 0 & 0 & 0 \end{bmatrix}$ is the 2×3 mull matrix.

(h) Unit Matrix: A square matrix having unity for its elements in the leading diagonal and all the other elements as zero is called an unit matrix.

For example $\begin{bmatrix} 1 & 0 & 0 & 0 \\ 0 & 0 & 0 & 0 \\ 0 & 0 & 1 & 0 \\ 0 & 0 & 0 & 1 \end{bmatrix}$ is unit matrix of order 4×4 denote it byI_4.

$\therefore$ an a-rowed square matrix $[a_{ij}]$ is called a unit matrix provided

$$a_{ij} = 1, \text{ whenever } i = j$$
$$= 0, \text{ whenever } i \neq j.$$

(i) Equal Matrix: Two matrices A $[(a_{ij})]$ are said to be equal if (a) they are of the same type *i.e.*, if they have same number of rows and columns and (b) the elements in the corresponding positions of the two matrices are equal.

From the definition given above it is evident that

1. If A = B, then B = A (Symmetry)
2. A = A, where A is any matrix. (Reflexivity
3. If A = B and B = C, then A = C (Transitivity)

i.e., the relation of equality in the set of all matrices is an equivalence relation.

(j) Diagonal Matrix: A square matrix in which all elements except those in the main (or leading) diagonal are zero is know as a diagonal matrix.

For example $\begin{bmatrix} 2 & 0 & 0 \\ 0 & 3 & 0 \\ 0 & 0 & 7 \end{bmatrix}$ is a 3-rowed diagonal matrix.

The um of the diagonal elements of a square matrix A (say) is called the trace of the matrix A.

(k) Sub-Matrix: A matrix which is obtained from a given matrix by deleting any number of rows and number of columns is called a sub-matrix of the given matrix.

For example $\begin{bmatrix} 1 & 2 \\ 3 & 4 \end{bmatrix}$ is a sub-matrix of $\begin{bmatrix} 5 & 3 & 2 \\ 1 & 1 & 2 \\ 7 & 3 & 4 \end{bmatrix}$

(l) Diagonal Element and Orinciple Diagonal: Those elements a_{ij} of any matrix $[a_{ij}]$ are called diagonal elements for which i = j.

The line along which the above elements lie is called the Principal diagonal or the Diagonal of the matrix.

1.4 TRIANGULAR MATRICES

It every element above or below the leading diagonal is zero, then the matrix is called a *Triangular Matrix.*

(a) Upper Triangular Matrix: A square matrix A whose elements $a_{ij} = 0$ for i < j is called an upper triangular matrix.

For example $\begin{bmatrix} a_{11} & a_{12} & a_{13}\ldots\ldots a_{1n} \\ 0 & a_{22} & a_{23}\ldots\ldots a_{2n} \\ 0 & 0 & a_{33}\ldots\ldots a_{3n} \\ \ldots & \ldots & \ldots\ldots\ldots \\ 0 & 0 & 0\ldots\ldots a_{nn} \end{bmatrix}$

(b) **Lower Triangular Matrix:** A square matrix A whose element $a_{ij} = 0$ for $i < j$ is called a lower triangular matrix.

For example
$$\begin{bmatrix} a_{11} & 0 & 0 \ldots\ldots 0 \\ a_{21} & a_{22} & 0 \ldots\ldots 0 \\ a_{31} & a_{32} & a_{33} \ldots\ldots 0 \\ \ldots & \ldots & \ldots \ldots\ldots 0 \\ a_{n1} & a_{n2} & a_{n3} \ldots\ldots a_{nn} \end{bmatrix}$$

1.5 DIAGONAL MATRIX

Defintion: A square matrix in which all element except those element there in the leading diagonal are zero is called a diagonal matrix. Thus, an n-rowed square matrix $[a_{ij}]$ is a diagonal matrix iff $a_{ij} = 0$ wherever $i \neq j$. If A $[a_{ij}]$ is a diagonal matrix of order n. It must be in the following form.

For eample
$$\begin{bmatrix} a_{11} & 0 & 0 \ldots\ldots 0 \\ 0 & a_{22} & 0 \ldots\ldots 0 \\ 0 & 0 & a_{33} \ldots\ldots \ldots \\ \ldots & \ldots & \ldots \ldots\ldots \ldots \\ 0 & 0 & 0 \ldots\ldots a_{nn} \end{bmatrix}$$

Theorem 1:

Any two diangonal matrices of the same order commute under mltiplication.

Proof:

Let any two diagonal matrices be

$$A = \begin{bmatrix} a_1 & 0 & 0 & \ldots & 0 \\ 0 & a_2 & 0 & \ldots & 0 \\ \ldots & \ldots & \ldots & \ldots & \ldots \\ 0 & 0 & 0 & \ldots & a_n \end{bmatrix} \text{ and } B = \begin{bmatrix} b_1 & 0 & 0 & \ldots & 0 \\ 0 & b_2 & 0 & \ldots & 0 \\ \ldots & \ldots & \ldots & \ldots & \ldots \\ 0 & 0 & 0 & \ldots & b_n \end{bmatrix}$$

Then we have

$$AB = \begin{bmatrix} a_1 & 0 & 0 & \ldots & 0 \\ 0 & a_2 & 0 & \ldots & 0 \\ \ldots & \ldots & \ldots & \ldots & \ldots \\ 0 & 0 & 0 & \ldots & a_n \end{bmatrix} \times \begin{bmatrix} b_1 & 0 & 0 & \ldots & 0 \\ 0 & b_2 & 0 & \ldots & 0 \\ \ldots & \ldots & \ldots & \ldots & \ldots \\ 0 & 0 & 0 & \ldots & b_n \end{bmatrix}$$

$$\Rightarrow \quad AB = \begin{bmatrix} a_1b_1 & 0 & 0 & \dots & 0 \\ 0 & a_2b_2 & 0 & \dots & 0 \\ \dots & 0 & \dots & \dots & \dots \\ 0 & 0 & 0 & \dots & a_nb_n \end{bmatrix} \qquad \dots(1)$$

$$\text{and} \quad BA = \begin{bmatrix} b_1 & 0 & 0 & \dots & 0 \\ 0 & b_2 & 0 & \dots & 0 \\ \dots & \dots & \dots & \dots & 0 \\ 0 & 0 & 0 & \dots & b_n \end{bmatrix} \times \begin{bmatrix} a_1 & 0 & 0 & \dots & 0 \\ 0 & a_2 & 0 & \dots & 0 \\ \dots & \dots & \dots & \dots & \dots \\ 0 & 0 & 0 & \dots & a_n \end{bmatrix}$$

$$= \begin{bmatrix} b_1a_1 & 0 & 0 & \dots & 0 \\ 0 & b_2a_1 & 0 & \dots & 0 \\ \dots & \dots & \dots & \dots & 0 \\ 0 & 0 & 0 & \dots & b_na_n \end{bmatrix} \qquad \dots(ii)$$

∴ From (1) and (2), we find that AB = BA and each one of them is a diagonal matrix. **Hence proved.**

Theorme 2:

Product of any two diagonal matrices of order n is a diagonal matrix of order n.

Proof:

The proof is same as of theorem 1 above. (Do yourself).

Theorem 3:

Sum of any two diagonal matrices of order n is a diagonal matrix of order n and commute under addition.

Proof:

Let any two diagonal matrices be

$$A = \begin{bmatrix} a_1 & 0 & 0 & \dots & 0 \\ 0 & a_2 & 0 & \dots & 0 \\ \dots & \dots & \dots & \dots & \dots \\ 0 & 0 & 0 & \dots & a_n \end{bmatrix} \text{ and } B = \begin{bmatrix} b_1 & 0 & 0 & \dots & 0 \\ 0 & b_2 & 0 & \dots & 0 \\ \dots & \dots & \dots & \dots & \dots \\ 0 & 0 & 0 & \dots & b_n \end{bmatrix}$$

$$\therefore A + B = \begin{bmatrix} a_1 + b_1 & 0 & 0 & \dots & 0 \\ 0 & a_2 + b_2 & 0 & \dots & \dots \\ \dots & \dots & \dots & \dots & 0 \\ 0 & 0 & 0 & \dots & a_n + b_n \end{bmatrix} \quad \dots(1)$$

$$\text{And } B + A = \begin{bmatrix} b_1 + a_1 & 0 & 0 & \dots & 0 \\ 0 & b_2 + a_2 & 0 & \dots & 0 \\ \dots & \dots & \dots & \dots & 0 \\ 0 & 0 & 0 & \dots & b_n + a_n \end{bmatrix} \quad \dots(2)$$

$\therefore$ From (1) and (2), we get A + B = B + A and each one of them is a diagonal matrix of odrer n.

1.6 SEALAR MATRIX

Definition: If in a square matrix A all the diagonal elements are equal to a (where a $\neq$ 0) and all the remaining elements are equal to zero then it is called a scalar matrix. Thus, n $\times$ n square matrix $[a_{ij}]$ is called a sealar matrix iff for some member a.

For example $\begin{bmatrix} a & 0 & 0 & 0 \\ 0 & a & 0 & 0 \\ 0 & 0 & a & 0 \\ 0 & 0 & 0 & a \end{bmatrix}$ is a scalar matrix of order 4 $\times$ 4.

Commutative Matrices

Definition: If A and B are two square matrices such that AB = BA, then A and B are called *commutative* or are said to *commute.*

If **AB** = – **BA,** the matrices **A** and **B** are said to *anti-commute.*

Example 1:

Show that the matrices A and B anti-commute, where

$$A = \begin{bmatrix} 1 & -1 \\ 2 & -1 \end{bmatrix} \text{ and } B = \begin{bmatrix} 1 & 1 \\ 4 & -1 \end{bmatrix}$$

Solution:

$$\text{Here } AB = \begin{bmatrix} 1 & -1 \\ 2 & -1 \end{bmatrix} \times \begin{bmatrix} 1 & 1 \\ 4 & -1 \end{bmatrix}$$

$$= \begin{bmatrix} 1.1 - 1.4 & 1.1 + (-1).(-1) \\ 2.1 - 1.4 & 2.1 + (-1).(-1) \end{bmatrix} = \begin{bmatrix} -3 & 2 \\ -2 & 3 \end{bmatrix}$$

$$\text{And } BA = \begin{bmatrix} 1 & 1 \\ 4 & -1 \end{bmatrix} \times \begin{bmatrix} 1 & -1 \\ 2 & -1 \end{bmatrix} \qquad ...(1)$$

$$= \begin{bmatrix} 1.1 + 1.2 & 1.(-1) + 1.(-1) \\ 4.1 + (-1).2 & 4.(-1) + (-1).(-1) \end{bmatrix} = \begin{bmatrix} 3 & -2 \\ 2 & -3 \end{bmatrix}$$

$$= - \begin{bmatrix} -3 & 2 \\ -2 & 3 \end{bmatrix} \qquad ...(2)$$

∴ From (1) and (2) we find that **AB** = – **BA**.

Hence **A** and **B** anti-commute.

Example 2:

If $A = \begin{bmatrix} a & 0 & 0 \\ 0 & a & 0 \\ 0 & 0 & a \end{bmatrix}$ *and* $B = \begin{bmatrix} a_{11} & a_{12} & a_{13} \\ a_{21} & a_{22} & a_{23} \\ a_{31} & a_{32} & a_{33} \end{bmatrix}$

Then prove that AB = BA = aB.

Solution:

$$AB = \begin{bmatrix} a & 0 & 0 \\ 0 & a & 0 \\ 0 & 0 & a \end{bmatrix} \times \begin{bmatrix} a_{11} & a_{12} & a_{13} \\ a_{21} & a_{22} & a_{23} \\ a_{31} & a_{32} & a_{33} \end{bmatrix}$$

$$= \begin{bmatrix} aa_{11} & aa_{12} & aa_{13} \\ aa_{21} & aa_{22} & aa_{23} \\ aa_{31} & aa_{32} & aa_{33} \end{bmatrix} = a \begin{bmatrix} a_{11} & a_{12} & a_{13} \\ a_{21} & a_{22} & a_{23} \\ a_{31} & a_{32} & a_{33} \end{bmatrix}$$

= aB.

$$\text{Similarly } BA = \begin{bmatrix} a_{11} & a_{12} & a_{13} \\ a_{21} & a_{22} & a_{23} \\ a_{31} & a_{32} & a_{33} \end{bmatrix} \times \begin{bmatrix} a & 0 & 0 \\ 0 & a & 0 \\ 0 & 0 & a \end{bmatrix}$$

$$= \begin{bmatrix} aa_{11} & aa_{12} & aa_{13} \\ aa_{21} & aa_{22} & aa_{23} \\ aa_{31} & aa_{32} & aa_{33} \end{bmatrix} = a \begin{bmatrix} a_{11} & a_{12} & aa_{13} \\ a_{21} & a_{22} & a_{23} \\ a_{31} & a_{32} & a_{33} \end{bmatrix}$$

= aB.

Hence AB = BA = aB.

1.7 UNIT MATRIX OR IDENTITY MATRIX

Definition: *If in a scalar martix the diagonal element a = 1, then the matrix is called the unit matrix or identity matrix and is denoted by* $\mathbf{I}_n$ *in the case of n × n matrix. Thus a square matrix* $A = [a_{ij}]$ *n × n is called an identity or unit matrix iff* $a_{ij}\begin{cases}1 & \text{when } i=j\\ 0 & \text{when } i\neq j\end{cases}$

$$\text{For example } I_4 = \begin{bmatrix}1 & 0 & 0 & 0\\ 0 & 1 & 0 & 0\\ 0 & 0 & 1 & 0\\ 0 & 0 & 0 & 1\end{bmatrix}$$

Example 1:

Prove that $I^m = I^{m-1} = ... = I^2 = I$, *where m is any positive integer and* I_n *is the unit matrix of order n × n.*

Solution:

Let A be any n × n matrix and **I** be the nuit martrix of order n × n *i.e.,* $\mathbf{I} = \mathbf{I}_n$.

Now we know that $A\mathbf{I}_n = \mathbf{I}_n A = A$

But $\mathbf{I}_n = \mathbf{I}$. ...(1)

$\therefore A\mathbf{I} = \mathbf{I}A = A$

Taking $A = \mathbf{I}$, we have $\mathbf{I}\cdot\mathbf{I} = \mathbf{I} \Rightarrow \mathbf{I}^2 = \mathbf{I}$ ...(2)

Again form (1), taking $A = \mathbf{I}^2$, where $\mathbf{I}^2 = \mathbf{I}$ (proved), we get

$\mathbf{I}^2\cdot\mathbf{I} = \mathbf{I}^3 \quad \Rightarrow \quad \mathbf{I}^3 = \mathbf{I}^2 = \mathbf{I}$, from (2).

Proceeding in this way, we can prove that

$\mathbf{I}^m = \mathbf{I}^{m-1} = ... = \mathbf{I}^3 = \mathbf{I}^3 = \mathbf{I}$, where m is any positive integer.

Example 2:

If A be any n × n matrix and I_n *is the identity martix of order n × n, then prove that* $AI_n = I_nA = A$.

Solution:

Let us suppose that

$$A = \begin{bmatrix} a_{11} & a_{12} & \dots & a_{1n} \\ a_{21} & a_{22} & \dots & a_{2n} \\ \dots & \dots & \dots & \dots \\ a_{n1} & a_{n2} & \dots & a_{nn} \end{bmatrix} \text{ and } I_n = \begin{bmatrix} 1 & 0 & \dots & 0 \\ 0 & 1 & \dots & 0 \\ \dots & \dots & \dots & 0 \\ 0 & 0 & \dots & 1 \end{bmatrix}$$

$$\therefore A \cdot \mathbf{I}_n = \begin{bmatrix} a_{11} & a_{12} & \dots & a_{1n} \\ a_{21} & a_{22} & \dots & a_{2n} \\ \dots & \dots & \dots & \dots \\ a_{n1} & a_{n2} & \dots & a_{nn} \end{bmatrix} \times \begin{bmatrix} 1 & 0 & \dots & 0 \\ 0 & 1 & \dots & 0 \\ \dots & \dots & \dots & 0 \\ 0 & 0 & \dots & 1 \end{bmatrix}$$

$$= \begin{bmatrix} a_{11}.1 + a_{12}.0 + \dots + a_{1n}.0 & a_{11}.0 + a_{12}.1 + \dots + a_{1n}.0 & \dots & \dots & a_{11}.0 + a_{12}.0 + \dots + a_{1n}.1 \\ a_{21}.1 + a_{22}.0 + \dots + a_{2n}.0 & a_{21}.0 + a_{22}.1 + \dots + a_{2n}.0 & \dots & \dots & a_{21}.0 + a_{22}.0 + \dots + a_{1n}.1 \\ \dots & \dots & \dots & \dots & \dots \quad \dots \\ a_{n1}.1 + a_{n2}.0 + \dots + a_{nn}.0 & a_{n1}.0 + a_{n2}.1 + \dots + a_{nn}.0 & \dots & \dots & a_{n1}.0 + a_{n2}.0 + \dots + a_{nn}.1 \end{bmatrix}$$

$$= \begin{bmatrix} a_{11} & a_{12} & \dots & a_{1n} \\ a_{21} & a_{22} & \dots & a_{2n} \\ \dots & \dots & \dots & \dots \\ a_{n1} & a_{n2} & \dots & a_{nn} \end{bmatrix} = A$$

Similarly we can show that $\mathbf{I}_n \cdot A = A$.

Hence we have $A \cdot \mathbf{I}_n = \mathbf{I}_n \cdot A = A$.

1.8 PERIODIC MATRIX

Definition: *A square matrix A is called periodic, if $A^{k+1} = A$, where k is a positive integer.*

If k is the least positive integer for which $A^{k+1} = A$, then A is said to be of **period k.**

Idempotent Matrix

Definition: A square matrix A is called idempotent provided it satisfies the relation $A^2 = A$.

Symmetric Idempotent Matrix

Definition: *A square matrix A is called symmetric idempotent if* $A = A'$ *and* $A^2 = A$, *where* A' *is the transposed matrix of A.*

Example 1:

Show that if A and B are matrices of order $n \times n$ *and such that* $AB = A$ *and* $BA = B$, *then A and B are idempotent martices.*

Solution:

We have $\mathbf{ABA} = \mathbf{(AB)\,A} = \mathbf{(A)\,A}$, $\because$ $\mathbf{AB} = \mathbf{A}$ (given)

$\Rightarrow \quad \mathbf{ABA} = \mathbf{A}^2 \quad ...(1)$

Also $\quad \mathbf{ABA} = \mathbf{A\,(BA)} = \mathbf{A\,(B)} \quad \because \mathbf{BA} = \mathbf{B}$ (given)

$= \mathbf{AB} = \mathbf{A}, \quad \because \mathbf{AB} = \mathbf{A}$ (given)

$\Rightarrow \quad \mathbf{ABA} = \mathbf{A} \quad ...(2)$

From (1) and (2), we have $A^2 = A$ *i.e.,* A is idempotent.

In a similar manner, we can prove that

$\mathbf{BAB} = \mathbf{B\,(AB)} = \mathbf{B\,(A)}, \quad \because \mathbf{AB} = \mathbf{A}$ (given)

$= \mathbf{BA} = \mathbf{B}, \quad \because \mathbf{BA} = \mathbf{B}$ (given)

$\Rightarrow \quad \mathbf{BAB} = \mathbf{B} \quad ...(3)$

Also $\quad \mathbf{BAB} = \mathbf{(BA)\,B} = \mathbf{(B)\,B}, \because \mathbf{BA} = \mathbf{B}$ (given)

$\Rightarrow \quad \mathbf{BAB} = \mathbf{B}^2 \quad ...(4)$

$\therefore$ From (3) and (4), we have $B^2 = B$ *i.e.,* B is idempotent.

Hence proved.

Example 2:

If a is an idempotent matrix, then the matrix $B = I - A$ *is idempotent and* $AB = O = BA$.

Solution:

We know $\mathbf{IA} = \mathbf{AI} = \mathbf{A}$.

Also **A** being an idempotent matrix, we have $\mathbf{A}^2 = \mathbf{A}$.

Since **I** and A are square matrices, so **I – A** is also a square matrix and therefore we have

$$\mathbf{(I - A)^2} = \mathbf{(I - A)\,(I - A)}$$

$= \mathbf{(I - A)\,I - (I - A)\,A}$, by distributive law

$= \mathbf{I^2 - AI - IA + A^2}$

$= \mathbf{I - A - A + A}$, from (1), (2) and $\mathbf{I^2 = I}$

$\Rightarrow$ $\mathbf{(I - A)^2 = I - A}$, *i.e.*, $\mathbf{I - A}$ or $\mathbf{B}$ is an idempotent matrix by definition.

Again $\mathbf{AB = A\,(I - A) = AI - A^2}$, by distributive law

$= \mathbf{A - A}$, from (1) and (2)

i.e., $\mathbf{AB = O}$.

And $\mathbf{BA = (I - A)\,A = IA - A^2}$, by distributive law

$= \mathbf{A - A}$, from (1) and (2)

$\Rightarrow$ $\mathbf{BA = O}$. **Hence proved.**

Example 3:

If A and B are idempotent martices, then show that AB is idempotent if A and B commute.

Solution:

If $\mathbf{A}$ is idempotent, then $\mathbf{A^2 = A}$ and if $\mathbf{B}$ is idempotent, then $\mathbf{B^2 = B}$. ...(1)

And if $\mathbf{A}$ and $\mathbf{B}$ commute, then $\mathbf{AB = BA}$ ...(2)

Now $\mathbf{(AB)^2 = (AB).(AB)}$

$= \mathbf{A\,(BA)\,B}$, by associative law

$= \mathbf{A\,(AB)\,B}$, from (2)

$= \mathbf{(AA)\,(BB)}$, by associative law

$= \mathbf{A^2B^2}$

$= \mathbf{AB}$, by (1)

Hence $\mathbf{AB}$ is idempotent.

Example 4:

Show that the matrix $A = \begin{bmatrix} 2 & -2 & -4 \\ -1 & 3 & 4 \\ 1 & -2 & -3 \end{bmatrix}$ *is idempotent.*

Solution:

$$A^2 = A \cdot A = \begin{bmatrix} 2 & -2 & -4 \\ -1 & 3 & 4 \\ 1 & -2 & -3 \end{bmatrix} \times \begin{bmatrix} 2 & -2 & -4 \\ -1 & 3 & 4 \\ 1 & -2 & -3 \end{bmatrix}$$

$$= \begin{bmatrix} 2.2 - 2(-1) - 4.1 & 2(-2) - 2.3 - 4(-2) & 2(-4) - 2.4 - 4(-3) \\ -1.2 + 3(-1) + 4.1 & -1(-2) + 3.3 + 4(-2) & -1(-4) + 3.4 + 4(-3) \\ 1.2 - 2(-1) - 3.1 & 1(-2) - 2.3 - 3(-2) & 1(-4) - 2.4 - 3(-3) \end{bmatrix}$$

$$= \begin{bmatrix} 2 & -2 & -4 \\ -1 & 3 & 4 \\ 1 & -2 & -3 \end{bmatrix} = A$$

Hence the matrix A is idempotent.

1.9 INVOLUTORY MATRIX

Definition: *A square matrix A is called Involutory provided it satisfies the relation $A^2 = \boldsymbol{I}$, where $\boldsymbol{I}$ is the identity matrix.*

For example, the matrix $A = \begin{bmatrix} 1 & 0 \\ 1 & -1 \end{bmatrix}$ is involutory matrix,

$$\text{since } A^2 = \begin{bmatrix} 1 & 0 \\ 1 & -1 \end{bmatrix} \times \begin{bmatrix} 1 & 0 \\ 1 & -1 \end{bmatrix}$$

$$= \begin{bmatrix} 1.1 + 0.0 & 1.0 + 0.(-1) \\ 0.1 + (-1).0 & 0.0 + (-1).(-1) \end{bmatrix} = \begin{bmatrix} 1 & 0 \\ 0 & 1 \end{bmatrix} = I$$

Example 1:

If A is any square martix of order n and I_n is the identity martix of order n, such that $(I_n - A) = O$, then show that A is an involutory martix.

Solution:

Given that $(I_n - A)(I_n + A) = O$

$\Rightarrow \quad I_n^2 + I_n \cdot A - A \cdot I_n - A^2 = O$

$\Rightarrow \quad I_n + A - A - A^2 = O, \qquad \because I_n^2 = I_n, I_n \cdot A = A = A \cdot I_n.$

$\Rightarrow \quad I_n - A^2 = O$

$\Rightarrow \quad A^2 = I_n$

i.e., **A** is involutory by definition.

Example 2:

Show that the matrix $A = \begin{bmatrix} -5 & -8 & 0 \\ 3 & 5 & 0 \\ 1 & 2 & -1 \end{bmatrix}$ *is involutory,*

Solution:

$$A^2 = \begin{bmatrix} -5 & -8 & 0 \\ 3 & 5 & 0 \\ 1 & 2 & -1 \end{bmatrix} \times \begin{bmatrix} -5 & -8 & 0 \\ 3 & 5 & 0 \\ 1 & 2 & -1 \end{bmatrix}$$

$$= \begin{bmatrix} (-5).(-5) + (-8).3 + 0.1 & (-5).(-8) + (-8).5 + 0.2 & (-5).0 + (-8).0 + 0\,(-1) \\ 3.(-5) + 5.3 + 0.1 & 3.(-8) + 5.5 + 0.2 & 3.0 + 5.0 + 0.(-1) \\ 1.(-5) + 2.3 + (-1).1 & 1.(-8) + 2.5 + (-1).2 & 1.0 + 2.0 + (-1)\,(-1) \end{bmatrix}$$

$$= \begin{bmatrix} 25 - 24 + 0 & -40 - 40 + 0 & 0 + 0 + 0 \\ -15 + 15 + 0 & -24 + 25 + 0 & 0 + 0 + 0 \\ -5 + 6 - 1 & -8 + 10 - 2 & 0 + 0 + 1 \end{bmatrix} = \begin{bmatrix} 1 & 0 & 0 \\ 0 & 1 & 0 \\ 0 & 0 & 1 \end{bmatrix} = I$$

Hence the given matrix a is involutory.

1.10 NILPOTENT MATRIX

Definition: *A square matrix **A** si called Nilpotent matrix of order **m**, provided it satisfies the relation* $\boldsymbol{A^m = O}$ *and* $\boldsymbol{A^{m-1} \neq O}$*, where m is a positive integer and **O** is the null matrix of order m.*

For example, the matrix $A = \begin{bmatrix} 0 & 1 \\ 0 & 0 \end{bmatrix}$ is a nilpotent matrix, since

$$A = \begin{bmatrix} 0 & 1 \\ 0 & 0 \end{bmatrix} \neq \mathbf{O},$$

$$A^2 = \begin{bmatrix} 0 & 1 \\ 0 & 0 \end{bmatrix} + \begin{bmatrix} 0 & 1 \\ 0 & 0 \end{bmatrix} = \begin{bmatrix} 0.0 + 1.0 & 0.1 + 1.0 \\ 0.0 + 0.0 & 0.1 + 0.0 \end{bmatrix}$$

$$= \begin{bmatrix} 0 & 0 \\ 0 & 0 \end{bmatrix} = \mathbf{O},$$

$$A^3 = \lambda^2 \cdot A = O \cdot A = O.$$

i.e., A is a matrix which is not itself a zero matrix though its powers are zero matrices and so it is a nilpotent matrix *(Another definition of nilpotent matrix).*

Example:

Show that $A = \begin{bmatrix} 1 & 2 & 3 \\ 1 & 2 & 3 \\ -1 & -2 & -3 \end{bmatrix}$ *is a nilpotent matrix of order 2.*

Solution:

Given $A = \begin{bmatrix} 1 & 2 & 3 \\ 1 & 2 & 3 \\ -1 & -2 & -3 \end{bmatrix} \neq O$

$$\therefore A^2 = \begin{bmatrix} 1 & 2 & 3 \\ 1 & 2 & 3 \\ -1 & -2 & -3 \end{bmatrix} \times \begin{bmatrix} 1 & 2 & 3 \\ 1 & 2 & 3 \\ -1 & -2 & -3 \end{bmatrix}$$

$$= \begin{bmatrix} 1.1 + 2.2 + 3(-1) & 1.2 + 2.2 + 3(-2) & 1.3 + 2.3 + 3(-3) \\ 1.1 + 2.1 + 3(-1) & 1.2 + 2.2 + 3(-2) & 1.3 + 2.3 + 3(-3) \\ -1.1 - 2.1 - 3(-1) & -1.2 - 2.2 - 3(-2) & -1.3 - 2.3 - 3(-3) \end{bmatrix}$$

$$= \begin{bmatrix} 0 & 0 & 0 \\ 0 & 0 & 0 \\ 0 & 0 & 0 \end{bmatrix} = \mathbf{O},$$ where **O** is the null matrix of order 3.

i.e., $\mathbf{A^2 = O}$ $\mathbf{A \neq O}$. Hence **A** is a nilpotent matrix of order 2.

1.11 SCALAR MULTIPLE OF A MATRIX

If A is a matrix and λ is a number then λA is defined as the matrix each element of which λ times the corresponding element of the matrix. A.

For example: $2\begin{bmatrix} 3 & 5 & 7 \\ 2 & 3 & 4 \end{bmatrix} = \begin{bmatrix} 6 & 10 & 14 \\ 4 & 6 & 8 \end{bmatrix}$

⇒ if A $[a_{ij}]$, then $\lambda A = [\lambda a_{ij}]$, where λ is a number.

1.12 ADDITION OF MATRICES

If there be two m × n matrices given by $A = [a_{ij}]$ and $B = [b_{ij}]$, then the matrix A + B is defined as the matrix each element of which is the sum of the corresponding elements of A and B *i.e.,*

$$A + B = [a_{ij} + b_{ij}],$$

where i = 1, 2, 3,..., m and j = 1, 2, 3,..., n.

For example: If $A = \begin{bmatrix} a_1 & b_1 & c_1 \\ a_2 & b_2 & c_2 \end{bmatrix}$ and $B = \begin{bmatrix} a_3 & b_3 & c_3 \\ a_4 & b_4 & c_4 \end{bmatrix}$

then $A = B = \begin{bmatrix} a_1 + a_3 & b_2 + b_3 & c_1 + c_3 \\ a_2 + a_4 & b_2 + b_4 & c_2 + c_4 \end{bmatrix}$

1.13 SUBTRACTION OF MATRICES

If there be two m × n matrices given by $A = [a_{ij}]$ and $B = [b_{ij}]$, then the matrix A – B is defined as the matrix each element of which is obtained by subtracting the element of B from the corresponding element A *i.e.,* $A - B = [a_{ij} - b_{ij}]$,

where i = 1, 2,...,m and j = 1, 2,...,n.

For example: If $A = \begin{bmatrix} a_1 & b_1 & c_1 \\ a_2 & b_2 & c_2 \end{bmatrix}$ and $B = \begin{bmatrix} a_3 & b_3 & c_3 \\ a_4 & b_4 & c_4 \end{bmatrix}$

then $A - B = \begin{bmatrix} a_1 - a_3 & b_2 - b_3 & c_1 - c_3 \\ a_2 - a_4 & b_2 - b_4 & c_2 - c_4 \end{bmatrix}$

Note: If the two matrices A and B are of the same order, then only their addition and subtraction is possible and these matrices are said to be conformable for addition or subtraction. On the other hand if the matrices A and B are of different orders, then their addition and subtraction is not possible and these matrices are called non-conformable for addition and subtraction.

1.14 PROPERTIES OF MATRIX ADDITION

Property I: Addition of Matrices is Commutative

$(A + B) = (B + A)$

i.e., $[a_{ij}] + [b_{ij}] = [b_{ij}] + [a_{ij}]$,

where $[a_{ij}]$ and $[b_{ij}]$ are any two m × n matrices *i.e.,* matrices of the same order.

Proof:

$[a_{ij}] + [b_{ij}] = [a_{ij} + b_{ij}]$, by definition of addition

$= [b_{ij} + a_{ij}]$, ∵ addition of numbers (elements) is commutative

$= [b_{ij} + a_{ij}]$

i.e., $[a_{ij}] + [b_{ij}] = [b_{ij}] + [a_{ij}]$. Hence the theorem.

Property II: Addition of Matrices is Associative

$[(A + B) + C] = A + [B + C]$

i.e., $\{[a_{ij}] + [b_{ij}]\} + [c_{ij}] = [a_{ij}] + [b_{ij}] + [c_{ij}]\}$,

where $[a_{ij}]$, $[b_{ij}]$ and $[c_{ij}]$ are any three matrices of the same order m × n, say.

Proof:

$\{[a_{ij}] + [b_{ij}]\} + [c_{ij}]$

$= [a_{ij} + b_{ij}] + [c_{ij}]$, by law of addition for matrices

$= [(a_{ij} + b_{ij}) + c_{ij}]$, by law of addition for matrices

$= [a_{ij} + (b_{ij} + c_{ij})]$, ∵ addition of numbers is associative

$= [a_{ij}] + [b_{ij} + c_{ij}]$

$= [a_{ij}] + \{[b_{ij}] + [c_{ij}]\}$. Hence the theorem.

Property III: Addition of Matrices Obey the Distributive Law

$k [A + B] = kA + kB$

i.e., $k ([a_{ij}] + [b_{ij}]) = k [a_{ij}] + 0 [b_{ij}]$

where $[a_{ij}]$ and $[b_{ij}]$ are any two matrices of the same order m × n, say.

Proof:

$k ([a_{ij}] + [b_{ij}] = k [a_{ij} + b_{ij}]$, by law of addition

$= [k (a_{ij} + b_{ij})]$, by law of scalar multiplication

$= (ka_{ij} + kb_{ij})]$, by distributive law for numbers

$= [ka_{ij}] + [kb_{ij}]$

$= k\ [a_{ij}] + k\ [b_{ij}]$. **Hencc the theorem.**

Property IV: Existence of Additive Identity:

If $A = [a_{ij}]$ be any $m \times n$ matrix and O be the $m \times n$ null matrix, then

$$A + O = A = O + A.$$

Proof:

Is obvious, since $a_{ij} + 0 = a_{ij} = 0 + a_{ij}$

Property V: Existence of Additive Inverse

If $A\ [a_{ij}]$ be any $m \times n$ matrix, then there exists another $m \times n$ matrix B such that

$$A + B + O = B = A,$$

where O is the $m \times n$ null matrix.

Here the matrix B is called the additive inverse of the matrix A or the negative of A.

Also the (i, j) the element of b is $-a_{ij}$ if $A = [a_{ij}]$

Example 1:

If $A = \begin{bmatrix} 1 & 5 & 6 \\ -6 & 7 & 0 \end{bmatrix}$ and $B = \begin{bmatrix} 1 & -5 & 7 \\ 8 & -7 & 7 \end{bmatrix}$ *then find A + B and A − B.*

Solution:

$$A + B = \begin{bmatrix} 1 & 5 & 6 \\ -6 & 7 & 0 \end{bmatrix} + \begin{bmatrix} 1 & -5 & 7 \\ 8 & -7 & 7 \end{bmatrix}$$

$$= \begin{bmatrix} 1+1 & 5-5 & 6+7 \\ -6+8 & 7-7 & 0+7 \end{bmatrix} = \begin{bmatrix} 2 & 0 & 13 \\ 2 & 0 & 7 \end{bmatrix}$$ **Ans.**

$$\text{and } A - B = \begin{bmatrix} 1 & 5 & 6 \\ -6 & 7 & 0 \end{bmatrix} - \begin{bmatrix} 1 & -5 & 7 \\ 8 & -7 & 7 \end{bmatrix}$$

$$= \begin{bmatrix} 1-1 & 5-(-5) & 6-7 \\ -6-7 & 8-7-(-7) & 0-7 \end{bmatrix} = \begin{bmatrix} 0 & 10 & -1 \\ -14 & 14 & -7 \end{bmatrix}$$ **Ans.**

Example 2:

Solve the following equations for A and B:

$$2A - B = \begin{bmatrix} 3 & -3 & 0 \\ 3 & 3 & 2 \end{bmatrix}, 2B + A = \begin{bmatrix} 4 & 1 & 5 \\ -1 & 4 & -4 \end{bmatrix}$$

Solution:

Given $2A - B = \begin{bmatrix} 3 & -3 & 0 \\ 3 & 3 & 2 \end{bmatrix}$

Multiplying both sides by 2, we get

$$4A - 2B = 2\begin{bmatrix} 3 & -3 & 0 \\ 3 & 3 & 2 \end{bmatrix} = \begin{bmatrix} 6 & -6 & 0 \\ 6 & 6 & 4 \end{bmatrix} \quad ...(1)$$

Also given that $2B + A = \begin{bmatrix} 4 & 1 & 5 \\ -1 & 4 & -4 \end{bmatrix}$ ...(2)

Adding (1) and (2) we get

$$5A = \begin{bmatrix} 6 & -6 & 0 \\ 6 & 6 & 4 \end{bmatrix} + \begin{bmatrix} 4 & 1 & 5 \\ -1 & 4 & -4 \end{bmatrix}$$

$$= \begin{bmatrix} 6+4 & -6+1 & 0+5 \\ 6-1 & 6+4 & 4+4 \end{bmatrix} = \begin{bmatrix} 10 & -5 & 5 \\ 5 & 10 & 0 \end{bmatrix}$$

$$\Rightarrow \quad A = \frac{1}{2}\begin{bmatrix} 10 & -5 & 5 \\ 5 & 10 & 0 \end{bmatrix} = \begin{bmatrix} 2 & -1 & 1 \\ 1 & 2 & 0 \end{bmatrix}$$

Again from (2) we get

$$2B = \begin{bmatrix} 4 & 1 & 5 \\ -1 & 4 & -4 \end{bmatrix} - A$$

$$\Rightarrow \quad 2B = \begin{bmatrix} 4 & 1 & 5 \\ -1 & 4 & -4 \end{bmatrix} - \begin{bmatrix} 2 & -1 & 1 \\ 1 & 2 & 0 \end{bmatrix}$$

$$= \begin{bmatrix} 4-2 & 1+1 & 5-1 \\ -1-1 & 4-2 & -4-0 \end{bmatrix} = \begin{bmatrix} 2 & 2 & 4 \\ -2 & 2 & -4 \end{bmatrix}$$

$$\Rightarrow \quad B = \frac{1}{2}\begin{bmatrix} 2 & 2 & 4 \\ -2 & 2 & -4 \end{bmatrix} = \begin{bmatrix} 1 & 1 & 2 \\ -1 & 1 & -2 \end{bmatrix}.$$

Ans.

Example 3:

Given $A = \begin{bmatrix} 1 & 2 & -3 \\ 5 & 0 & 2 \\ 1 & -1 & 1 \end{bmatrix}$ and $B = \begin{bmatrix} 3 & -1 & 2 \\ 4 & 2 & 5 \\ 2 & 0 & 3 \end{bmatrix}$, *find the matrix C such that A + 2C = B.*

Solution:

Given that A + 2C = B or 2= B – A

$$\Rightarrow \quad 2C = \begin{bmatrix} 3 & -1 & 2 \\ 4 & 2 & 5 \\ 2 & 0 & 3 \end{bmatrix} - \begin{bmatrix} 1 & 2 & -3 \\ 5 & 0 & 2 \\ 1 & -1 & 1 \end{bmatrix}$$

$$= \begin{bmatrix} 3-1 & -1-2 & 2-(-3) \\ 4-5 & 2-0 & 5-2 \\ 2-1 & 0-(-1) & 3-1 \end{bmatrix} = \begin{bmatrix} 2 & -3 & 5 \\ -1 & 2 & 3 \\ 1 & 1 & 2 \end{bmatrix}$$

$$\Rightarrow \quad C = \frac{1}{2}\begin{bmatrix} 2 & -3 & 0 \\ -1 & 2 & 3 \\ 1 & 1 & 2 \end{bmatrix} = \begin{bmatrix} 1 & -\frac{3}{2} & \frac{5}{2} \\ -\frac{1}{2} & 1 & \frac{3}{2} \\ \frac{1}{2} & \frac{1}{2} & 1 \end{bmatrix}$$

Example 4:

If $A = \begin{bmatrix} 2 & 3 & 1 \\ 0 & -1 & 5 \end{bmatrix}$ and $B = \begin{bmatrix} 1 & 2 & -6 \\ 0 & -1 & 3 \end{bmatrix}$ *evaluate 3A – 4B.*

Solution:

$$2A - 4B = 3\begin{bmatrix} 2 & 3 & 1 \\ 0 & -1 & 5 \end{bmatrix} - 4\begin{bmatrix} 1 & 2 & -6 \\ 0 & -1 & 3 \end{bmatrix}$$

$$= \begin{bmatrix} 6 & 9 & 3 \\ 0 & -3 & 15 \end{bmatrix} - \begin{bmatrix} 4 & 8 & -24 \\ 0 & -4 & 12 \end{bmatrix}$$

$$= \begin{bmatrix} 6-4 & 9-8 & 3-(-24) \\ 0-0 & -3-(-4) & 15-12 \end{bmatrix}$$

$$= \begin{bmatrix} 2 & 1 & 27 \\ 0 & 1 & 3 \end{bmatrix}.$$

1.15 MULTIPLICATION OF MATRICES

Definition: *Let* $A = [a_{ij}]$ $m \times n$ *and* $B = B = [b_{lk}]$ $n \times p$ *be two matrices such that the number of column in A is equal to the number of row in B.*

Then the matrix $[c_{ik}]$, $m \times p$ such that $c_{ik} = \sum_{j=1}^{n} a_{ij} b_{ik}$ *is called the product of the matrices A ande B in that order and we write = AB.*

As an example, consider the matrices

$$A = \begin{bmatrix} 1 & 2 & 3 \\ 4 & 5 & 6 \end{bmatrix} \text{ and } B \begin{bmatrix} 7 & 8 \\ 9 & 10 \\ 11 & 12 \end{bmatrix}$$

Here the number of columns in A = 3 = then number of rows in B and thus we can evaluate AB.

Let Ab = $[c_{ij}]$, where $[c_{ij}]$ is 2×2 matrix.

Now to write c_{11}, we take the elements of the first row of A *viz.*, 1, 2, 3 in this order and the elements of the first column of B *viz.*, 7, 9, 11 in this order and form the products 1.7, 2.9, 3.11 and finally add them.

i.e., $c_{11} = 1.7 + 2.9 + 3.11 = 58$

Similarly $c_{12} = 1.8 + 2.10 + 3.12 = 64;$

$c_{21} = 4.7 + 5.9 + 6.11 = 139$

and $c_{22} = 4.8 + 5.10 + 6.12 = 154$

Hence $AB = [c_{ij}] = \begin{bmatrix} c_{11} & c_{12} \\ c_{21} & c_{12} \end{bmatrix} = \begin{bmatrix} 58 & 64 \\ 139 & 154 \end{bmatrix}.$

Note: The product AB can be calculated only if the number of columns in A be equal to the number of rows in b. The two matrices A and B satisfying this condition are called conformable to multiplication.

Post-Multiplication and Pre-Multiplication of Matrices

The matrix AB is the matrix A post-multiplied by B whereas the matrix BA is the matrix A pre-multiplied by B..

In the product AB, the matrix A is know as the pre-factor and the matrix B is know as the post-factor.

The product in both the above the above cases viz. AB and BA may or may not exist and may be equal or different.

i.e., we say $AB \neq BA$ in general.

The same is discussed on the next page:

Case 1: If the matrix A is $m \times n$ and the matrix B is $n \times k$, then the product AB exists whereas BA does not exist, since we know that AB can be calculated only if the numbers of columns in A is equal to the number of rows in B.

Case 2: If the matrix A is $m \times n$ and the matrix B is $n \times m$, then both AB and BA exist, but the matrix AB is $m \times m$ while the matrix BA is $n \times n$.

Hence $AB \neq BA$ thought AB and BA exist.

Case 3: If both A and B are square matrices of the same order, then AB as well as BA exist but are not necessarily equal *i.e.,* if

$$A = \begin{bmatrix} 1 & 2 \\ 3 & 4 \end{bmatrix} \text{ and } B = \begin{bmatrix} 3 & 1 \\ 4 & 7 \end{bmatrix}$$

then $$AB = \begin{bmatrix} 1 & 2 \\ 3 & 4 \end{bmatrix} \times \begin{bmatrix} 3 & 1 \\ 4 & 7 \end{bmatrix} = \begin{bmatrix} 1.3 + 2.4 & 1.1 + 2.7 \\ 3.3 + 4.4 & 3.1 + 4.7 \end{bmatrix}$$

$$= \begin{bmatrix} 11 & 25 \\ 25 & 31 \end{bmatrix}$$

and $$BA = \begin{bmatrix} 3 & 1 \\ 4 & 7 \end{bmatrix} \times \begin{bmatrix} 1 & 2 \\ 3 & 4 \end{bmatrix} = \begin{bmatrix} 3.1 + 1.3 & 3.2 + 1.4 \\ 4.1 + 7.3 & 4.2 + 7.4 \end{bmatrix}$$

$$= \begin{bmatrix} 6 & 10 \\ 25 & 36 \end{bmatrix}$$

$\therefore \quad AB \neq BA.$

But if $$A = \begin{bmatrix} 1 & 0 \\ 0 & -2 \end{bmatrix} \text{ and } B = \begin{bmatrix} 1 & 0 \\ 0 & 4 \end{bmatrix}$$

then $$AB = \begin{bmatrix} 1 & 0 \\ 0 & -2 \end{bmatrix} \times \begin{bmatrix} 1 & 0 \\ 0 & 4 \end{bmatrix} = \begin{bmatrix} 1.1 + 0.0 & 1.0 + 0.4 \\ 0.1 - 2.0 & 0.0 - 2.4 \end{bmatrix}$$

$$= \begin{bmatrix} 1 & 0 \\ 0 & -8 \end{bmatrix}$$

and $$BA = \begin{bmatrix} 1 & 0 \\ 0 & 4 \end{bmatrix} \times \begin{bmatrix} 1 & 0 \\ 0 & -2 \end{bmatrix} = \begin{bmatrix} 1.1 + 0.0 & 1.0 + 0\,(-2) \\ 0.1 + 4.0 & 0.0 + 4\,(-2) \end{bmatrix}$$

$$= \begin{bmatrix} 1 & 0 \\ 0 & -8 \end{bmatrix}$$

$\therefore \quad AB = BA.$

Hence in general AB $\neq$ BA.

Note 1: If AB = BA, then matrices A and B are said to commit. If AB = – BA, the matrices A and B are said to anticommute.

Note 2: If $A = \begin{bmatrix} 1 & 1 \\ 1 & 1 \end{bmatrix}$ and $B = \begin{bmatrix} 1 & 0 \\ -1 & 0 \end{bmatrix}$,

then $$AB = \begin{bmatrix} 1 & 1 \\ 1 & 1 \end{bmatrix} \times \begin{bmatrix} 1 & 0 \\ -1 & 0 \end{bmatrix}$$

$$= \begin{bmatrix} 1.1 + 1.(-1) & 1.0 + 1.0 \\ 1.1 + 1.(-1) & 1.0 + 1.0 \end{bmatrix} = \begin{bmatrix} 0 & 0 \\ 0 & 0 \end{bmatrix}$$

i.e., AB is zero matrix (or null matrix) whereas neither A nor B is a zero matrix.

$\therefore$ AB = O does not imply that either A = 0 or B = O.

Here $$BA = \begin{bmatrix} 1 & 0 \\ -1 & 0 \end{bmatrix} \times \begin{bmatrix} 1 & 1 \\ 1 & 1 \end{bmatrix}$$

$$= \begin{bmatrix} 1.1 + 0.1 & 1.1 + 0.1 \\ -1.1 + 0.1 & -1.1 + 0.1 \end{bmatrix} = \begin{bmatrix} 1 & 1 \\ -1 & -1 \end{bmatrix}$$

i.e., BA $\neq$ O

If $A = \begin{bmatrix} 4 & 4 \\ 3 & 3 \end{bmatrix}$ and $B = \begin{bmatrix} -1 & 1 \\ 1 & -1 \end{bmatrix}$, then

$$AB = \begin{bmatrix} 4 & 4 \\ 3 & 3 \end{bmatrix} \times \begin{bmatrix} -1 & 1 \\ 1 & -1 \end{bmatrix}$$

$$= \begin{bmatrix} 4(-1) + 4(1) & 4(1) + 4(-1) \\ 3(-1) + 3(1) & 3(1) + 3(-1) \end{bmatrix} = \begin{bmatrix} 0 & 0 \\ 0 & 0 \end{bmatrix} = 0$$

i.e., the product of two non-zero square matrices can be a zero matrix.

and $$BA = \begin{bmatrix} -1 & 1 \\ 1 & -1 \end{bmatrix} \times \begin{bmatrix} 4 & 4 \\ 3 & 3 \end{bmatrix}$$

$$= \begin{bmatrix} (-1).4 + 1.3 & (-1).4 + 1.3 \\ 1.4 + (-1).3 & 1.4 + (-1).3 \end{bmatrix} = \begin{bmatrix} -1 & -1 \\ 1 & 1 \end{bmatrix} \neq 0$$

Example 1:

Calculate AB and BA if

$$A = \begin{bmatrix} 1 & -1 & 1 \\ -3 & 2 & -1 \\ -2 & 1 & 0 \end{bmatrix}, B = \begin{bmatrix} 1 & 2 & 3 \\ 2 & 4 & 6 \\ 1 & 2 & 3 \end{bmatrix}$$

Solution:

$$AB = \begin{bmatrix} 1 & -1 & 1 \\ -3 & 2 & -1 \\ -2 & 1 & 0 \end{bmatrix} \times \begin{bmatrix} 1 & 2 & 3 \\ 2 & 4 & 6 \\ 1 & 2 & 3 \end{bmatrix}$$

$$= \begin{bmatrix} 1.1 + (-1).2 + 1.1 & 1.2 \ (-1).4 + 1.2 & 1.3 + (-1).6 + 1.3 \\ (-3).1 + 2.2 + (-1).1 & (-3).2 + 2.4 + (-1).2 & (-3).3 + 2.6 + (-1).3 \\ (-2).1 + 1.2 + 0.1 & (-2).2 + 1.4 + 0.2 & (-2).3 + 1.6 + 0.3 \end{bmatrix}$$

$$= \begin{bmatrix} 0 & 0 & 0 \\ 0 & 0 & 0 \\ 0 & 0 & 0 \end{bmatrix} = O, \text{ where O is } 3 \times 3 \text{ null matrix.}$$ **Ans.**

$$\text{And } BA = \begin{bmatrix} 1 & 2 & 3 \\ 2 & 4 & 6 \\ 1 & 2 & 3 \end{bmatrix} \times \begin{bmatrix} 1 & -1 & 1 \\ -3 & 2 & -1 \\ -2 & 1 & 0 \end{bmatrix}$$

$$= \begin{bmatrix} 1.1 + 2.(-3) + 3(-2) & 1(-1) + 2.2 + 3.1 & 1.1 + 2.(-1) + 3.0 \\ 2.1 + 4.(-3) + 6(-2) & 2(-1) + 4.2 + 6.1 & 2.1 + 4.(-1) + 6.0 \\ 1.1 + 2.(-3) + 3(-2) & 1(-1) + 2.2 + 3.1 & 1.1 + 2.(-1) + 30 \end{bmatrix}$$

$$= \begin{bmatrix} -11 & 6 & -1 \\ -22 & 12 & -2 \\ -11 & 6 & -1 \end{bmatrix}$$ **Ans.**

[**Note:** $AB \neq BA$].

Example 2:

If $A = \begin{bmatrix} 2 & 3 & 4 \\ 1 & 2 & 3 \\ -1 & 1 & 2 \end{bmatrix}$ and $B - \begin{bmatrix} 1 & 3 & 0 \\ -1 & 2 & 1 \\ 0 & 0 & 2 \end{bmatrix}$ *evaluate AB, BA or whichever exists.*

Solution:

$$AB = \begin{bmatrix} 2 & 3 & 4 \\ 1 & 2 & 3 \\ -1 & 1 & 2 \end{bmatrix} \times \begin{bmatrix} 1 & 3 & 0 \\ -1 & 2 & 1 \\ 0 & 0 & 2 \end{bmatrix}$$

$$= \begin{bmatrix} 2.1 + 3(-1) + 4.0 & 2.3 + 3.2 + 4.0 & 2.0 + 3.1 + 4.2 \\ 1.1 + 2(-1) + 3.0 & 1.3 + 2.2 + 3.0 & 1.0 + 2.1 + 3.2 \\ -1.1 + 1(-1) + 2.0 & 1.3 + 1.2 + 2.0 & -1.0 + 1.1 + 2.2 \end{bmatrix}$$

$$= \begin{bmatrix} -1 & 12 & 11 \\ -1 & 7 & 8 \\ -2 & -1 & 5 \end{bmatrix}$$

$$\text{And } BA = \begin{bmatrix} 1 & 3 & 0 \\ -1 & 2 & 1 \\ 0 & 0 & 2 \end{bmatrix} \times \begin{bmatrix} 2 & 3 & 4 \\ 1 & 2 & 3 \\ -1 & 1 & 2 \end{bmatrix}$$

$$= \begin{bmatrix} 1.2 + 3.1 + 0(-1) & 1.3 + 3.2 + 0.1 & 1.4 + 3.3 + 0.2 \\ -1.2 + 2.1 + 1(-1) & -1.3 + 2.2 + 1.1 & -1.4 + 2.3 + 1.2 \\ 0.2 + 0.1 + 2(-1) & 0.3 + 0.2 + 2.1 & 0.4 + 0.3 + 2.2 \end{bmatrix}$$

$$= \begin{bmatrix} 5 & 9 & 13 \\ -1 & 2 & 4 \\ -2 & 2 & 4 \end{bmatrix}$$

Example 3:

Find the product matrix of the matrices

$$\begin{bmatrix} 2 & 1 & 2 & 1 \\ 1 & 1 & 1 & 1 \end{bmatrix} \textit{ and } \begin{bmatrix} 2 & -1 & 0 \\ 0 & 4 & 1 \\ -2 & 1 & 0 \\ 1 & -3 & 2 \end{bmatrix}$$

Solution:

The required matrix

$$= \begin{bmatrix} 2 & 1 & 2 & 1 \\ 1 & 1 & 1 & 1 \end{bmatrix} \times \begin{bmatrix} 2 & -1 & 0 \\ 0 & 4 & 1 \\ -2 & 1 & 0 \\ 1 & -3 & 2 \end{bmatrix}$$

$$= \begin{bmatrix} 2.2 + 1.0 + 2(-2) + 1.1 & 2(-1) + 1.4 + 2.1 + 1(-3) & 2.0 + 1.1 + 2.0 + 1.2 \\ 1.2 + 1.0 + 1(-2) + 1.1 & 1(-1) + 1.4 + 1.1 + 1(-3) & 1.0 + 1.1 + 1.0 + 1.2 \end{bmatrix}$$

$$= \begin{bmatrix} 1 & 1 & 3 \\ 1 & 1 & 3 \end{bmatrix}.$$ **Ans.**

Example 4:

If $\begin{bmatrix} 4 \\ 1 \\ 3 \end{bmatrix} A = \begin{bmatrix} -4 & 8 & 4 \\ -1 & 2 & 1 \\ -3 & 6 & 3 \end{bmatrix}$, *find A.*

Solution:

We know that if X is an m × n matrix, Y is an n × k matrix, then the product XY is an m × k matrix.

Hence $\begin{bmatrix} 4 \\ 1 \\ 3 \end{bmatrix}$ is 3 × 1 matrix and $\begin{bmatrix} -4 & 8 & 4 \\ -1 & 2 & 1 \\ -3 & 6 & 3 \end{bmatrix}$.

is 3 × 3 matrix, os A must be a 1 × 3 matrix *i.e.,* a row matrix. (**Note**)

∴ Let A = [a, b, c]

Then $\begin{bmatrix} 4 \\ 1 \\ 3 \end{bmatrix} \times [a\ b\ c] = \begin{bmatrix} -4 & 8 & 4 \\ -1 & 2 & 1 \\ -3 & 6 & 3 \end{bmatrix}$

Which gives $\begin{bmatrix} 4a & 4b & 4c \\ a & b & c \\ 3a & 3b & 3c \end{bmatrix} = \begin{bmatrix} -4 & 8 & 4 \\ -1 & 2 & 1 \\ -3 & 6 & 3 \end{bmatrix}$

Comparing corresponding elements we have

$4a = -4,\ a = -1,\ 3a = -3,\ 4b = 8,\ b = 2,\ 3b = 6$

and $4c - 4,\ c = 1,\ 3c = 3.$

All these are satisfied by $a = -1$, $b = 2$, $c = 1$.

Hence from (1) we have

A = [a b c] = [–1, 2, 1]. **Ans.**

Example 5:

$$A = \begin{bmatrix} 1 & 1 & -1 \\ 2 & -3 & 4 \\ 3 & -2 & 3 \end{bmatrix};\ B = \begin{bmatrix} -1 & -2 & -1 \\ 6 & 12 & 6 \\ 5 & 10 & 5 \end{bmatrix},\ C = \begin{bmatrix} -1 & -1 & 1 \\ 2 & 2 & -2 \\ -3 & -3 & 3 \end{bmatrix}$$

show that AB and CA are null matrices but AB ≠ O, AC ≠ O.

Solution:

$$AB = \begin{bmatrix} 1 & 1 & -1 \\ 2 & -3 & 4 \\ 3 & -2 & 3 \end{bmatrix} \times \begin{bmatrix} -1 & -2 & -1 \\ 6 & 12 & 6 \\ 5 & 10 & 5 \end{bmatrix}$$

$$= \begin{bmatrix} 1(-1) + 1.6 + (-1).5 & 1(-2) + 1.12 + (-1).10 & 1(-1) + 1.6 + (-1).5 \\ 2(-1) - 3.6 + 4.5 & 2(-2) - 3.12 + 4.10 & 2(-1) - 3.6 + 4.5 \\ 3(-1) - 2.6 + 3.5 & 3(-2) - 2.12 + 3.10 & 3(-1) - 2.6 + 3.5 \end{bmatrix}$$

$$= \begin{bmatrix} 0 & 0 & 0 \\ 0 & 0 & 0 \\ 0 & 0 & 0 \end{bmatrix}, \text{ which is null matrix.}$$

This is known as 'unusual property' of matrix multiplication

$$CA = \begin{bmatrix} -1 & -1 & 1 \\ 2 & 2 & -2 \\ -3 & -3 & 3 \end{bmatrix} \times \begin{bmatrix} 1 & 1 & -1 \\ 2 & -3 & 4 \\ 3 & -2 & 3 \end{bmatrix}$$

$$= \begin{bmatrix} 1.1 - 1.2 + 1.3 & -1.1 - 1(-3) + 1(-2) & -1(-1) - 1.4 + 1.3 \\ 2.1 + 2.2 - 2.3 & 2.1 + 2(-3) - 2(-2) & 2(-1) + 2.4 - 2.3 \\ -3.1 - 3.2 + 3.3 & -3.1 - 3(-3) + 3(-2) & -3(-1) - 3.4 + 3.3 \end{bmatrix}$$

$$= \begin{bmatrix} 0 & 0 & 0 \\ 0 & 0 & 0 \\ 0 & 0 & 0 \end{bmatrix}, \text{ which is a null matrix.}$$

Hence proved.

We can prove in a similar way that BA ≠ O and AC ≠ O.

Example 6:

If $A = \begin{bmatrix} 1 & -2 & 3 \\ -4 & 2 & 5 \end{bmatrix}$ and $B = \begin{bmatrix} 2 & 3 \\ 4 & 5 \\ 2 & 1 \end{bmatrix}$ *find AB and show that* $AB \neq BA$.

Solution:

$$AB = \begin{bmatrix} 1 & -2 & 3 \\ -4 & 2 & 5 \end{bmatrix} \times \begin{bmatrix} 2 & 3 \\ 4 & 5 \\ 2 & 1 \end{bmatrix}$$

$$= \begin{bmatrix} 1.2 + (-2).\,4 + 3.2 & 1.3 + (-2).\,5 + 3.1 \\ -4.2 + \quad 2.4 + 5.2 & -4{,}3 + \;2.5 + 5.1 \end{bmatrix}$$

$$= \begin{bmatrix} 0 & -4 \\ 10 & 3 \end{bmatrix}$$

and $\quad BA = \begin{bmatrix} 2 & 3 \\ 4 & 5 \\ 2 & 1 \end{bmatrix} \times \begin{bmatrix} 1 & -2 & 3 \\ -4 & 2 & 5 \end{bmatrix}$

$$= \begin{bmatrix} 2.1 + 3\,(-4) & 2\,(-2) + 3\,(2) & 2\,(3) + 3\,(5) \\ 4.1 + 5\,(-4) & 4\,(-2) + 5\,(2) & 4\,(3) + 5\,(5) \\ 2.1 + 1\,(-4) & 2\,(-2) + 1\,(2) & 2\,(3) + 1\,(5) \end{bmatrix}$$

$$= \begin{bmatrix} -10 & 2 & 21 \\ -16 & 2 & 37 \\ -2 & -2 & 11 \end{bmatrix}$$

Hence $AB \neq BA$.

Example 7:

If $A = \begin{bmatrix} 1 & 2 \\ 3 & 0 \\ 4 & 1 \end{bmatrix}$ and $B = \begin{bmatrix} 0 & 1 & 0 \\ 0 & 2 & 1 \\ 2 & 3 & 0 \end{bmatrix}$, *find BA.*

Solution:

$$BA = \begin{bmatrix} 0 & 1 & 0 \\ 0 & 2 & 1 \\ 2 & 3 & 0 \end{bmatrix} \times \begin{bmatrix} 1 & 2 \\ 3 & 0 \\ 4 & 1 \end{bmatrix}$$

$$= \begin{bmatrix} 0.1 + 1.3 + 0.4 & 0.2 + 1.0 + 0.1 \\ 0.1 + 2.3 + 1.4 & 0.2 + 2.0 + 1.1 \\ 2.1 + 3.3 + 0.4 & 2.2 + 3.0 + 0.1 \end{bmatrix}$$

$$= \begin{bmatrix} 3 & 0 \\ 10 & 1 \\ 11 & 4 \end{bmatrix}$$

Example 8:

Multiply [3 – 1 4] and $\begin{bmatrix} -2 \\ 6 \\ 3 \end{bmatrix}$

Solution:

$$[3 \; -1 \; 4] \times \begin{bmatrix} -2 \\ 6 \\ 3 \end{bmatrix}$$

$$= [3\,(-2) + (-1) \,.\, 6 + 4.3] = [0]$$ **Ans.**

Example 9:

If $A = \begin{bmatrix} 0 & 1 \\ 1 & 2 \end{bmatrix}$ and $B = \begin{bmatrix} 1 \\ 2 \end{bmatrix}$, *find AB and BA, if they exist.*

Solution:

$$AB = \begin{bmatrix} 0.1 + 1.2 \\ 1.1 + 2.2 \end{bmatrix} = \begin{bmatrix} 2 \\ 5 \end{bmatrix}$$ **Ans.**

Also BA does not exist as the number of columns in B and number of rows in A are not equal.

Example 10:

If $A = \begin{bmatrix} 1 & 4 & 0 \\ 2 & 5 & 0 \\ 3 & 6 & 0 \end{bmatrix}$, $B = \begin{bmatrix} 3 & 2 & 1 \\ 1 & 2 & 3 \\ 4 & 5 & 6 \end{bmatrix}$ and $C = \begin{bmatrix} 3 & 2 & 1 \\ 1 & 2 & 3 \\ 7 & 8 & 9 \end{bmatrix}$

then evaluate AB – AC.

Solution:

$$AB = \begin{bmatrix} 1 & 4 & 0 \\ 2 & 5 & 0 \\ 3 & 6 & 0 \end{bmatrix} \times \begin{bmatrix} 3 & 2 & 1 \\ 1 & 2 & 3 \\ 4 & 5 & 6 \end{bmatrix}$$

$$= \begin{bmatrix} 1.3 + 4.1 + 0.4 & 1.2 + 4.2 + 0.5 & 1.1 + 4.3 + 0.6 \\ 2.3 + 5.1 + 0.4 & 2.2 + 5.2 + 0.5 & 2.1 + 5.3 + 0.6 \\ 3.3 + 6.1 + 0.4 & 3.2 + 6.2 + 0.5 & 3.1 + 6.3 + 0.6 \end{bmatrix}$$

$$= \begin{bmatrix} 7 & 10 & 13 \\ 11 & 14 & 17 \\ 15 & 18 & 21 \end{bmatrix} \qquad ...(1)$$

And

$$AC = \begin{bmatrix} 1 & 4 & 0 \\ 2 & 5 & 0 \\ 3 & 6 & 0 \end{bmatrix} \times \begin{bmatrix} 3 & 2 & 1 \\ 1 & 2 & 3 \\ 7 & 8 & 9 \end{bmatrix}$$

$$= \begin{bmatrix} 1.3 + 4.1 + 0.7 & 1.2 + 4.2 + 0.8 & 1.1 + 4.3 + 0.9 \\ 2.3 + 5.1 + 0.7 & 2.2 + 5.2 + 0.8 & 2.1 + 5.3 + 0.2 \\ 3.3 + 6.1 + 0.7 & 3.2 + 6.2 + 0.8 & 3.1 + 6.3 + 0.9 \end{bmatrix}$$

$$= \begin{bmatrix} 7 & 10 & 13 \\ 11 & 14 & 17 \\ 15 & 18 & 21 \end{bmatrix} \qquad ...(2)$$

$\therefore$ From (1) and (2) we get AB – AC

$$= \begin{bmatrix} 7 & 10 & 13 \\ 11 & 14 & 17 \\ 15 & 18 & 21 \end{bmatrix} - \begin{bmatrix} 7 & 10 & 13 \\ 11 & 14 & 17 \\ 15 & 18 & 21 \end{bmatrix} = \begin{bmatrix} 0 & 0 & 0 \\ 0 & 0 & 0 \\ 0 & 0 & 0 \end{bmatrix}$$ **Ans.**

Example 11:

Find the square of the matrix

$$\begin{bmatrix} -1 & 1 & 1 & 1 \\ 1 & -1 & 1 & 1 \\ 1 & 1 & -1 & 1 \\ 1 & 1 & 1 & -1 \end{bmatrix}$$

Solution:

$$\begin{bmatrix} -1 & 1 & 1 & 1 \\ 1 & -1 & 1 & 1 \\ 1 & 1 & -1 & 1 \\ 1 & 1 & 1 & -1 \end{bmatrix}^2$$

$$= \begin{bmatrix} -1 & 1 & 1 & 1 \\ 1 & -1 & 1 & 1 \\ 1 & 1 & -1 & 1 \\ 1 & 1 & 1 & -1 \end{bmatrix} \times \begin{bmatrix} -1 & 1 & 1 & 1 \\ 1 & -1 & 1 & 1 \\ 1 & 1 & -1 & 1 \\ 1 & 1 & 1 & -1 \end{bmatrix}$$

$$= \begin{bmatrix} (-1)(-1) + 1.1 + 1.1 + 1.1 & (-1).1 + 1(-1) + 1,1 + 1.1 & (-1).1 + 1.1 + 1(-1) + 1.1 & (-1).1 + 1.1 + 1.1 + 1(-1) \\ 1.(-1) + (-1).1 + 1.1 + 1.1 & 1.1 + (-1)(-1) + 1.1 + 1.1 & 1.1 + (-1).1 + 1(-1) + 1.1 & 1.1 + (-1).1 + 1.1 + 1(-1) \\ 1.(-1) + 1.1 + (-1).1 + 1.1 & 1.1 + 1(-1) + (-1).1 + 1.1 & 1.1 + 1.1 + (-1).(-1) + 1.1 & 1.1 + 1.1 + (-1).1 + 1(-1) \\ 1.(-1) + 1.1 + 1.1 + (-1).1 & 1.1 + 1.(-1) + 1.1 + (-1).1 & 1.1 + 1.1 + 1,(-1) + (-1).1 & 1.1 + 1.1 + 1.1 + (-1)(-1) \end{bmatrix}$$

$$= \begin{bmatrix} 4 & 0 & 0 & 0 \\ 0 & 4 & 0 & 0 \\ 0 & 0 & 4 & 0 \\ 0 & 0 & 0 & 4 \end{bmatrix} = 4 \begin{bmatrix} 1 & 0 & 0 & 0 \\ 0 & 1 & 0 & 0 \\ 0 & 0 & 1 & 0 \\ 0 & 0 & 0 & 1 \end{bmatrix}$$ **Ans.**

Example 12:

If $A = \begin{bmatrix} 1 & 1 & 3 \\ 2 & 2 & 6 \\ -1 & -1 & -3 \end{bmatrix}$, *show that* $A^2 = O$.

Solution:

$$A^2 = \begin{bmatrix} 1 & 1 & 3 \\ 2 & 2 & 6 \\ -1 & -1 & -3 \end{bmatrix} \times \begin{bmatrix} 1 & 1 & 3 \\ 2 & 2 & 6 \\ -1 & -1 & -3 \end{bmatrix}$$

$$= \begin{bmatrix} 1.1 + 1.2 + 3.(-1) & 1.1 + 1.2 + 3.(-1) & 1.3 + 1.6 + 3.(-3) \\ 2.1 + 2.2 + 6.(-1) & 2.1 + 2.2 + 6.(-1) & 2.3 + 2.6 + 6.(-3) \\ -1.1 - 1.2 - 3.(-1) & -1.1 + 1.2 - 3(-1) & -1.3 + 1.6 - 3,(-3) \end{bmatrix}$$

$$= \begin{bmatrix} 0 & 0 & 0 & 0 \\ 0 & 0 & 0 & 0 \\ 0 & 0 & 0 & 0 \end{bmatrix}$$ = O, where O is 3 × 3 null matrix. **Hence proved**

Example 13:

Evaluate A^3 if $A = \begin{bmatrix} \cosh\theta & \sinh\theta \\ \sinh\theta & \cosh 0 \end{bmatrix}$

Solution:

$$A^2 = \begin{bmatrix} \cosh\theta & \sinh\theta \\ \sinh\theta & \cosh\theta \end{bmatrix} \times \begin{bmatrix} \cosh\theta & \sinh\theta \\ \sinh\theta & \cosh\theta \end{bmatrix}$$

$$= \begin{bmatrix} \cosh^2\theta + \sinh^2\theta & \cosh\theta\sinh\theta + \sinh\theta\cosh\theta \\ \sinh\theta\cosh\theta + \cosh\theta\sinh\theta & \sinh^2\theta + \cosh\theta \end{bmatrix}$$

$$= \begin{bmatrix} \cosh 2\theta & \sinh 2\theta \\ \sinh 2\theta & \cosh 2\theta \end{bmatrix} \because \begin{matrix} \cosh^2\theta + \sinh^2\theta = \cosh 2\theta \\ 2\cosh\theta\cosh\theta = \sinh 2\theta \end{matrix}$$

$$\therefore \quad A^3 = A^2A = \begin{bmatrix} \cosh 2\theta & \sinh 2\theta \\ \sinh 2\theta & \cosh 2\theta \end{bmatrix} \begin{bmatrix} \cosh 2\theta & \sinh 2\theta \\ \sinh 2\theta & \cosh 2\theta \end{bmatrix}$$

$$= \begin{bmatrix} \cosh 2\theta\cosh\theta + \sinh 2\theta\sinh\theta & \cosh 2\theta\sinh\theta + \sinh 2\theta\cosh\theta \\ \sinh 2\theta\cosh\theta + \cosh 2\theta\sinh\theta & \sinh 2\theta\sinh\theta + \cosh 2\theta\cosh\theta \end{bmatrix}$$

$$= \begin{bmatrix} \cosh(2\theta+\theta) & \sinh(2\theta+\theta) \\ \sinh(2\theta+\theta) & \cosh(2\theta+\theta) \end{bmatrix},$$

$\because$ sinh (A + B) = sinh A cosh B + cosh A sin B

sinh (A + B) = cosh A cosh B + sinh A sin B

$$= \begin{bmatrix} \cosh 3\theta & \sinh 2\theta \\ \sinh 3\theta & \cosh 2\theta \end{bmatrix}$$

Example 14:

If A, B, C are three matrices such that

$A = [x, y, z]$, $B = \begin{bmatrix} a & h & g \\ h & b & f \\ g & f & c \end{bmatrix}$, $C = \begin{bmatrix} x \\ y \\ z \end{bmatrix}$, *evaluate ABC.*

Solution:

$$B = [x, y, z] \times \begin{bmatrix} a & h & g \\ h & b & f \\ g & f & c \end{bmatrix}$$

$$= [x.a + y.h + z.g \quad x.h + y.b + z.f \quad x.g + y.f + z.c]$$

$$\Rightarrow \quad ABC = [ax + hy + gz \quad hx + by + fz \quad gx + fy + cz \times \begin{bmatrix} x \\ y \\ z \end{bmatrix}$$

$$= [x\,(ax + hy + gz) + y\,(hx + by + fz) + z\,(gx + fy + cz)] \quad \textbf{(Note)}$$

$$= [ax^2 + by^2 + cz^2 + 2hxy + 2gzx + 2fyz]. \qquad \textbf{Ans.}$$

Example 15:

Find the product of the following two matrices

$$\begin{bmatrix} 0 & c & -b \\ -c & 0 & a \\ b & -a & 0 \end{bmatrix} \times \begin{bmatrix} a^2 & ab & ac \\ ab & b^2 & bc \\ ac & bc & c^2 \end{bmatrix}$$

Solution:

The required product

$$= \begin{bmatrix} 0 & c & -b \\ -c & 0 & a \\ b & -a & 0 \end{bmatrix} \times \begin{bmatrix} a^2 & ab & ac \\ ab & b^2 & bc \\ ac & bc & c^2 \end{bmatrix}$$

$$= \begin{bmatrix} 0.a^2 + c.ab - b.ac & 0.ab + c.b^2 - b.bc & 0ac + c.bc - b.c^2 \\ -ca^2 + 0.ab\ a.ac & -c.ab + 0b^2 + a.bc & -c.ac + 0.bc + a.c^2 \\ b.a^2 - a.ab + 0.ac & b.ab - a.b^2 + 0.bc & b.ac - a.bc + 0.c^2 \end{bmatrix}$$

$$= \begin{bmatrix} 0 & 0 & 0 \\ 0 & 0 & 0 \\ 0 & 0 & 0 \end{bmatrix}$$

Example 16:

Find the values of x, y, z in the following equation

$$\begin{bmatrix} 1 & 2 & 3 \\ 3 & 1 & 2 \\ 2 & 3 & 1 \end{bmatrix} \times \begin{bmatrix} x \\ y \\ z \end{bmatrix} = \begin{bmatrix} 4 & -2 \\ 0 & -6 \\ -1 & 2 \end{bmatrix} \times \begin{bmatrix} 2 \\ 1 \end{bmatrix}$$

Solution:

$$\begin{bmatrix} 1 & 2 & 3 \\ 3 & 1 & 2 \\ 2 & 3 & 1 \end{bmatrix} \times \begin{bmatrix} x \\ y \\ z \end{bmatrix} = \begin{bmatrix} 1.x + 2y + 2z \\ 3x + 1.y + 2z \\ 2x + 3y + 1.z \end{bmatrix} \qquad ...(1)$$

And $\begin{bmatrix} 4 & -2 \\ 0 & -6 \\ -1 & 2 \end{bmatrix} \times \begin{bmatrix} 2 \\ 1 \end{bmatrix} = \begin{bmatrix} 4.2 + (-2).1 \\ 0.2 + (-6).1 \\ -1.2 + \quad 2.1 \end{bmatrix} = \begin{bmatrix} 6 \\ -6 \\ 0 \end{bmatrix} \qquad .(2)$

With the help of (1.) and (2), the given equation reduces to

$$\begin{bmatrix} x + 2y + 3z \\ 3x + y + 2z \\ 2x + 3y + z \end{bmatrix} = \begin{bmatrix} 6 \\ -6 \\ 0 \end{bmatrix}$$

From this on comparing the corresponding elements on both sides we get $x + 2y + 3z = 6$; $3x + y + 2z = -6$ and $2x + 3y + z = 0$.

Solving these we get $x = -4$, $y = 2$, $z = 2$. **Ans.**

Example 17:

If $A = \begin{bmatrix} 1 & 2 & 3 \\ 0 & 1 & 2 \\ 0 & 0 & 1 \end{bmatrix}$; $B = \begin{bmatrix} x \\ y \\ z \end{bmatrix}$ and $AB = \begin{bmatrix} 6 \\ 3 \\ 1 \end{bmatrix}$ *find the values of x, y, z.*

Solution:

$$AB = \begin{bmatrix} 1 & 2 & 3 \\ 0 & 1 & 2 \\ 0 & 0 & 1 \end{bmatrix} \times \begin{bmatrix} x \\ y \\ z \end{bmatrix}$$

$$\Rightarrow \quad \begin{bmatrix} 6 \\ 3 \\ 1 \end{bmatrix} = \begin{bmatrix} 1.x + 2.y + 3z \\ 0.x + 1.y + 2z \\ 0.x + 0.y + 1.z \end{bmatrix}$$

$\Rightarrow \quad 6 = x + 2y + 3z,\ 3 = y + 2z,\ 1 = z,$ **(Note)**

comparing the corresponding elements of the matrices on both sides. Solving these we get $x = 1$, $y = 1$, $z = 1$. **Ans.**

Example 18:

If $A = \begin{bmatrix} i & 0 \\ 0 & -i \end{bmatrix}$, $B = \begin{bmatrix} 0 & -1 \\ 1 & 0 \end{bmatrix}$, $C = \begin{bmatrix} 0 & i \\ i & 0 \end{bmatrix}$, *prove that*

$A^2 = B^2 = C^2 = -I$ *and* $AB = -C = -BA$, *where* $I = \begin{bmatrix} 1 & 1 \\ 0 & 1 \end{bmatrix}$

Solution:

$$A^2 = \begin{bmatrix} i & 0 \\ 0 & -i \end{bmatrix} \times \begin{bmatrix} i & 0 \\ 0 & -i \end{bmatrix}$$

$$= \begin{bmatrix} i.i + 0.0 & i.0 + (-1) \\ 0.i - i.0 & 0.0 + (-i)(-i) \end{bmatrix} = \begin{bmatrix} -1 & 0 \\ 0 & -1 \end{bmatrix}$$

$$= \begin{bmatrix} 1 & 0 \\ 0 & 1 \end{bmatrix}$$

$$= -I.$$

and $B^2 = \begin{bmatrix} 1 & -1 \\ 0 & 0 \end{bmatrix} \times \begin{bmatrix} 1 & -1 \\ 0 & 0 \end{bmatrix} = \begin{bmatrix} 0.0 - 1.1 & 0.(-1) + (-1).0 \\ 1.0 + 0.1 & 1(-1) + 0.0 \end{bmatrix}$

$$= \begin{bmatrix} -1 & 0 \\ 0 & -1 \end{bmatrix} = -\begin{bmatrix} 1 & 0 \\ 0 & 1 \end{bmatrix} = -1$$

Similarly we can prove that $C^2 = -I$. Hence $A^2 = B^2 = C^2 = -I$.

Again $AB = \begin{bmatrix} i & 0 \\ 0 & -i \end{bmatrix} \times \begin{bmatrix} 0 & -1 \\ 1 & 0 \end{bmatrix}$

$$= \begin{bmatrix} i.0 + 0.1 & i(-1) + 0.0 \\ 0.0 - i(1) & 0(-1) - i.0 \end{bmatrix}$$

$$= \begin{bmatrix} 0 & -i \\ -i & 0 \end{bmatrix} = -\begin{bmatrix} 0 & i \\ i & 0 \end{bmatrix} = -C$$

and $BA = \begin{bmatrix} 0 & -1 \\ 1 & 0 \end{bmatrix} \times \begin{bmatrix} i & 0 \\ 0 & -i \end{bmatrix}$

$$= \begin{bmatrix} 0.i - i.o & 0.0 - 1(-i) \\ 1.i + 0.0 & 1.0 + 0(-1) \end{bmatrix}$$

$$= \begin{bmatrix} 0 & i \\ i & 0 \end{bmatrix} = C$$

Hence $AB = -C = -BA$.

Example 19:

Given $A_i = \begin{bmatrix} 0 & 0 & 0 & 1 \\ 0 & 0 & 1 & 0 \\ 0 & 1 & 0 & 0 \\ 1 & 0 & 0 & 0 \end{bmatrix}$; $A_2 = \begin{bmatrix} 0 & 0 & 0 & i \\ 0 & 0 & -i & 0 \\ 0 & i & 0 & 0 \\ -i & 0 & 0 & 0 \end{bmatrix}$

$A_2 = \begin{bmatrix} 0 & 0 & 1 & 0 \\ 0 & 0 & 0 & -1 \\ 1 & 0 & 0 & 0 \\ 0 & -1 & 0 & 0 \end{bmatrix}$ and $A_4 = \begin{bmatrix} 1 & 0 & 0 & 0 \\ 0 & 1 & 0 & 0 \\ 0 & 0 & -1 & 0 \\ 0 & 0 & 0 & -1 \end{bmatrix}$

Show that $A_i A_k + A_k A_i = 2I$ *or O according as* $i = k$ *or* $i \neq k$ *and I is the matrix of order 4 and i and k take the values 1, 2, 3 and 4.*

Solution:

Let i = k = 1 (say). Then

$A_iA_k = A_1A_2 = A_kA_i$

$\therefore A_iA_k = A_1A_1 = \begin{bmatrix} 0 & 0 & 0 & 1 \\ 0 & 0 & 1 & 0 \\ 0 & 1 & 0 & 0 \\ 1 & 0 & 0 & 0 \end{bmatrix} \times \begin{bmatrix} 0 & 0 & 0 & 1 \\ 0 & 0 & 1 & 0 \\ 0 & 1 & 0 & 0 \\ 1 & 0 & 0 & 0 \end{bmatrix}$

$= \begin{bmatrix} 0+0+0+1 & 0+0+0+0 & 0+0+0+0 & 0+0+0+0 \\ 0+0+0+0 & 0+0+1+0 & 0+0+0+0 & 0+0+0+0 \\ 0+0+0+0 & 0+0+0+0 & 0+1+0+0 & 0+0+0+0 \\ 0+0+0+0 & 0+0+0+0 & 0+0+0+0 & 1+0+0+0 \end{bmatrix}$

$= \begin{bmatrix} 1 & 0 & 0 & 0 \\ 0 & 1 & 0 & 0 \\ 0 & 0 & 1 & 0 \\ 0 & 0 & 0 & 1 \end{bmatrix} = I$

$\therefore \quad A_iA_k + A_kA_i = I + I = 2I$ **Hence proved.**

If $i \neq k$, let i = 3 and k = 2

Then $A_iA_k = A_3A_2 = \begin{bmatrix} 0 & 0 & 1 & 0 \\ 0 & 1 & 0 & -1 \\ 1 & 0 & 0 & 0 \\ 0 & -1 & 0 & 0 \end{bmatrix} \times \begin{bmatrix} 0 & 0 & 0 & i \\ 0 & 0 & -i & 0 \\ 0 & i & 0 & 0 \\ -i & 0 & 0 & 0 \end{bmatrix}$

$$= \begin{bmatrix} 0+0+0+0 & 0+0+i+0 & 0+0+0+0 & 0+0+0+0 \\ 0+0+0+i & 0+0+0+0 & 0+0+0+0 & 0+0+0+0 \\ 0+0+0+0 & 0+0+0+0 & 0+0+0+0 & i+0+0+0 \\ 0+0+0+0 & 0+0+0+0 & 0+i+0+0 & 0+0+0+0 \end{bmatrix}$$

$$= \begin{bmatrix} 0 & i & 0 & 0 \\ i & 0 & 0 & 0 \\ 0 & 0 & 0 & i \\ 0 & 0 & i & 0 \end{bmatrix} = i \begin{bmatrix} 0 & 1 & 0 & 0 \\ 1 & 0 & 0 & 0 \\ 0 & 0 & 0 & 1 \\ 0 & 0 & 1 & 0 \end{bmatrix}$$

And $A_k A_i = A_2 A_3 = \begin{bmatrix} 0 & 0 & 0 & i \\ 0 & 0 & -i & 0 \\ 0 & i & 0 & 0 \\ -i & 0 & 0 & 0 \end{bmatrix} \times \begin{bmatrix} 0 & 0 & 1 & 0 \\ 0 & 0 & 0 & -1 \\ 1 & 0 & 0 & 0 \\ 0 & -1 & 0 & 0 \end{bmatrix}$

$= \begin{bmatrix} 0 & -i & 0 & 0 \\ -i & 0 & 0 & 0 \\ 0 & 0 & 0 & -i \\ 0 & 0 & -i & 0 \end{bmatrix}$ Multiplying in the usual way

$$= -i \begin{bmatrix} 0 & 1 & 0 & 0 \\ 1 & 0 & 0 & 0 \\ 0 & 0 & 0 & 1 \\ 0 & 0 & 1 & 0 \end{bmatrix}$$

$$\therefore A_i A_k + A_k A_i = i \begin{bmatrix} 0 & 1 & 0 & 0 \\ 1 & 0 & 0 & 0 \\ 0 & 0 & 0 & 1 \\ 0 & 0 & 1 & 0 \end{bmatrix} - i \begin{bmatrix} 0 & 1 & 0 & 0 \\ 1 & 0 & 0 & 0 \\ 0 & 0 & 0 & 1 \\ 0 & 0 & 1 & 0 \end{bmatrix}$$

$= O$ **Hence proved.**

We can in a similar way prove the above result by giving i and k other values also.

Example 20:

If $A = \begin{bmatrix} 2 & 3 & 1 \\ 0 & -1 & 5 \end{bmatrix}$ and $B = \begin{bmatrix} 1 & 2 & -6 \\ 0 & -1 & 3 \end{bmatrix}$ *evaluate (a)* $A^2 - B^2$ *and (b) AB and BA.*

Solution:

(a) $A^2 = \begin{bmatrix} 2 & 3 & 1 \\ 0 & -1 & 5 \end{bmatrix} \times \begin{bmatrix} 2 & 3 & 1 \\ 0 & -1 & 5 \end{bmatrix}$, which does not exist as number of columns in the first matrix is not equal to number of rows in the second matrix.

Similarly B^2 does not exist.

(b) AB and BA both do not exist, the reason being the same as in part (a) above.

Example 21:

If $A = \begin{bmatrix} \cos\theta & -\sin\theta \\ \sin\theta & \cos\theta \end{bmatrix}$, $B = \begin{bmatrix} \cos\phi & -\sin\phi \\ \sin\phi & \cos\phi \end{bmatrix}$ *show that AB = BA.*

Solution:

$$AB = \begin{bmatrix} \cos\theta & -\sin\theta \\ \sin\theta & \cos\theta \end{bmatrix} \times \begin{bmatrix} \cos\phi & -\sin\phi \\ \sin\phi & \cos\phi \end{bmatrix}$$

$$= \begin{bmatrix} \cos\theta\cos\phi - \sin\theta\sin\phi & -\cos\theta\sin\phi - \sin\theta\cos\phi \\ \sin\theta\cos\phi + \cos\theta\sin\phi & -\sin\theta\sin\phi + \cos\theta\cos\phi \end{bmatrix}$$

$$= \begin{bmatrix} \cos(\theta+\phi) & -\sin(\theta+\phi) \\ \sin(\theta+\phi) & \cos(\theta+\phi) \end{bmatrix} \qquad ...(1)$$

$$\text{And } BA = \begin{bmatrix} \cos\phi & -\sin\phi \\ \sin\phi & \cos\phi \end{bmatrix} \times \begin{bmatrix} \cos\theta & -\sin\theta \\ \sin\theta & \cos\theta \end{bmatrix}$$

$$= \begin{bmatrix} \cos\phi\cos\theta - \sin\phi\sin\theta & -\cos\phi\sin\theta - \sin\phi\cos\theta \\ \sin\phi\cos\theta + \cos\phi\sin\theta & -\sin\phi\sin\theta + \cos\phi\cos\theta \end{bmatrix}$$

$$= \begin{bmatrix} \cos(\theta+\phi) & -\sin(\theta+\phi) \\ \sin(\theta+\phi) & \cos(\theta+\phi) \end{bmatrix}$$

∴ From (1) and (2) we get AB = BA. **Hence proved.**

Example 22:

Prove that the product of two matrices

$$\begin{bmatrix} \cos^2\theta & \cos\theta\sin\theta \\ \cos\theta\sin\theta & \sin^2\theta \end{bmatrix} \text{ and } \begin{bmatrix} \cos^2\phi & \cos\phi\sin\phi \\ \cos\phi\sin\phi & \sin^2\phi \end{bmatrix}$$

is zero when θ aznd ϕ differ by an odd multiple of $\frac{1}{2}\pi$.

Solution:

The required product

$$= \begin{bmatrix} \cos^2\theta\cos^2\phi + \cos\theta\sin\theta\cos\phi\sin\phi & \cos^2\theta\cos\phi + \cos\theta\sin\theta\cos\phi\sin^2\phi \\ \cos\theta\sin\theta\cos^2\phi + \sin^2\theta\cos\phi\sin\phi & \cos\theta\sin\theta\cos\phi\sin\phi + \sin^2\theta\sin\phi \end{bmatrix}$$

$$= \begin{bmatrix} \cos\theta\cos\phi(\cos\theta\cos\phi + \sin\theta\sin\phi) & \cos\theta\sin\phi(\cos\theta\cos\phi + \sin\theta\sin\phi) \\ \sin\theta\cos\phi(\cos\theta\cos\phi + \sin\theta\sin\phi) & \sin\theta\sin\phi(\cos\theta\cos\phi + \sin\theta\sin\phi) \end{bmatrix}$$

$$= \begin{bmatrix} \cos\theta\cos\phi\cos(\theta \sim \phi) & \cos\theta\sin\phi\cos(\theta \sim \phi) \\ \sin\theta\cos\phi\cos(\theta \sim \phi) & \sin\theta\sin\phi\cos(\theta \sim \phi) \end{bmatrix}$$

If $\theta \sim \phi$ = an odd multiple of $\frac{1}{2}\pi$, then $\cos(\theta \sim \phi) = 0$ and consequently the above product is zero (*i.e.*, the null matrix of order 2×2).

1.16 PROPERTIES OF MULTIPLICATION OF MATRICES

Property I: Multiplication of Matrices is Associative

Let $A = [a_{ij}]$, $B = [b_{jk}]$ and $C = [c_{kr}]$ be three $m \times n$, $n \times p$ and $p \times 1$ matrices respectively, then (AB). C = A. (BC).

Proof:

Let $AB = [d_{ik}]$, where $d_{ik} = \sum_{j=1}^{n} a_{ij} b_{jk}$ ...(1)

Then $(AB).\ C = [d_{ik}] \times [c_{kr}] = [e_{ir}]$,

Where $$e_{ir} = \sum_{k=1}^{p} d_{ik} b_{kr}$$

$$= \sum_{k=1}^{p}\left(\sum_{j=1}^{p} a_{ij}\, b_{jk}\right) . C_{jr}, \text{ from (1)}$$

i.e., (i, r)th element of (AB). C $= \sum_{k=1}^{p}\sum_{j=1}^{n} a_{ij}\, b_{jk}\, c_{kr}$ ...(2)

And let BC $= [g_{jr}]$, $g_{jr} = \sum_{k=1}^{p} b_{jk}\, c_{kr}$ ...(3)

Then A. BC $= [a_{ij}] \times [g_{jr}] = [h_{ir}]$,

Where $h_{ir} = \sum_{j=1}^{n} a_{ij}\, g_{jr}$

$$= \sum_{j=1}^{n} a_{ij}\left(\sum_{k=1}^{p} b_{jk}\, c_{kr}\right), \text{ form (3)}$$

i.e., (i, r)th element of A. (BC) $= \sum_{k=1}^{p}\sum_{j=1}^{n} a_{ij}\, b_{jk}\, c_{kr}$, ...(4)

since the summation can be interchanged.

A From (3) and (4) we can conclude the (i, r)th elements of (AB)· C and A· (BC) are the same and their orders are also m × *l*.

Hence (AB)· C = A· (BC).

Property II: Multiplication of Matrices is Disributive with Respect to Matrix Addition

(a) Let A $= [a_{ij}]$, B $= [b_{jk}]$ and C $= [c_{jk}]$ be three m × n, n × p and n × p matrices respectively, then

A (B + C) = AB + AC

Proof:

A (B + C) $= [a_{ij}] \times \{[b_{jk}] + [c_{jk}]\}$

$= [a_{ij}]\, [b_{jk} + c_{jk}] = [b_{ik}]$, say,

Where $d_{ik} = \sum_{j=1}^{n} a_{ij}\, (b_{jk} + c_{jk})$

⇒ (i, k)th element of A (B + C) $= \sum_{j=1}^{n} a_{ij}\, b_{jk} + \sum_{j=1}^{n} a_{ij}\, c_{jk}$ ...(1)

Again AB $= [a_{ij}]\, [b_{jk}] = [e_{ik}]$ say,

Where $e_{ik} = \sum_{j=1}^{n} a_{ij}\, b_{jk}$ *i.e.,*)i, k)th element of AB $= \sum_{i=1}^{n} a_{ij}\, b_{jk}$ ...(2)

Similarly we can prove that

$$\text{(i, k)th element of AC} = \sum_{j=1}^{n} a_{ij}\, c_{jk} \qquad ...(3)$$

$\therefore$ From (2) and (3) we have

$$\text{(i, k)th element o AB + AC} = \sum_{j=1}^{n} a_{ij}\, b_{jk} + \sum_{j=1}^{n} a_{ij}\, c_{jk} \qquad ...(4)$$

Hence from (1) and (4) we conclude that A (B + C) = AB + AC.

(b) Let $A = [a_{ij}]$, $B = [b_{jk}]$ and $C = [c_{jk}]$ be three $n \times p$, $m \times n$ and $m \times n$ matrices respectively.

The (B + C) = A = BA + CA.

(Note: If A and be $m \times n$ and $n \times p$ matrices then BA can not exist whereas AB exists).

Proof:

Its proof is simlar to that of part (a)

Positive Integral Power of a Square Matrix

We find that if A is square matrix, then only the product AA is defined and we write A^2 for AA. Also by associative law

$$A^2A = (AA)\,A = A\,(AA) = AA^2$$

So A^2A or AA^2 is written as A^3.

In general AAA...A is denoted by A^n if there are n factors.

Definition: *In general AAA...A upto m factors is denoted by* A^m.

Theorem 1:

If A be a square matrix (n × n say), then

$A^p \cdot A^q = A^{p \cdot q}$ *, for any pair of positive integers p and q.*

Proof:

We shall prove that by the method of induction.

From definition we know that $A^p \cdot A = A^{p+1}$, where p is any positive integer.

$\therefore \quad A^p \cdot A^q = A^{p+q}$ holds when $q = 1$, whatever p may be.

We shall now prove that if it holds for a particular value m say of q, for all values of p.

Now $A^p \cdot A^{m+1} = A^p \cdot (A^m \cdot A)$, by definition given above

$= (A^p \cdot A^m) \cdot A$, by associative law

$= (A^{p+m}) \cdot A$, by hypothesis

$= A^{p+m+1}$, by definition given above

$= A^{p+(m+1)}$, by associative law of addition of numbers.

i.e., $\quad A^p \cdot A^q = A^{p+q}$ holds for the value $m + 1$ of q, whatever p may be if it holds for $q = m$.

Hence the result is established by mathematical induction.

Theorem 2:

If A be a square matrix, then

$(A^p)^q = A^{pq}$, *for every pair of positive integers p and q.*

Proof:

The proof is similar to the above theorem.

Example 1:

If $I = \begin{bmatrix} 1 & 0 \\ 0 & 1 \end{bmatrix}$ *and* $E = \begin{bmatrix} 0 & 1 \\ 0 & 0 \end{bmatrix}$, *Prove that*

$(aI + bE)^3 = a^3I + 3a^2bE.$

Solution:

$$aI + bE = a\begin{bmatrix} 1 & 0 \\ 0 & 1 \end{bmatrix} + b\begin{bmatrix} 0 & 1 \\ 0 & 0 \end{bmatrix}$$

$$= \begin{bmatrix} a & 0 \\ 0 & a \end{bmatrix} + \begin{bmatrix} 0 & b \\ 0 & 0 \end{bmatrix} = \begin{bmatrix} a+0 & 0+b \\ 0+0 & a+0 \end{bmatrix}$$

$$= \begin{bmatrix} a & b \\ 0 & a \end{bmatrix} = B \text{ (say)}$$

$\therefore \quad (aI + bE)^2 = B^2 = \begin{bmatrix} a & b \\ 0 & a \end{bmatrix} \times \begin{bmatrix} a & b \\ 0 & a \end{bmatrix}$

$$= \begin{bmatrix} a.a + b.0 & a.b + b.a \\ 0.a + a.0 & 0.b + a.a \end{bmatrix} = \begin{bmatrix} a^2 & 2ab \\ 0 & a^2 \end{bmatrix}$$

$\therefore (aI + BE)^3 = B^3 = B^2B$ **(Note)**

$$= \begin{bmatrix} a^2 & 2ab \\ 0 & a^2 \end{bmatrix} \times \begin{bmatrix} a & b \\ 0 & a \end{bmatrix}$$

$$= \begin{bmatrix} a^2.a + 2ab.0 & a^2.b + 2ab.a \\ 0.a + a^2.0 & 0.b + a^2.a \end{bmatrix} = \begin{bmatrix} a^3 & 3a^2b \\ 0 & a^3 \end{bmatrix} \quad ...(1)$$

Now $a^3I + 3a^2bE = a^2 \begin{bmatrix} 1 & 0 \\ 0 & 1 \end{bmatrix} + 3a^2b \begin{bmatrix} 0 & 1 \\ 0 & 0 \end{bmatrix}$

$$= \begin{bmatrix} a^3 & 0 \\ 0 & a^3 \end{bmatrix} + \begin{bmatrix} 0 & 3a^2b \\ 0 & 0 \end{bmatrix}$$

$$= \begin{bmatrix} a^3 + 0 & 0 + 3a^2b \\ 0 + 0 & a^3 + 0 \end{bmatrix} = \begin{bmatrix} a^3 & 3a^2b \\ 0 & a^2 \end{bmatrix} \quad ...(2)$$

$\therefore$ From (1) and (2) we get $(aI + bE)^3 = a^3I + 3a^2bE$.

Example 2:

If $A = \begin{bmatrix} 1 & -1 \\ -1 & 1 \end{bmatrix}$, *then show that* $A^2 = 2A$ *and* $A^3 = 4A$.

Solution:

Given $A = \begin{bmatrix} 1 & -1 \\ -1 & 1 \end{bmatrix}$ ...(1)

$\therefore A^2 = A \cdot A = \begin{bmatrix} 1 & -1 \\ -1 & 1 \end{bmatrix} \times \begin{bmatrix} 1 & -1 \\ -1 & 1 \end{bmatrix}$

$$= \begin{bmatrix} 1.1 + (-1).(-1) & 1.(-1) + (-1).1 \\ (-1).1 + 1.(-1) & (-1).(-1) + 1.1 \end{bmatrix} = \begin{bmatrix} 2 & -2 \\ -2 & 2 \end{bmatrix}$$

$$= 2 \begin{bmatrix} 1 & -1 \\ -1 & 1 \end{bmatrix} = 2A, \text{ from (1)} \qquad ...(2)$$

Again $A^3 = A \cdot A^2 = A \cdot (2A)$, from (2)

$= 2A \cdot A = 2A^2 = 2\ (2A)$, from (2)

$= 4A$. **Hence proved.**

Example 3:

If A denoteds the matrix $\begin{bmatrix} a & b \\ c & d \end{bmatrix}$, *prove that* $A^2-(a+d)A+(ad-bc)\,I = O$.

Solution:

$$A^2 = \begin{bmatrix} a & b \\ c & d \end{bmatrix} \times \begin{bmatrix} a & b \\ c & d \end{bmatrix}$$

$$= \begin{bmatrix} a.a + b.c & a.b + b.d \\ c.a + d.c & c.d + d.d \end{bmatrix} = \begin{bmatrix} a^2 + bc & b(a + d) \\ c(a + d) & cb + d^2 \end{bmatrix}$$

$\therefore \quad A^2 - (a + d)\ A + (ad - dc)\ I$

$$= \begin{bmatrix} a^2 + bc & b(a+d) \\ c(a+d) & cb + d^2 \end{bmatrix} - (a + d) \begin{bmatrix} a & b \\ c & d \end{bmatrix} + (ad - bc) \begin{bmatrix} 1 & 0 \\ 0 & 1 \end{bmatrix}$$

$$= \begin{bmatrix} a^2 + bc & b(a+d) \\ c(a+d) & cb + d^2 \end{bmatrix} + \begin{bmatrix} -a(a+d) & -b(a+d) \\ -c(a+d) & -d(a+d) \end{bmatrix}$$

$$+ \begin{bmatrix} ad - bc & 0 \\ 0 & ad - bc \end{bmatrix}$$

$$= \begin{bmatrix} a^2 + bc - a(a+d) + ad - bc & b(a+d) - b(a+d) + 0 \\ c(c+d) - c(a+d) + 0 & cb + d^2 - d(a+d) + ad - bc \end{bmatrix}$$

$$= \begin{bmatrix} 0 & 0 \\ 0 & 0 \end{bmatrix} = O, \text{ where O is the } 2 \times 2 \text{ null matrix.}$$

Example 4:

Show that $\begin{bmatrix} \cos\theta & -\sin\theta \\ \sin\theta & \cos\theta \end{bmatrix}^n = \begin{bmatrix} \cos n\theta & -\sin n\theta \\ \sin n\theta & \cos n\theta \end{bmatrix}$ *where n is a positive integer.*

Solution:

Let $$A = \begin{bmatrix} \cos\theta & -\sin\theta \\ \sin\theta & \cos\theta \end{bmatrix} \quad ...(1)$$

Then $$(A^2) = A.A = \begin{bmatrix} \cos\theta & -\sin\theta \\ \sin\theta & \cos\theta \end{bmatrix} \times \begin{bmatrix} \cos\theta & -\sin\theta \\ \sin\theta & \cos\theta \end{bmatrix}$$

$$= \begin{bmatrix} \cos^2\theta - \sin^2\theta & -\sin\theta\cos\theta - \sin\theta\cos\theta \\ \sin\theta\cos\theta + \sin\theta\cos\theta & -\sin^2\theta + \cos^2\theta \end{bmatrix}$$

$$\Rightarrow \quad A^2 = \begin{bmatrix} \cos 2\theta & -\sin 2\theta \\ \sin 2\theta & \cos 2\theta \end{bmatrix} \quad ...(2)$$

Similarly $(A)^3 = (A)^2.\ A$

$$= \begin{bmatrix} \cos 2\theta & -\sin 2\theta \\ \sin 2\theta & \cos 2\theta \end{bmatrix} \times \begin{bmatrix} \cos\theta & -\sin\theta \\ \sin\theta & \cos\theta \end{bmatrix}, \text{ from (1)}$$

$$= \begin{bmatrix} \cos 2\theta\cos\theta - \sin 2\theta\sin\theta & -\cos 2\theta\sin\theta - \sin 2\theta\cos\theta \\ \sin 2\theta\cos\theta + \cos 2\theta\sin\theta & -\sin 2\theta\sin\theta + \cos 2\theta\cos\theta \end{bmatrix}$$

$$= \begin{bmatrix} \cos(2\theta+\theta) & -\sin(2\theta+\theta) \\ \sin(2\theta+\theta) & \cos(2\theta+\theta) \end{bmatrix}$$

$$\Rightarrow \quad (A)^3 = \begin{bmatrix} \cos 3\theta & -\sin 3\theta \\ \sin 3\theta & \cos 3\theta \end{bmatrix} \quad ...(3)$$

In the light of (1), (2) and (3) let us assume that

$$(A)^n = \begin{bmatrix} \cos n\theta & -\sin n\theta \\ \sin n\theta & \cos n\theta \end{bmatrix} \quad ...(4)$$

Now $(A)^{n+1} = (A)^n \cdot (A)$

$$= \begin{bmatrix} \cos n\theta & -\sin n\theta \\ \sin n\theta & \cos n\theta \end{bmatrix} \times \begin{bmatrix} \cos \theta & -\sin \theta \\ \sin \theta & \cos \theta \end{bmatrix}$$

$$= \begin{bmatrix} \cos n\theta \cos \theta - \sin n\theta \sin \theta & -\cos n\theta \sin \theta - \sin n\theta \cos \theta \\ \sin n\theta \cos \theta + \cos n\theta \sin \theta & -\sin n\theta \sin \theta + \cos n\theta \cos \theta \end{bmatrix}$$

$$= \begin{bmatrix} \cos (n\theta + \theta) & -\sin (n\theta + \theta) \\ \sin (n\theta + \theta) & \cos (n\theta + \theta) \end{bmatrix} = \begin{bmatrix} \cos (n+1)\theta & -\sin (n+1)\theta \\ \sin (n+1)\theta & \sin (n+1)\theta \end{bmatrix}$$

i.e., (4) holds for n + 1 if it is true for n.

We have already proved in (2) and (3) that (4) holds for n = 2 and 3. Hence (4) holds for all positive integral values of n.

i.e., $(A)^n = \begin{bmatrix} \cos \theta & -\sin \theta \\ \sin \theta & \cos \theta \end{bmatrix}^n \begin{bmatrix} \cos n\theta & -\sin n\theta \\ \sin n\theta & \sin n\theta \end{bmatrix}$ **Hence proved.**

Example 5:

If $A = \begin{bmatrix} 3 & -4 \\ 1 & -1 \end{bmatrix}$, *show that* $A^n = \begin{bmatrix} 1+2n & -4n \\ n & 1-2n \end{bmatrix}$

Solution:

$$A^2 = A \cdot A = \begin{bmatrix} 3 & -4 \\ 1 & -1 \end{bmatrix} \times \begin{bmatrix} 3 & -4 \\ 1 & -1 \end{bmatrix}$$

$$= \begin{bmatrix} 3.3 - 4(1) & 3.(-4) - 4.(-1) \\ 1.3 - 1.(1) & 1.(-4) - 1.(-1) \end{bmatrix} = \begin{bmatrix} 5 & -8 \\ 2 & -3 \end{bmatrix}$$

$$= \begin{bmatrix} 1 + 2(2) & -4n(2) \\ (2) & 1 - 2(2) \end{bmatrix}$$

$= A^n$, when n = 2 **(Note)**

$\therefore A^n = \begin{bmatrix} 1+2n & -4n \\ n & 1-2n \end{bmatrix}$ holds when n = 2

Now $A^{n+1} = A^n.A$

$$= \begin{bmatrix} 1+2n & -4n \\ n & 1-2n \end{bmatrix} \cdot \begin{bmatrix} 3 & -4 \\ 1 & -1 \end{bmatrix}$$

$$= \begin{bmatrix} (1+2n).3 - 4n(1) & (1+2n)(-4) - 4n(-1) \\ n.3 + (1-2n)(1) & n(-4) + (1-2n)(-1) \end{bmatrix}$$

$$= \begin{bmatrix} 3 + 2n & -4 - 4n \\ 1 + n & -1 - 2n \end{bmatrix} = \begin{bmatrix} 1 + 2(n+1) & -4(n+1) \\ (n+1) & 1-2(n+1) \end{bmatrix}$$

i.e., $A^n = \begin{bmatrix} 1 + 2n & -4n \\ n & 1 - 2n \end{bmatrix}$ holds for n + 1.

Also we have shown above that it holds for n = 2.

Hence it is true for all positive integral values of n.

Example 6:

Let $A = \begin{bmatrix} a & b \\ 0 & 1 \end{bmatrix}$, *where* $a \neq 0$. *Show that for* $n \geq 0$,

$$A^n = \begin{bmatrix} a^n & \dfrac{b(a^n - 1)}{(a-1)} \\ 0 & 1 \end{bmatrix}$$

Solution:

$$A^2 = A.A = \begin{bmatrix} a & b \\ 0 & 1 \end{bmatrix} \times \begin{bmatrix} a & b \\ 0 & 1 \end{bmatrix}$$

$$= \begin{bmatrix} a.a + b.0 & a.b + b.1 \\ 0.a + 1.0 & 0.b + 1.1 \end{bmatrix} = \begin{bmatrix} a^2 & b(a+1) \\ 0 & 1 \end{bmatrix}$$

$$= \begin{bmatrix} a^2 & \dfrac{b(a^2 - 1)}{(a-1)} \\ 0 & 1 \end{bmatrix} = A^n, \text{ when } n = 2.$$ **(Note)**

$\therefore$ $A^n = \begin{bmatrix} a^n & b(a^n - 1)/(a-1) \\ 0 & 1 \end{bmatrix}$ holds when n = 2.

Now $A^{n+1} = A^n. A$.

$$= \begin{bmatrix} a^n & b(a^n - 1)/(a-1) \\ 0 & 1 \end{bmatrix} \times \begin{bmatrix} a & b \\ 0 & 1 \end{bmatrix}$$

$$= \begin{bmatrix} a^n.a + 0 & a^n.b + 1.\{b(a^n - 1)/(a-1) \\ 0 + 0 & 0 + 1 \end{bmatrix}$$

$$= \begin{bmatrix} a^{n+1} & b\{a^n(a-1) + (a-1)\}/(a-1) \\ 0 & 1 \end{bmatrix}$$

$$= \begin{bmatrix} a^{n+1} & b(a^{n+1} - 1)/(a-1) \\ 0 & 1 \end{bmatrix}$$

$\therefore \quad A^n = \begin{bmatrix} a^n & b(a^n - 1)/(a-1) \\ 0 & 1 \end{bmatrix}$ holds for n = 1.

Also we have shown above that it holds for n = 2.

Hence it is true for all positive integral values of $n \geq 0$.

Hence proved.

Example 7:

Show that if $A = \begin{bmatrix} \cosh\theta & \sinh\theta \\ \sinh\theta & \cosh\theta \end{bmatrix}$, *then* $A^n = \begin{bmatrix} \cosh n\theta & \sinh n\theta \\ \sinh n\theta & \cosh n\theta \end{bmatrix}$

Solution:

Here $A^2 = A \cdot A$

$$= \begin{bmatrix} \cosh\theta & \sinh\theta \\ \sinh\theta & \cosh\theta \end{bmatrix} \times \begin{bmatrix} \cosh\theta & \sinh\theta \\ \sinh\theta & \cosh\theta \end{bmatrix}$$

$$= \begin{bmatrix} \cosh^2\theta + \sinh^2\theta & \cosh\theta\sinh\theta + \sinh\theta\cosh\theta \\ \text{Sinh}\,\theta\cosh\theta + \cosh\theta\sinh\theta & \sinh^2\theta + \cosh^2\theta \end{bmatrix}$$

$$\Rightarrow \quad A^2 = \begin{bmatrix} \cosh 2\theta & \sinh 2\theta \\ \sinh 2\theta & \cosh 2\theta \end{bmatrix} \quad ...(1)$$

Similarly $A^3 = A^2 \cdot A$

$$= \begin{bmatrix} \cosh 2\theta & \sinh 2\theta \\ \sinh 2\theta & \cosh 2\theta \end{bmatrix} \times \begin{bmatrix} \cosh \theta & \sinh \theta \\ \sinh \theta & \cosh \theta \end{bmatrix}, \text{ from (1)}$$

$$= \begin{bmatrix} \cosh 2\theta \cosh \theta + \sinh 2\theta \sinh \theta & \cosh 2\theta \sinh \theta + \sinh 2\theta \cosh \theta \\ \sinh 2\theta \cosh \theta + \cosh 2\theta \sinh \theta & \sinh 2\theta \sinh \theta + \cosh 2\theta \cosh \theta \end{bmatrix}$$

$$= \begin{bmatrix} \cosh (2\theta+\theta) & \sinh (2\theta+\theta) \\ \sinh (2\theta+\theta) & \cosh (2\theta+\theta) \end{bmatrix}$$

$$\Rightarrow \quad A^3 = \begin{bmatrix} \cosh 3\theta & \sinh 3\theta \\ \sinh 3\theta & \cosh 3\theta \end{bmatrix} \qquad \text{....(1)}$$

In the light of (1), (2) and the given value of A, let us assume that

$$A^n = \begin{bmatrix} \cosh n\theta & \sinh n\theta \\ \sinh n\theta & \cosh n\theta \end{bmatrix} \qquad \text{...(2)}$$

Now $A^{n+1} = A^n \cdot A$

$$= \begin{bmatrix} \cosh n\theta & \sinh n\theta \\ \sinh n\theta & \cosh n\theta \end{bmatrix} \times \begin{bmatrix} \cosh \theta & \sinh \theta \\ \sinh \theta & \cosh \theta \end{bmatrix}$$

$$= \begin{bmatrix} \text{csoh } n\theta \cosh \theta + \sinh n\theta \sinh \theta & \cosh n\theta \sinh \theta + \sinh n\theta \cosh \theta \\ \sinh n\theta \cosh \theta + \cosh n\theta \sinh \theta & \sinh n\theta \sinh \theta + \cosh n\theta \cosh \theta \end{bmatrix}$$

$$= \begin{bmatrix} \cosh (n\theta+\theta) & \sinh (n\theta+\theta) \\ \sinh (n\theta+\theta) & \cosh (n\theta+\theta) \end{bmatrix}$$

$$= \begin{bmatrix} \cosh (n+1)\theta & \sinh (n+1)\theta \\ \sinh (n+1)\theta & \cosh (n+1)\theta \end{bmatrix}$$

i.e., (3) holds for n + 1 if it is true for n.

Also from (1) and (2) we know that (3) holds for n = 2 and n = 3. Hence (3) holds for all positive integral values of n.

i.e., $A^n = \begin{bmatrix} \cosh n\theta & \sinh n\theta \\ \sinh n\theta & \cosh n\theta \end{bmatrix}$ **Hence proved**

Example 8:

If $A = \begin{bmatrix} 2 & -1 \\ 0 & 1 \end{bmatrix}$ and $B = \begin{bmatrix} 1 & 0 \\ -1 & -1 \end{bmatrix}$, *show that*

$(A + B)^2 = A^2 + AB + BA + B^2 \; ' \; A^2 + 2AB + B^2$

Solution:

$$A^2 = \begin{bmatrix} 2 & -1 \\ 0 & 1 \end{bmatrix} \times \begin{bmatrix} 2 & -1 \\ 0 & 1 \end{bmatrix}$$

$$= \begin{bmatrix} 2.2 + (-1).0 & 2.(-1) - 1.1 \\ 0.2 + 1.0 & 0.(-1) + 1.1 \end{bmatrix} = \begin{bmatrix} 4 & -3 \\ 0 & -1 \end{bmatrix}$$

$$AB = \begin{bmatrix} 2 & -1 \\ 0 & 1 \end{bmatrix} \times \begin{bmatrix} 1 & 0 \\ -1 & -1 \end{bmatrix}$$

$$= \begin{bmatrix} 2.2 - 1(-1) & 2.0 - 1(-1) \\ 0.1 + 1(-1) & 0.0 + 1(-1) \end{bmatrix} = \begin{bmatrix} 3 & 1 \\ -1 & -1 \end{bmatrix};$$

$$BA = \begin{bmatrix} 1 & 0 \\ -1 & -1 \end{bmatrix} \times \begin{bmatrix} 0 & -1 \\ 0 & 1 \end{bmatrix}$$

$$= \begin{bmatrix} 1.2 + 0.0 & 1(-1) + 0.1 \\ -1.2 - 1.0 & -1(-1) - 1.1 \end{bmatrix} = \begin{bmatrix} 2 & -1 \\ -2 & 0 \end{bmatrix}$$

$$B^2 = \begin{bmatrix} 1 & 0 \\ -1 & -1 \end{bmatrix} \times \begin{bmatrix} 1 & 0 \\ -1 & -1 \end{bmatrix}$$

$$= \begin{bmatrix} 1.1 + 0(-1) & 1.0 + 0(-1) \\ -1.1 - 1(-1) & -1.0 - 1(-1) \end{bmatrix} = \begin{bmatrix} 1 & 0 \\ 0 & 1 \end{bmatrix}$$

$$A + B = \begin{bmatrix} 2 & -1 \\ 0 & 1 \end{bmatrix} + \begin{bmatrix} 1 & 0 \\ -1 & -1 \end{bmatrix}$$

$$= \begin{bmatrix} 2+1 & -1+0 \\ 0-1 & 1-1 \end{bmatrix} = \begin{bmatrix} 3 & -1 \\ -1 & 0 \end{bmatrix}$$

$$(A + B)^2 = \begin{bmatrix} 3 & -1 \\ -1 & 0 \end{bmatrix} \times \begin{bmatrix} 3 & -1 \\ -1 & 0 \end{bmatrix}$$

$$= \begin{bmatrix} 3.3-1(-1) & 3(-1)-1.0 \\ -1.3+0(-1) & -1(-1)+0.0 \end{bmatrix} = \begin{bmatrix} 10 & -3 \\ -3 & 1 \end{bmatrix} \quad ...(1)$$

Now $A^2 + AB + BA + B^2$

$$= \begin{bmatrix} 4 & -3 \\ 0 & 1 \end{bmatrix} + \begin{bmatrix} 3 & 1 \\ -1 & -1 \end{bmatrix} + \begin{bmatrix} 2 & -1 \\ -2 & 0 \end{bmatrix} + \begin{bmatrix} 1 & 0 \\ 0 & 1 \end{bmatrix}$$

$$= \begin{bmatrix} 4+3+2+1 & -3+1-1+0 \\ 0-1-2+0 & 1-1+0+1 \end{bmatrix} = \begin{bmatrix} 10 & -3 \\ -3 & 1 \end{bmatrix}$$

$= (A + B)^2$, from (i) **Hence proved.**

Also $A^2 + 2AB + B^2$

$$= \begin{bmatrix} 4 & -3 \\ 0 & 1 \end{bmatrix} + 2\begin{bmatrix} 3 & 1 \\ -1 & -1 \end{bmatrix} + \begin{bmatrix} 1 & 0 \\ 0 & 1 \end{bmatrix}$$

$$= \begin{bmatrix} 4 & -3 \\ 0 & 1 \end{bmatrix} + \begin{bmatrix} 6 & 2 \\ -2 & -2 \end{bmatrix} + \begin{bmatrix} 1 & 0 \\ 0 & 1 \end{bmatrix}$$

$$= \begin{bmatrix} 4+6+1 & -3+2+0 \\ 0-2+0 & 1-2+1 \end{bmatrix} = \begin{bmatrix} 11 & -1 \\ -2 & 0 \end{bmatrix} \neq (A + B)^2$$

Hence proved

Example 9:

Evaluate $A^2 - 4A - 5I$, where

$$A = \begin{bmatrix} 1 & 2 & 2 \\ 2 & 1 & 2 \\ 2 & 2 & 1 \end{bmatrix} \text{ and } I = \begin{bmatrix} 1 & 0 & 0 \\ 0 & 1 & 0 \\ 0 & 0 & 1 \end{bmatrix}$$

Solution:

$$A^2 = \begin{bmatrix} 1 & 2 & 2 \\ 2 & 1 & 2 \\ 2 & 2 & 1 \end{bmatrix} \times \begin{bmatrix} 1 & 2 & 2 \\ 2 & 1 & 2 \\ 2 & 2 & 1 \end{bmatrix}$$

$$= \begin{bmatrix} 1.1 + 2.2 + 2.2 & 1.2 + 2.1 + 2.2. & 1.2 + 2.2 + 2.1 \\ 2.1 + 1.2 + 2.2 & 2.2 + 1.1 + 2.2 & 2.2 + 1.2 + 2.1 \\ 2.1 + 2.2 + 1.2 & 2.2 + 2.1 + 1.2 & 2.2 + 2.2 + 1.1 \end{bmatrix}$$

$$= \begin{bmatrix} 1+4+4 & 2+2+4 & 2+4+2 \\ 2+2+4 & 4+1+4 & 4+2+2 \\ 2+4+2 & 4+2+2 & 4+4+2 \end{bmatrix} = \begin{bmatrix} 9 & 8 & 8 \\ 8 & 9 & 8 \\ 8 & 8 & 9 \end{bmatrix}$$

$\therefore A^2 - 4A - 5I$

$$= \begin{bmatrix} 9 & 8 & 8 \\ 8 & 9 & 8 \\ 8 & 8 & 9 \end{bmatrix} - 4 \begin{bmatrix} 1 & 2 & 2 \\ 2 & 1 & 2 \\ 2 & 2 & 1 \end{bmatrix} - 5 \begin{bmatrix} 1 & 0 & 0 \\ 0 & 1 & 0 \\ 0 & 0 & 1 \end{bmatrix}$$

$$= \begin{bmatrix} 9 & 8 & 8 \\ 8 & 9 & 8 \\ 8 & 8 & 9 \end{bmatrix} + \begin{bmatrix} -4 & -8 & -8 \\ -8 & -4 & -8 \\ -8 & -8 & -4 \end{bmatrix} + \begin{bmatrix} -5 & 0 & 0 \\ 0 & -5 & 0 \\ 0 & 0 & -5 \end{bmatrix}$$

$$= \begin{bmatrix} 9-4-5 & 8-8+0 & 8-8+0 \\ 8-8+0 & 9-4-5 & 8-8+0 \\ 8-8+0 & 8-8+0 & 9-4-5 \end{bmatrix} = \begin{bmatrix} 0 & 0 & 0 \\ 0 & 0 & 0 \\ 0 & 0 & 0 \end{bmatrix} = O,$$

where O is the null matrix. **Ans.**

Example 10:

Let $f(x) = x^2 - 5x + 6$, find $f(A)$ if

$$A = \begin{bmatrix} 2 & 0 & 1 \\ 2 & 1 & 3 \\ 1 & -1 & 0 \end{bmatrix}$$

Solution:

$$f(A) = A^2 - 5A + 6$$

$$= A^2 - 5A + 6I, \text{ where } I = \begin{bmatrix} 1 & 0 & 0 \\ 0 & 1 & 0 \\ 0 & 0 & 1 \end{bmatrix}$$

Now proceed $\begin{bmatrix} 1 & -1 & -3 \\ -1 & -1 & -10 \\ -5 & 4 & 4 \end{bmatrix}$ **Ans.**

Example 11:

Show that if A is the matrix $\begin{bmatrix} 2 & -1 & 1 \\ -1 & 2 & -1 \\ 1 & -1 & 2 \end{bmatrix}$, *then* $A^3 - 6A^2 + 9A - 4I$ $= O$, *I being the* 3×3 *unit matrix and O being the* 3×3 *null matrix.*

Solution:

$$A^2 = \begin{bmatrix} 2 & -1 & 1 \\ -1 & 2 & -1 \\ 1 & -1 & 2 \end{bmatrix} \times \begin{bmatrix} 2 & -1 & 1 \\ -1 & 2 & -1 \\ 1 & -1 & 2 \end{bmatrix}$$

$$= \begin{bmatrix} 4+1+1 & -2-2-1 & 2+1+2 \\ -2-2-1 & 1+4+1 & -1-2-2 \\ 2+1+2 & -1-2-2 & 1+1+4 \end{bmatrix} = \begin{bmatrix} 6 & -5 & 5 \\ -5 & 6 & -5 \\ 5 & -5 & 6 \end{bmatrix}$$

$$\therefore A^3 = A^2 \times A = \begin{bmatrix} 6 & -5 & 5 \\ -5 & 6 & -5 \\ 5 & -5 & 6 \end{bmatrix} \times \begin{bmatrix} 2 & -1 & 1 \\ -1 & 2 & -1 \\ 1 & -1 & 2 \end{bmatrix}$$

$$= \begin{bmatrix} 12+5+5 & -6-10-5 & 6+5+10 \\ -10-6-5 & 5+12+5 & -5-6-10 \\ 10+5+6 & -5-10-6 & 5+5+12 \end{bmatrix}$$

$$= \begin{bmatrix} 22 & -21 & 21 \\ -21 & 22 & -21 \\ 21 & -21 & 22 \end{bmatrix}$$

$$\therefore A^3 - 6A^2 + 9A - 4I$$

$$= \begin{bmatrix} 22 & -21 & 21 \\ -21 & 22 & -21 \\ 21 & -21 & 22 \end{bmatrix} - 6 \begin{bmatrix} 6 & -5 & 5 \\ -5 & 6 & -5 \\ 5 & -5 & 6 \end{bmatrix}$$

$$+ 9 \begin{bmatrix} 2 & -1 & 1 \\ -1 & 2 & -1 \\ 1 & -1 & 2 \end{bmatrix} - 4 \begin{bmatrix} 1 & 0 & 0 \\ 0 & 0 & 0 \\ 0 & 0 & 1 \end{bmatrix}$$

$$= \begin{bmatrix} 22 & -21 & 21 \\ -21 & 22 & -21 \\ 21 & -21 & 22 \end{bmatrix} + \begin{bmatrix} -36 & 30 & -30 \\ 30 & -36 & 30 \\ -30 & 30 & -36 \end{bmatrix}$$

$$+ \begin{bmatrix} 18 & -9 & 9 \\ -9 & 18 & -9 \\ 9 & -9 & 18 \end{bmatrix} + \begin{bmatrix} -4 & 0 & 0 \\ 0 & -4 & 0 \\ 0 & 0 & -4 \end{bmatrix}$$

$$= \begin{bmatrix} 22+36+18-4 & -21+30-9+0 & 21-30+9+0 \\ -21+30-9+0 & 22-36+18-4 & -21+30-9+0 \\ 21-30+9+0 & -21+30-9+0 & 22+36+18-4 \end{bmatrix}$$

$$= \begin{bmatrix} 0 & 0 & 0 \\ 0 & 0 & 0 \\ 0 & 0 & 0 \end{bmatrix} = O$$

Hence proved

Example 12:

If $A = \begin{bmatrix} 0 & 1 \\ 1 & 1 \end{bmatrix}$ *and* $B = \begin{bmatrix} 0 & -1 \\ 1 & 0 \end{bmatrix}$, *show that* $(A + B)(A - B) \neq A^2 - B^2$

Solution:

$$A + B = \begin{bmatrix} 0 & 1 \\ 1 & 1 \end{bmatrix} - \begin{bmatrix} 0 & -1 \\ 1 & 0 \end{bmatrix} = \begin{bmatrix} 0+0 & 1-1 \\ 1+1 & 1+0 \end{bmatrix}$$

$$= \begin{bmatrix} 0 & 0 \\ 2 & 1 \end{bmatrix}$$

$$A - B = \begin{bmatrix} 0 & 1 \\ 1 & 1 \end{bmatrix} - \begin{bmatrix} 0 & -1 \\ 1 & 0 \end{bmatrix} = \begin{bmatrix} 0-0 & 1-(-1) \\ 1-1 & 1-0 \end{bmatrix}$$

$$= \begin{bmatrix} 0 & 2 \\ 0 & 1 \end{bmatrix}$$

$$\therefore \quad (A + B)(A - B) = \begin{bmatrix} 0 & 0 \\ 2 & 1 \end{bmatrix} \times \begin{bmatrix} 0 & 2 \\ 0 & 1 \end{bmatrix}$$

$$= \begin{bmatrix} 0.0 + 0.0 & 0.2 + 0.1 \\ 2.0 + 1.0 & 2.2 + 1.1 \end{bmatrix} = \begin{bmatrix} 0 & 0 \\ 0 & 5 \end{bmatrix}$$

$$A^2 = \begin{bmatrix} 0 & 1 \\ 1 & 1 \end{bmatrix} \times \begin{bmatrix} 0 & 1 \\ 1 & 1 \end{bmatrix} = \begin{bmatrix} 0.0 + 1.1 & 0.1 + 1.1 \\ 1.0 + 1.1 & 1.1 + 1.1 \end{bmatrix}$$

$$= \begin{bmatrix} 1 & 1 \\ 1 & 2 \end{bmatrix}$$

$$B^2 = \begin{bmatrix} 0 & -1 \\ 1 & 0 \end{bmatrix} \times \begin{bmatrix} 0 & -1 \\ 1 & 0 \end{bmatrix} = \begin{bmatrix} 0.0 - 1.1 & 0.1 - 1.0 \\ 1.0 - 0.1 & 1.1 + 0.0 \end{bmatrix}$$

$$= \begin{bmatrix} -1 & 0 \\ 0 & -1 \end{bmatrix}$$

$$\therefore \quad A^2 - B^2 = \begin{bmatrix} 1 & 1 \\ 1 & 2 \end{bmatrix} - \begin{bmatrix} -1 & 0 \\ 0 & -1 \end{bmatrix} = \begin{bmatrix} 1 + 1 & 1 - 0 \\ 1 - 0 & 2 + 1 \end{bmatrix}$$

$$= \begin{bmatrix} 2 & 1 \\ 1 & 3 \end{bmatrix}$$

Hence $(A + B)(A - B) \neq A^2 - B^2$.

Example 13:

If $A = \begin{bmatrix} 0 & -\tan \frac{1}{2}\alpha \\ \tan \frac{1}{2}\alpha & 0 \end{bmatrix}$ *and I is a unit matrix, then prove that* $I + A = (I - A) \begin{bmatrix} \cos\alpha & -\sin\alpha \\ \sin\alpha & \cos\alpha \end{bmatrix}$

Solution:

$$I + A = \begin{bmatrix} 1 & 0 \\ 0 & 1 \end{bmatrix} + \begin{bmatrix} 0 & -\tan\frac{1}{2}\alpha \\ \tan\frac{1}{2}\alpha & 0 \end{bmatrix}$$

$$= \begin{bmatrix} 1+0 & 0-\tan\frac{1}{2}\alpha \\ 0+\tan\frac{1}{2}\alpha & 1+0 \end{bmatrix} = \begin{bmatrix} 1 & -\tan\frac{1}{2}\alpha \\ \tan\frac{1}{2}\alpha & 1 \end{bmatrix} \quad ...(1)$$

$$I - A = \begin{bmatrix} 1 & 0 \\ 0 & 2 \end{bmatrix} - \begin{bmatrix} 0 & -\tan\frac{1}{2}\alpha \\ \tan\frac{1}{2}\alpha & 0 \end{bmatrix}$$

$$= \begin{bmatrix} 1-0 & 0+\tan\frac{1}{2}\alpha \\ 0-\tan\frac{1}{2}\alpha & 1-0 \end{bmatrix} = \begin{bmatrix} 1 & \tan\frac{1}{2}\alpha \\ -\tan\frac{1}{2}\alpha & 1 \end{bmatrix}$$

$$\therefore (I - A) \begin{bmatrix} \cos\alpha & -\sin\alpha \\ \sin\alpha & \cos\alpha \end{bmatrix}$$

$$= \begin{bmatrix} 1 & \tan\frac{1}{2}\alpha \\ -\tan\frac{1}{2}\alpha & 1 \end{bmatrix} \begin{bmatrix} \cos\alpha & -\sin\alpha \\ \sin\alpha & \cos\alpha \end{bmatrix}$$

$$= \begin{bmatrix} 1.\cos\alpha + \tan\frac{1}{2}\alpha.\sin\alpha & 1.(-\sin\alpha) + \tan\frac{1}{2}\alpha\cos\alpha \\ -\tan\frac{1}{2}\alpha\cos\alpha + 1.\sin\alpha & (\sin\alpha).\tan\frac{1}{2}\alpha + 1.\cos\alpha \end{bmatrix},$$

$$= \begin{bmatrix} \left(1-2\sin^2\frac{1}{2}a\right) + 2\sin^2\frac{1}{2}\alpha & -2\sin\frac{1}{2}\alpha\cos\frac{1}{2}\alpha \\ & +\tan\frac{1}{2}\alpha\cos\alpha \\ -\tan\frac{1}{2}\alpha\cos\alpha + 2\sin\frac{1}{2}\alpha\cos\frac{1}{?}\alpha & 2\sin^2\frac{1}{2}\alpha \\ +\left(1+2\sin^2\frac{1}{2}\alpha\right) & \end{bmatrix}$$

writing $\cos\alpha = 1 - 2\sin^2\frac{1}{2}\alpha$

$$= \begin{bmatrix} 1 & -2\tan\frac{1}{2}\alpha\cos^2\frac{1}{2}\alpha + \tan\frac{1}{2}\alpha\cos\alpha \\ -\tan\frac{1}{2}\alpha\cos\alpha + 2\tan\frac{1}{2}\alpha\cos^2\frac{1}{2}\alpha & 1 \end{bmatrix}.$$

writiting $\sin\frac{1}{2}\alpha$ as $\tan\frac{1}{2}\alpha\ \cos\frac{1}{2}\alpha$

$$= \begin{bmatrix} 1 & -\tan\frac{1}{2}\alpha\ 2\cos^2\frac{1}{2}\alpha - \left(2\cos^2\frac{1}{2}\alpha - 1\right) \\ \tan\frac{1}{2}\alpha\left\{-\left(2\cos^2\frac{1}{2}\alpha - 1\right) + 2\cos^2\frac{1}{2}\alpha\right\} & 1 \end{bmatrix}$$

writing $\cos\alpha = 2\cos^2\frac{1}{2}\alpha - 1$

$$= \begin{bmatrix} 1 & -\tan\frac{1}{2}\alpha \\ \tan\frac{1}{2}\alpha & 1 \end{bmatrix} = I + A, \text{ from } (1)$$

Hence proved

1.17 APPLICATION OF MATRIX

Example 1:

In a development plan of a city, a contractor has taken a contract to construct certain houses for which he needs building materials like stones, sand etc. There are three firms A, B, C that can supply him these materials. At one time these firms A, B, C supplied him 40, 35 and 25 truck loads of stones and 10, 5 and 8 truck loads of sand respectively. If the cost of one truck load of stone and sand are Rs. 1,200 and Rs. 500 respectively, then find the total amount paid by the contractor to each of these firms A, B, C separately.

Solution:

The truck-loads of stone and sand supplied by the firms A, B, and C can be written in the form of a matrix A (say) given by

$$A = \begin{array}{c} \\ \text{Stone} \\ \text{Sand} \end{array} \begin{array}{c} \begin{array}{ccc} A & B & C \end{array} \\ \begin{bmatrix} 40 & 35 & 25 \\ 10 & 5 & 8 \end{bmatrix} \end{array}$$, which is a 2×3 matrix.

And the cost per truck of stone and sand can be given in the form of a matrix B (say), given by

$$B = \begin{array}{c} \begin{array}{cc} \text{Stone} & \text{Sand} \end{array} \\ [1200 \quad 500] \end{array}$$

The required total amount paid to each of the firms A, B and C are given by the product matrix BA. (Note here AB can not be calculated).

$$\text{Now } BA = [1200 \quad 500] \times \begin{bmatrix} 40 & 35 & 25 \\ 10 & 5 & 8 \end{bmatrix}$$

$$= [(1200 \times 40) + (500 \times 10) \;\; (1200 \times 35) + (500 \times 5) \;\; (1200 \times 25) + (500 \times 8)]$$

$$= [48000 + 5000 \quad 42000 + 2500 \quad 30000 + 4000]$$

$$= [53{,}000 \quad 44,500 \quad 34{,}000]$$

$\therefore$ The amounts paid to the firms A, B and C by the contractor are Rs. 53,000, Rs. 44,500 and Rs. 34,000 respectively. **Ans.**

Example 2:

A man buys 8 dozens of mangoes, 10 dozens of ampples and 4 dozens of bananas. Mangoes cost Rs. 18 per dozen, apples Rs. 9 per dozen and bananas Rs. 6 per dozen. Represent the quantities bought by a row matrix and the prices by a column matrix and hence obtain the total cost.

Solution:

The quantities bought are represented by 3×1 row matrix $[8 \quad 10 \quad 4]$ and the prices are represented by 3×1 column matrix

$$\begin{bmatrix} 18 \\ 9 \\ 6 \end{bmatrix}$$

$\therefore$ The cost of fruits is a single number *i.e.*, 1×1 matrix given by the product matrix

$$[8 \quad 10 \quad 4] \times \begin{bmatrix} 18 \\ 9 \\ 6 \end{bmatrix}$$

i.e., [(8 × 18) + (10 × 9) + (4 × 6) *i.e.,* [144 + 90 + 24] *i.e.,* [258]

∴ The required total cost = Rs. 258. **Ans.**

Example 3:

A manufacturer produces three products A, B, C which he sells in the market. Annual sale volumes are indicated as follows:

Markets	***Products***		
	A	*B*	*C*
I	*8,000*	*10,000*	*15,000*
II	*10,000*	*2,000*	*20,000*

(1) If unit sale prices of A, B and C are Rs. 2.25, Rs. 1.50 and Rs. 1.25 respectively find the total revenue in each market with the help os matrices, (2) if the unit costs of the above three products are Rs. 1.60, Rs 1.20 and Rs. 0. 90 respectively, find the gross profit with the help of matrices.

Solution:

(1) The total revenue in each market is given by the product matrix.

$$[2.25 \quad 1.50 \quad 1.25] \times \begin{bmatrix} 8,00 & 10,000 \\ 10,000 & 2,000 \\ 15,000 & 20,000 \end{bmatrix}$$ **(Note)**

= [(2, 25 × 8,000) + (1.50 × 10,000) + (1.25 × 15,000)
(2.25 × 10,000) + (1.50 × 2,000) + (1.25 × 20,000)]

= [18,000 + 15,000 + 18,750 22,500 + 3,000 + 25,000]

= [51750 50500]

∴ Total revenue from the market I = Rs. 51.750 and total revenue from the market II = Rs. 50,500. **Ans.**

(2) similarly the total cost of products which the manufacturer sells in the markets are given by the product matrix.

$$[1.60 \quad 1.20 \quad 0.90] \times \begin{bmatrix} 8,00 & 10,000 \\ 10,000 & 2,000 \\ 15,000 & 20,000 \end{bmatrix}$$

= [(1.60 × 8,000) + (1.20 × 10,000) + (0.90 × 15,000)
(1.60 × 10,000) + (1,20 × 2,000) + (0.90 × 20,000)]

= [12,800 + 12,000 + 13,500   16,500 + 2,400 + 18,000]

= [38,300   36,400]

∴ Total cost of products which the manufacturer sells in the markets I and I are Rs. 38,300 and Rs. 36,400 respectively.

∴ Requited gross profit = (Total revenue received from both the markets) – (Total cost of products which the manufacturer sold in both the markets)

= (Rs. 51,750 + Rs. 50,500) – (Rs. 38,300 + Rs. 36,400)

= Rs. 102,250 – Rs. 74,700 = Rs. 27, 550. **Ans.**

Example 4:

A store has in stock 20 dozen shirts, 15 dozen trousers and 25 dozen pairs of socks. If the selling prices are Rs. 50 per shirt, Rs. 90 per trouser and Rs. 12 per pair of socks, then find the total amount the stove owner will get after selling all the items in the stock.

Solution:

The stock in the store can be written in the form of a row matrix A given by $A = [20 \times 12 \quad 15 \times 12 \quad 25 \times 12]$

⇒ $A = [240 \quad 180 \quad 300]$, which is a 1 × 3 matrix.

The prices can be written in the form of a column matrix B given by

$B = \begin{bmatrix} 50 \\ 90 \\ 12 \end{bmatrix}$, which is a 3 × 1 matrix.

The required amount is a single number *i.e.*, a matrix of order 1 × 1 and so the same can be obtained by multiplying the matrices A and B, since their product would be a 1 × 1 matrix.

Now $AB = [240 \quad 180 \quad 300] \times \begin{bmatrix} 50 \\ 90 \\ 12 \end{bmatrix}$

$= [(240 \times 50) + (180 \times 90) + (300 \times 12)]$

$= [12000 + 16200 + 3600] = [31800]$

∴ The required amount received by the store owner

= Rs. 31800.

Example 5:

A finance company has offices located in every division, every district and every taluka in a certain state in India. Assume that there are five

divisions, thirty districts and 200 talukas in the state. Each office has one heeadclerk, one cashier, one clerk and one person. A divisional office has, in addition, one office superintendent, two clerks, one typist and one peon. A district office, has in addition, one clerk and one peon. The basic monthly salaries are as follows: office superintendent Rs. 500, Head clerk Rs. 200, cashier Rs. 175, clerks and typists Rs. 150 and person Rs 100. Using matrix notation find:

(1) the total number of posts of each kind in all the offices taken together,

(2) the total basic monthly salary bill of all the offices taken together.

Solution:

Let us use the symbols Div, Dis, Tal for division, district, taluka respectively and O, H, C, CL, T and P for office superintendent, Head clerk, cashier, clerk, typist and peon respectively.

Then the number of offices can be arranged as elements of a row matrix A (say) given by

$$\begin{matrix} & \text{Div.} & \text{Dis.} & \text{Tal.} \\ A = (& 5 & 30 & 200) \end{matrix}$$

The composition of staff in various offices can be arranged in a 3 × 6 matrix B (say) given by

$$\begin{matrix} & \text{O} & \text{H} & \text{C} & \text{Cl} & \text{T} & \text{P} \end{matrix}$$

$$B = \begin{bmatrix} 1 & 1 & 1 & 2+1 & 1 & 1+1 \\ 0 & 1 & 1 & 1+1 & 0 & 1+1 \\ 0 & 1 & 1 & 1 & 0 & 1 \end{bmatrix}$$

The basic monthly salaries of various types of employees of these offices correspond to the elements of the column matrix C (say) given by

$$C = \begin{matrix} \text{O} & 500 \\ \text{H} & 200 \\ \text{C} & 175 \\ \text{Cl} & 150 \\ \text{T} & 150 \\ \text{P} & 100 \end{matrix}$$

(1) Total number of posts of each kind in all the offices are the elements of the product matrix AB

i.e., $[5 \quad 30 \quad 200] \times \begin{bmatrix} 1 & 1 & 1 & 3 & 1 & 2 \\ 0 & 1 & 1 & 2 & 0 & 2 \\ 0 & 1 & 1 & 1 & 0 & 1 \end{bmatrix}$ **(Note)** *i.e.,*[5 + 0 + 0, 5 + 30 + 200, 5 + 30 + 200, 15 + 60 + 200, 5 + 0 + 0, 10 + 60 + 200]

$$\begin{matrix} & O & H & C & Cl & T & P \\ i.e., & [5 & 235 & 235 & 275 & 5 & 270 \end{matrix}$$ **Ans.**

[*i.e.,* required number of posts in all the offices taken together are 5 office supdts, 235. Head clerks, 235 cashiers, 275 cashiers, 275 clerks, 5 typists and 270 persons.

(2) Total basic monthly salary bill of each kind of office are the elements of the product matrix BC

$$i.e., \quad \begin{bmatrix} 1 & 1 & 1 & 3 & 1 & 2 \\ 0 & 1 & 1 & 2 & 0 & 2 \\ 0 & 1 & 1 & 1 & 0 & 1 \end{bmatrix} \times \begin{pmatrix} 500 \\ 200 \\ 175 \\ 150 \\ 150 \\ 100 \end{pmatrix}$$

$$= \begin{bmatrix} (1\times 500) + (1\times 200) + (1\times 175) + (3\times 150) + (1\times 150) + (2\times 150) \\ (0\times 500) + (1\times 200) + (1\times 175) + (2\times 150) + (0\times 150) + (2\times 150) \\ (0\times 500) + (1\times 200) + (1\times 175) + (1\times 150) + (0\times 150) + (1\times 150) \end{bmatrix}$$

$$= \begin{bmatrix} 500 + 200 + 175 + 450 + 150 + 200 \\ 0 + 200 + 175 + 300 + \quad 0 + 200 \\ 0 + 200 + 175 + 150 + \quad 0 + 100 \end{bmatrix} = \begin{bmatrix} 1675 \\ 875 \\ 625 \end{bmatrix}$$

[*i.e.,* The total basic monthly salary bill of each divisional, district and taluka offices are Rs. 1675, Rs. 875 and Rs. 625 respectively].

(3) Total basic monthly salary bill of all the offices (*i.e.,* of five divisional, 30 district and 200 taluka offices) is the element of the product matrix ABC

$$i.e., \quad [5 \quad 30 \quad 200] \times \begin{bmatrix} 1675 \\ 875 \\ 625 \end{bmatrix}$$ **(Note)**

i.e., [(5 × 1675) + (30 × 875) + (200 × 625)

i.e., [8375 + 26250 + 125000] *i.e.,* [159625] **Ans.**

i.e., [total basic monthly salary bill of all the offices taken together is Rs. 159,625]. **Ans.**

Example 6:

A trust fund has Rs. 50.000 that is to be invested into two types of bonds. The first bond pays 5% interest per year and the second bond pays 6%

interest per year. Using matrix multiplication, determine how to divide Rs. 50.000 among the two types of bonds so as to obtain an annual total interest of Rs. 2780.

Solution:

Let Rs. 50,000 be divided into two parts Rs. x and Rs. (50,000 – x) out of which first part is invested in first type of bonds and the second part is invested in second type of bonds.

The values of these bonds can be written in the form of a row matrix A given by A = [x 50,000 – x], which is a 1 × 2 matrix.

And the amounts received as interest per rupee annually from these two types of bonds can be written in the form of a column matrix B given by

$B = \begin{bmatrix} 5/100 \\ 6/100 \end{bmatrix}$, which is a 2 × 1 matrix.

Here the interest has been calculated per rupee annually.

Now, the interest to be obtained annually is a single number *i.e.,* a matrix of order 1 × 1 and the same can be obtained by the product matrix AB, since this product matrix would be a 1 × 1 matrix. **(Note)**

$$\text{Here AB} = [x \quad 50{,}000 -] \times \begin{bmatrix} 5/100 \\ 6/100 \end{bmatrix}$$

$$= \left[x.\frac{5}{100} + (50{,}000 - x).\frac{6}{100}\right]$$

$$= \left[300 - \frac{x}{100}\right]$$

Also we are given that the annual interest = Rs. 2,780.

$\therefore$ We must have $\left[300 - \dfrac{x}{100}\right] = [2780]$ **(Note)**

$$\Rightarrow \quad 3000 - \frac{x}{100} = 2780$$

$$\Rightarrow \quad x = (3000 - 2780) \times 100$$

$$\Rightarrow \quad x = 220 \times 100 = 22{,}000$$

Hence the required amounts are

Rs. 22,000 and Rs. (50,000 – 22,000) *i.e.,*

Rs. 22, 000 and Rs. 28,000. **Ans.**

Example 7:

If A and B are both skew-symmetric matrices of same ofer such that AB = Ba, then show that AB is symmetric.

Solution:

If **A** and **B** are both skew-symmetric matrices,

then $\mathbf{A} = -\mathbf{A}'$ and $\mathbf{B} = -\mathbf{B}'$ ...(1)

Also given that $\mathbf{AB} = \mathbf{BA}$

$= (-\mathbf{B}')(-\mathbf{A}')$, from (1)

$= \mathbf{B}'\,\mathbf{A} = (\mathbf{AB})'$

$\Rightarrow$ $\mathbf{AB} = (\mathbf{AB})'$ *i.e.*, 1 **AB** is a symmetric matrix. **Hence proved.**

Example 8:

If A is a symmetic matrix, then whow that kA is also symmetric for any scalar k.

Solution:

Here $(k\mathbf{A})' = k\mathbf{A}', = k\mathbf{A}$, $\mathbf{A}' = \mathbf{A}$, **A** being symmetic

Hence k**A** is symmetric, if **A** is so.

Example 9:

Find the symmetric and skew-symmetric parts of the matrix

$$A = \begin{bmatrix} 1 & 2 & 4 \\ 6 & 8 & 1 \\ 3 & 5 & 7 \end{bmatrix}$$

Solution:

Here A' = transpose of A

$$= \begin{bmatrix} 1 & 6 & 3 \\ 2 & 8 & 5 \\ 4 & 1 & 7 \end{bmatrix}$$

The symmetric part of $\mathbf{A} = \frac{1}{2}(\mathbf{A} + \mathbf{A}')$

$$= \frac{1}{2}\begin{bmatrix} 1 & 2 & 4 \\ 6 & 8 & 1 \\ 3 & 5 & 7 \end{bmatrix} + \begin{bmatrix} 1 & 6 & 3 \\ 2 & 8 & 5 \\ 4 & 1 & 7 \end{bmatrix}$$

$$= \frac{1}{2}\begin{bmatrix} 1+1 & 2+6 & 4+3 \\ 6+2 & 8+8 & 1+5 \\ 3+4 & 5+1 & 7+7 \end{bmatrix} = \frac{1}{2}\begin{bmatrix} 2 & 8 & 7 \\ 8 & 16 & 6 \\ 7 & 6 & 14 \end{bmatrix}$$

$$= \begin{bmatrix} 1 & 4 & \frac{7}{8} \\ 4 & 8 & 3 \\ \frac{7}{8} & 3 & 7 \end{bmatrix}$$ **Ans.**

And the skew-symmetric part of A = $\frac{1}{2}$ (A – A')

$$= \frac{1}{2}\begin{bmatrix} 1 & 2 & 4 \\ 6 & 8 & 1 \\ 3 & 5 & 7 \end{bmatrix} - \begin{bmatrix} 1 & 6 & 3 \\ 2 & 8 & 5 \\ 4 & 1 & 7 \end{bmatrix}$$

$$= \frac{1}{2}\begin{bmatrix} 1-1 & 2-6 & 4-3 \\ 6-2 & 8-8 & 1-5 \\ 3-4 & 5-1 & 7-7 \end{bmatrix} = \frac{1}{2}\begin{bmatrix} 0 & -4 & 1 \\ 4 & 0 & -4 \\ -1 & 4 & 0 \end{bmatrix}$$

$$= \begin{bmatrix} 0 & -2 & \frac{1}{2} \\ 2 & 0 & -2 \\ -\frac{1}{2} & 2 & 0 \end{bmatrix}$$ **Ans.**

Example 10:

If A is any square matrix, show that AA' is a symmetric matrix.

Solution:

(**A'A**) = transpose of **AA'**

= (**A'**)' **A'**

= **AA'**

i.e., **AA** = (**AA'**). Hence **AA'** is a symmetric matrix by definition.

Example 11:

If A be a square matrix, show that A + A' is symmetric and A – A' is a skew-symmetric matrix.

Solution:

If **A** is a square matrix, then

$$(\mathbf{A} + \mathbf{A}') = \mathbf{A}' + (\mathbf{A}')'$$
$$= \mathbf{A}' + \mathbf{A}$$
$$= \mathbf{A} + \mathbf{A}', \text{ by commutative law of addition}$$

Hence by definition **A** + **A'** is symmetric.

Again $(\mathbf{A} - \mathbf{A}')' = \mathbf{A}' - (\mathbf{A}')',$

$$= \mathbf{A}' - \mathbf{A},$$
$$= -(\mathbf{A} - \mathbf{A})'$$

Hence by definintion **A** – **A'** is skew-symmetric.

Example 12:

If A is a skew-symmetric matrix, then show that AA' = A'A and A^2 symmetric.

Solution:

If **A** is a skew-symmetric matrix, then we know that

$$\mathbf{A}' = -\mathbf{A} \qquad ...(1)$$

Pre-multiplying both sides of (1) by **A**, we get

$$\mathbf{AA}' = -\mathbf{AA} = -\mathbf{A}^2 \qquad ...(2)$$

Post-multiplying both sides of (1) by **A**, we get

$$\mathbf{A'A} = -\mathbf{AA} = -\mathbf{A}^2 \qquad ...(3)$$

From (2) and (3) we conclude that **AA'** = **A'A**.

Further we can prove that **AA'** and **A'A** are symmetric matrices Hence from (2) and (3) we find that $-\mathbf{A}^2$ is a symmetic matrix or $\mathbf{A}^2$ is a symmetric matrix, as we know is a symmetric matrix or $\mathbf{A}^2$ is a symmetric matrix, as we know that k**A** is also symmetric if k is scalar and **A** is symmetric.

Hence proved.

1.18 HERMITIAN AND SKEW-HERMITIAN MATRICES

(a) Hermitian Matrix

Definition: *A square matrix A such that* $\overline{\mathbf{A}}'$ *= A is called Hermitian i.e., the matrix* $[a_{ij}]$ *is Hermitian provided* $a_{ij} = a_{ji}$, *for all values of i and j.*

For example: $A = \begin{bmatrix} 1 & \alpha + i\beta & \gamma + i\delta \\ \alpha - i\beta & m & x + iy \\ \gamma - i\delta & x - iy & n \end{bmatrix}$

It A is Hermitian Matrix then $a_{ij} = \bar{a}_{ij}$ (by definition)

$\therefore$ a_{ij} is real for a_{ij}. Thus, every diogonal element of a Hermitian Matrix must be real. A Hermitian Matrix over the field of real numbers is nothing but a real symmetric matrix.

(b) Skew-Hermitian Matrix

Definition: *A square matrix A such that* $\overline{\mathbf{A}}' = -A$ *is called Skew-Hermitian i.e., the matrix* $[a_{ij}]$ *is skew Hermitian provided* $a_{ij} = -a_{ji}$ *for all values of i and j.*

For example: $A = \begin{bmatrix} 2i & \alpha + i\beta & 3 \\ \alpha - i\beta & -i & \gamma + i\delta \\ 3 & \gamma - i\delta & 0 \end{bmatrix}.$

It A is Skew Hermitian Matrix then

$$a_{ij} = -\bar{a}_{ij}$$

$$\therefore \quad a_{ij} = -\bar{a}_{ij} = 0$$

i.e., a_{ij} must be either a pure imaginary must be number or zero. Thus, the diagonal elements of a Skew Hermitian Matrix must be pure imaginary number or zero

1.19 THEOREMS ON HERMITIAN AND SKEW-HERMITIAN MATRICES

Theorem 1:

The diagonal elements of a skew-hermition matrix are either purely imaginary or zero.

Proof:

Let $[a_{ij}]$ be an $n \times n$ skew-Hermitian matrix, then according to definition we have

$$a_{ij} = \bar{a}_{ij}, \text{ for all } 1 \le i \le n,\ 1 \le j \le n \qquad \text{...(1)}$$

Now the diagonal elements are a_{ij}, where $1 \le i \le n$.

$\therefore$ From (1), we have $a_{ij} = \bar{a}_{ij}$, for all $1 \le i \le n$ ...(2)

If $a_{ij} = \alpha + i\beta$, where α and β are real,

then $\bar{a}_{ij} = \alpha - i\beta$.

$\Rightarrow$ From (2), we get $\alpha + i\beta = (\alpha - i\beta)$

$\Rightarrow$ $\alpha + i\beta = -\alpha + i\beta \Rightarrow 2\alpha = 0 \Rightarrow \alpha = 0$

$\therefore a_{ij} = 0 + i\beta = i\beta$, which is purely imaginary and can be zero if$\beta = 0$

Hence, the diagonal elements of a Skew Hermitan Matrix are either purely imaginarly or zero. **Hence proved.**

Theorem 2:

Every square matrix (with complex elements) can be uniquely expressed as the sum of a Hermitian and a skew-hermitian matrices.

Proof:

Let **A** be a square matrix. Then we can write

$$A = \frac{1}{2}(A + A^{\Theta}) + \frac{1}{2}(A - A^{\Theta}) \qquad ...(1)$$

Now $\left(\overline{A + A^{\Theta}}\right) = \bar{A} + \overline{A^{\Theta}}$

$$\therefore \left\{\left(\overline{A + A^{\Theta}}\right)\right\} = \left\{\bar{A} + \overline{A^{\Theta}}\right\} = (\bar{A})' - (\overline{A^{\Theta}})',$$

$$= A^{\Theta} + (\overline{A^{\Theta}})', \text{ by def } (\bar{A}) = A^{\Theta},$$

$$= A^{\Theta} + (\overline{A^{\Theta}})' \qquad ...(2)$$

Now $(\overline{\mathbf{A}^{\Theta}})'$ = transposed conjugate of $\mathbf{A}^{\Theta}$

= transposed conjugate of $(\bar{\mathbf{A}})'$

= transposed matrix of $(\mathbf{A})'$

since conjugate of $\bar{\mathbf{A}}$ is **A**

= **A**, $\because$ $(\mathbf{A}')' = \mathbf{A}$

$\therefore$ From (2) we get, $\left\{\left(\overline{\mathbf{A} + \mathbf{A}^{\Theta}}\right)\right\} = \mathbf{A}^{\Theta} + \mathbf{A} = \mathbf{A} + \mathbf{A}^{\Theta}$

as addition If matrices obey commutative law.

$\therefore$ By definition we find that $\mathbf{A} + \mathbf{A}^{\Theta}$ is a Hermitian matrix.

Again $\left\{\left(\overline{A + A^{\Theta}}\right)\right\}' = \left\{\overline{A} - \overline{A}^{\Theta}\right\}' = (A)' - \left(A^{\Theta}\right)$

$= \mathbf{A}^{\Theta} - A$, as above

$= -(A - \mathbf{A}^{\Theta})$.

$\therefore$ By definition we find that $\mathbf{A} - \mathbf{A}^{\Theta}$ is a Skew-Hermitian matrix

$\therefore$ From we conclude that the square matrix A is the sum of a Hermitian and a Skew-Hermitian Matrices.

Theorem 3:

The diagonal elements of a Hermitian matrix are necessary real.

Proof:

Let $[a_{ij}]$ be an $n \times n$ Hermitian matrix, then according to definition we have

$a_{ij} = \bar{a}_{ij}$, for all $1 \leq i \leq n$, $1 \leq j \leq n$ ...(1)

Now the diagonal elements are a_{ij}, where $1 \leq i \leq n$.

$\therefore$ From (1), we have $a_{ij} = \bar{a}_{ij}$, for all $1 \leq i \leq n$ ...(2)

If $a_{ij} = \alpha + i\beta$, where α and β are real,

then $\bar{a}_{ij} = \alpha - i\beta$

$\Rightarrow$ From (2), we get $\alpha + i\beta = \alpha - i\beta$

$\Rightarrow$ $2i\beta = 0 \Rightarrow \beta = 0$

$\therefore a_{ij} = \alpha + i\,(0) = \alpha$, which is purely real.

Hence the diagonal elements of a Hermitian matrix, are necessarily real

Hence proved.

Example 1:

If A is a Skew-Hermitian Matrix, then shwo that ia is Hermitian.

Solution:

If **A** is a Skew-Hermitian Matrix, then

we have $-\mathbf{A} = \overline{\mathbf{A}}$

Also $\quad \overline{\mathbf{A}} = \mathbf{A}^{\Theta}$

$\therefore \quad -\mathbf{A} = \overline{\mathbf{A}}' = \mathbf{A}^{\Theta}$

Now $\quad (\mathbf{iA})^{\Theta} = -\,\mathbf{iA}^{\Theta} \qquad \because \ \bar{i} = -i$

$= -\,i\,(-\mathbf{A})$, from (i)

$\Rightarrow \quad (\mathbf{iA})^{\Theta} = \mathbf{iA} \qquad ...(2)$

We know that if **A** is a Hermitian matrix, then $\overline{\mathbf{A}'} = \mathbf{A} = \mathbf{A}^{\Theta}$, from (1)

And from (2) we find that (i**A**) hence i**A** is a Hermitian matrix.

Example 2:

If A is any square matrix, show that $\mathbf{AA}^{\Theta}$ *and* $\mathbf{A}^{\Theta}\mathrm{A}$ *are Hermitian.*

Solution:

$$(\mathbf{AA}^{\Theta})^{\Theta} = (\mathbf{A}^{\Theta})^{\Theta}\ \mathbf{A}^{\Theta}$$

$$= \mathbf{AA}^{\Theta}$$

$\therefore$ By definition $\mathbf{AA}^{\Theta}$ Hermitian.

Similarly $(\mathrm{A}^{\Theta}\mathrm{A})^{\Theta} = \mathrm{A}^{\Theta}\ (\mathrm{A}^{\Theta})^{\Theta}$

$$= \mathrm{A}^{\Theta}\ \mathrm{A}$$

$\therefore$ By definition A^{Θ} A is Hermitian.

Example 3:

Show that A is Hermitan iff $\overline{\mathrm{A}}$ *is Hermitian.*

Solution:

Let **A** be Hermitian, then $\mathbf{A} = \mathbf{A}^{\varnothing} \qquad ...(1)$

Now $(\overline{\mathrm{A}}^{\Theta})$ = transposed conjugate of $\overline{\mathrm{A}}$

= Transposed matrix of **A**, since $(\overline{\overline{\mathrm{A}}}) = \mathbf{A}$

$= \mathrm{A} = (\mathrm{A}^{\Theta})'$, by (1)

= thanspose of transposed conjugate of A

= conjugate of **A**, $\because$ (**B**')', = **B**

i.e., $(\overline{A})^{\Theta} = \overline{A}$

Hence by defintion, $\overline{A}$ is a Hermitian matrix,

Again if $\overline{A}$ is Hermitian, then we have

$\overline{A} = (\overline{A})^{\Theta}$

= transposed conjugate of $\overline{A}$

= transpose of **A**

$\Rightarrow \quad \overline{A} = \mathbf{A}'$

Now $\mathbf{A}^{\varnothing} = (\overline{A})'$ by definition

= (A')' by (ii)

= $\mathbf{A}^{\Theta} = \mathbf{A}$

Hence by defintion A is Hermitian. **Hence proved.**

Example 4:

If $A = \begin{bmatrix} 3 & 2-3i & 3+5i \\ 2+3i & 5 & i \\ 3-5i & -i & 7 \end{bmatrix}$, *then prove that* $\overline{A}$ *is Hermitian*

Solution:

$$\overline{A} = \begin{bmatrix} 3 & 2+3i & 3+5i \\ 2-3i & 5 & -i \\ 3+5i & i & 7 \end{bmatrix} = B \text{ (say)}$$

Then $B' = \begin{bmatrix} 3 & 2-3i & 3+5i \\ 2+3i & 5 & i \\ 3-5i & -i & 7 \end{bmatrix}$

$$\therefore \quad \overline{B'} = \begin{bmatrix} 3 & 2-3i & 3-5i \\ 2-3i & 5 & -i \\ 3+5i & -i & 7 \end{bmatrix} = B$$

$\therefore$ B *i.e.,* $\overline{A}$ is Hermitian Matrix.

Example 5:

Prove that the matrix $A = \begin{bmatrix} 1 & 1-i & 2 \\ 1+i & 3 & i \\ 2 & -i & 0 \end{bmatrix}$ *is Hermitian.*

Solution:

$$A' = \begin{bmatrix} 1 & 1+i & 2 \\ 1-i & 3 & -i \\ 2 & i & 0 \end{bmatrix}$$

$$\overline{A'} = \begin{bmatrix} 1 & 1-i & 2 \\ 1+i & 3 & i \\ 2 & -i & 0 \end{bmatrix} = A$$

$\therefore$ A is Hermitian Matrix.

Example 6:

If A is a Hermitian matrix, then show that iA is Skew-Hermitian Matrix.

Solution:

If **A** is a Hermitian matrix, then

we heve $\mathbf{A} = \overline{\mathbf{A'}}$

Also $\overline{\mathbf{A'}} = \mathbf{A}^{\Theta}$

$\therefore$ Here $\mathbf{A} = \overline{\mathbf{A'}} = \mathbf{A}^{\Theta}$ ...(1)

Now $(i\mathbf{A})^{\Theta} = -i\mathbf{A}^{\Theta}$, $\because \bar{i} = -i$

$= -\left(i\mathbf{A}^{\Theta}\right)$

$\Rightarrow$ $(i\mathbf{A})^{\Theta} = -(i\mathbf{A})$, from (1) ...(2)

We know that if **A** is a Skew Hermitian Matrix, then $\overline{\mathbf{A}}' = -A = \mathbf{A}^{\Theta}$, from (1).

And from (2), we find that $-(i\mathbf{A}) = (i\mathbf{A})^{\Theta}$, hence (i**A**) is a Skew Hermitian Matrix.

Example 7:

If A and B are Hermitian, then show that AB is Hermitian if and only if A and B commute.

Solution:

If **A** and **B** are Hermitian matrices, then we have

$$\mathbf{A} = (\overline{\mathbf{A}})' = \mathbf{A}^{\Theta} \text{ and } \mathbf{B} = (\overline{\mathbf{B}})' = \mathbf{B}^{\Theta} \quad ...(1)$$

Then $(\mathbf{AB})^{\Theta} = \mathbf{B}^{\Theta}\ \mathbf{A}^{\Theta}$,

$= \mathbf{BA}$, by (1) above

$= \mathbf{AB}$, if **A** and **B** commute

i.e., $(\mathbf{AB})^{\Theta} = \mathbf{AB} \Rightarrow (\overline{\mathbf{AB}})' = \mathbf{AB}$, $\quad \because A^{\Theta} = (\overline{A})'$

Hence, by definition **AB** is Hermitian.

Converse of this can be proved to be true by reversing the above calculations.

1.20 ADJOINT MATRIX

Let A $[a_{ij}]$ n × n be any n × n matrix. The transpose B' of the matrix B = $[A_i]$ n × n, where a_{ij} denoted the cofactor of the element a_{ij} in the determinant |A| is called the adjoint of the matrix A ad is denoted by the symbol ads.

Thus the adjoint o A matrix A is the transpose o the matrix formed by the cofactor of A if

$$A = \begin{bmatrix} a_{11} & a_{12} & a_{1n} \\ a_{21} & a_{22} & a_{2n} \\ \cdots\cdots & & \\ a_n & a_{n2} & a_{nn} \end{bmatrix}$$

Then Adj A = the transpose of the matrix $\begin{bmatrix} A_{11} & A_{22} & A_{1n} \\ A_{21} & A_{22} & A_{2n} \\ \cdots\cdots & \cdots\cdots & \cdots\cdots \\ A_{n1} & A_{n2} & A_{nn} \end{bmatrix}$

1.21 THE INVERSE OF A MATRIX

If for a given square matrix **A**, there exists a matrix **B** such that **AB** = **BA** = **I**, where **I** is a nuit matrix, then **A** is called **non-singular** and **B** is called **inverse of A** and we write **B** = $\mathbf{A}^{-1}$ (read as **B** equals **A** inverse).

Here **A** is the inverse of **B** and we can write $\mathbf{A} = \mathbf{B}^{-1}$.

If B *i.e.,* $\mathbf{A}^{-1}$ does not exist, then **A** is called **singular.**

Note 1: Non-square matrix has no inverse.

For example: $\begin{bmatrix} 1 & 2 \\ 3 & 4 \end{bmatrix} \begin{bmatrix} -2 & 1 \\ \dfrac{3}{2} & -\dfrac{1}{2} \end{bmatrix} = \begin{bmatrix} 1 & 0 \\ 0 & 1 \end{bmatrix} = I$

Each matrix in the product is the inverse of the other.

Note 2: If **AB** and **BA** are both defined and equal then the matrices **A** and **B** should both be square matrices of the same order.

1.22 THEOREMS ON INVERSE OF A MATRIX

Theorem 1:

If **A** *and* **B** *be two non-singular matrices of the same order then* **AB** *is also non-singular and* $(\mathbf{AB})^{-1} = \mathbf{B}^{-1}\,\mathbf{A}^{-1}$.

Or

The inverse of a product is the product of the inverse taken in the reverse order. This is also known as the **Reciprocal lawe for the inverse of a product.**

Proof:

$\mathbf{A}^{-1}$ and $\mathbf{B}^{-1}$ exisxts since **A** and **B** are non-singular.

$\therefore (\mathbf{AB})\,(\mathbf{B}^{-1}\,\mathbf{A}^{-1}) = \mathbf{A}\,(\mathbf{BB}^{-1})\,\mathbf{A}^{-1}$, by associative law

$= \mathbf{AIA}^{-1} = \mathbf{AA}^{-1}$

$= \mathbf{I}.$

And $(\mathbf{B}^{-1}\,\mathbf{A}^{-1})\,(\mathbf{AB}) = \mathbf{B}^{-1}\,(\mathbf{A}^{-1}\,\mathbf{A})\,\mathbf{B}$, by associative law

$= \mathbf{B}^{-1}\,(\mathbf{I})\,\mathbf{B},$ $\qquad \because \mathbf{A}^{-1}\,\mathbf{A} = \mathbf{I}$

$= \mathbf{B}^{-1}\,(\mathbf{IB}) = \mathbf{B}^{-1}\,\mathbf{B},$

$= \mathbf{I}.$

$\therefore (\mathbf{B}^{-1}\,\mathbf{A}^{-1})\,(\mathbf{AB}) = (\mathbf{AB})\,(\mathbf{B}^{-1}\,\mathbf{A}^{-1}) = \mathbf{I}$

i.e., $\mathbf{B}^{-1}\,\mathbf{A}^{-1}$ is the inverse of **AB** or $(\mathbf{AB})^{-1} = \mathbf{B}^{-1}\,\mathbf{A}^{-1}$, and as such **AB** is also non-singualr.

Theorem 2:

If a given square matrix A has an inverse, then it is unique or there exists one and only one inverse matrix to a given matrix.

Proof:

Let us suppose that **B** and **C** are two possible inverse of **A.** Then we must have

$$\mathbf{AB} = \mathbf{BA} = \mathbf{I} \qquad ...(1)$$

and $$\mathbf{AC} = \mathbf{CA} = \mathbf{I} \qquad ...(2)$$

$\therefore$ From (1) and (2), we get **AB** = **AC**, each being equal to **I**

$\Rightarrow$ $\mathbf{B\,(AB)} = \mathbf{B\,(AC)}$

$\Rightarrow$ $\mathbf{(BA)\,B} = \mathbf{(BA)\,C}$

$\Rightarrow$ $\mathbf{IB} = \mathbf{IC}$, from (1)

$\Rightarrow$ $\mathbf{B} = \mathbf{C}$

Hence there cannot be two inverses of **A**.

1.23 ORTHOGONAL MATRIX

Definition: *A square matrix **A** is called an orthogonal matrix if **AA'** = **I**, where **I** is an identity matrix and **A'** is the transposed matrix of **A**.*

Theorems on Orthogonal Matrices

Theorem 1:

For any two orthogonal matrices A and B, show that AB is an orthogonal matrix.

Proof:

If A and B are orthogonal matrices, then by definition we have

$$AA' = A'A = I \qquad ...(1)$$

and $$BB' = B'B = I \qquad ...(2)$$

$$\therefore (AB)\,(AB)' = (AB)\,(B'A),$$
$$= AB\,B'A' = A'\,(BB')\,A'$$
$$= AIA', \text{ from } (2)$$
$$= AA' = I, \text{ from } (1).$$

Similarly we can prove that

$$(AB)'\,(AB) = B'A'\,AB,$$
$$= B'IB, \text{ from } (1)$$
$$= B'B = I, \text{ from } (2)$$

Hence AB is an orthogonal matrix by definition.

Theorem 2:

For any square matrix **A**, *if* **AA'** = **I**, *then* **A'A** = **I**.

Proof:

Since **AA'** = **I**, so **A** is invertible (*i.e.*, **A** possesses an inverse) and there exists another matrix **B** such that

$$\mathbf{AB} = \mathbf{BA} = \mathbf{I} \qquad ...(1)$$

Now $\mathbf{B} = \mathbf{BI} = \mathbf{B}\,(\mathbf{AA}')$, $\quad \because \ \mathbf{AA}' = \mathbf{I}$ (given)

$= (\mathbf{BA})\,\mathbf{A}' = \mathbf{IA}'$, from (1)

i.e., $\mathbf{B} = \mathbf{A}'$

∴ From (1), we get $\mathbf{AA}' = \mathbf{A}'\mathbf{A} = \mathbf{I}$. **Hence proved.**

Theorem 3:

If **A** *is an orthogonal matrix, then* **A'** *is also orthogonal.*

Proof:

By definition if **A** is an orthogonal matrix, then

$$\mathbf{AA}' = \mathbf{A}'\mathbf{A} = \mathbf{I}$$

⇒ $(\mathbf{AA}')' = (\mathbf{A}'\mathbf{A})' = \mathbf{I}$, transposing and remembering $\mathbf{I}' = \mathbf{I}$

⇒ $(\mathbf{A}')'\,\mathbf{A}' = \mathbf{A}'\,(\mathbf{A}')' = \mathbf{I}$.

⇒ **A'** is othogonal by definition. **Hence proved.**

i.e., Transpose of an orthogonal matrix is also orthogonal

Theorem 4:

If **A** *is an othogonal matrix, then* $\mathbf{A}^{-1}$ *is also orthogonal*

Poof:

By definition if **A** is orthogonal, then

$$\mathbf{AA}' = \mathbf{A}'\mathbf{A} = \mathbf{I}$$

⇒ $(\mathbf{AA}')^{-1} = (\mathbf{A}'\mathbf{A})^{-1} = \mathbf{I}$,

taking inverse and remembering $\mathbf{I}^{-1} = \mathbf{I}$

⇒ $(\mathbf{A}')^{-1}\,\mathbf{A}^{-1} = \mathbf{A}^{-1}\,(\mathbf{A}')^{-1} = \mathbf{I}$,

⇒ $(\mathbf{A}^{-1})'\,\mathbf{A}^{-1} = \mathbf{A}^{-1}\,(\mathbf{A}^{-1})' = \mathbf{I}$ **(Note)**

⇒ $\mathbf{A}^{-1}$ is orthogonal by definition. **Hence proved.**

i.e., Inverse of an orthogonal matrix is also orthogonal matrix is also orthogonal.

1.24 UNITARY MATRIX

Definition: *A square matrix A is called an unitary matrix if* $A^{\Theta} = I$, *where I is an identity matrix and* A^{Θ} *is the transposed conjugate of A.*

Theormes on Unitary Matrices

Theorem 1:

For any two unitary matrices A and B show that Ab is an unitary matrix.

Proof:

If **A** and **B** are unitary matrices, then by definition we have

$$AA^{\Theta} = A^{\Theta}\ \mathbf{A} = \mathbf{I} \qquad ...(1)$$

and $$BB^{\Theta} = B^{\Theta}\ \mathbf{B} = \mathbf{I} \qquad ...(2)$$

$$\therefore (AB)\ (AB)^{\Theta} = (AB)\ (B^{\Theta}\ A^{\Theta}),$$

$$= \mathbf{A}\ (BB^{\Theta})\ A^{\Theta} = AIA^{\Theta}, \text{ from (2)}$$

$$= AA^{\Theta} = \mathbf{I}, \text{ from (1)}$$

Similarly $(AB)^{\Theta}\ (AB) = B^{\Theta}\ A^{\Theta}\ \mathbf{AB},$

$$= B^{\Theta}\ \mathbf{IB}, \text{ from (1)}$$

$$= B^{\Theta}\ \mathbf{B} = \mathbf{I}, \text{ from (2)}$$

Hence AB is an Unitary Matrix. **Hence proved.**

Theorem 2:

If A is an unitary matrix, then A^{-1} *is also unitary.*

Proof:

By definition if **A** is an unitary matrix, then

$$AA^{\Theta} = \mathbf{A}^{\varnothing}\mathbf{A} = \mathbf{I}$$

$$\Rightarrow \quad (AA^{\Theta})^{-1} = (A^{\Theta}\ A)^{-1} = \mathbf{I}, \text{ taking inverse}$$

$\Rightarrow \quad (A^{\Theta})^{-1}\ \mathbf{A^{-1}} = \mathbf{A^{-1}}\ (A^{\Theta})^{-1} = \mathbf{I}$

$\Rightarrow \quad (A^{-1})^{\Theta}\ \mathbf{A^{-1}} = \mathbf{A^{-1}}\ (A^{-1})^{\Theta} = \mathbf{I}$ **(Note)**

$\Rightarrow$ $\mathbf{A^{-1}}$ is an unitary matrix by definition. **Hence proved.**

Theorem 3:

If **A** *is an unitary matrix, then* **A'** *is also unitary.*

Proof:

By definition if A is an unitary matrix, then

$$\mathbf{AA^{\Theta} = A^{\Theta}A = I}.$$

$\Rightarrow \quad \mathbf{(AA^{\Theta})^{\Theta} = (A^{\Theta}A)^{\Theta} = I}$ taking transposed conjugate and remembering that $I^{\Theta} = \mathbf{I}$ **(Note)**

$\Rightarrow \quad (A^{\Theta})^{\Theta}\ A^{\Theta} = A^{\Theta}\ (A^{\Theta})^{\Theta} = \mathbf{I},$

$\Rightarrow \quad AA^{\Theta} = A^{\Theta}\ \mathbf{A} = \mathbf{I}$, since $(A^{\Theta})^{\Theta} = \mathbf{A}$

$\Rightarrow \quad (AA^{\Theta})' = (A^{\Theta}A)' = \mathbf{I}$, taking transpose of each side

$\Rightarrow \quad (A^{\Theta})'\ A' = A'\ (A^{\Theta})' = \mathbf{I}$ using

$\Rightarrow \quad (A')^{\Theta}\ \mathbf{A'} = \mathbf{A'}\ (A')^{\Theta} = \mathbf{I}$ **(Note)**

$\Rightarrow$ **A'** is an uniary matrix. **Hence proved.**

Theorem 4:

For any square matrix **A,** *if* $AA^{\Theta} = \mathbf{I}$, *then* $A^{\Theta}\ \mathbf{A} = \mathbf{I}$.

Proof:

Since $AA^{\Theta} = \mathbf{I}$, where **I** is the unit matrix, so we find that **A** is invertible and there exists another matrix **B** such that

$$\mathbf{AB = BA = I} \qquad ...(1)$$

Now $\mathbf{B = BI = B}\ (AA^{\Theta})$, $\quad \because AA^{\Theta} = \mathbf{I}$ (given)

$= \mathbf{(BA)}\ A^{\Theta} = \mathbf{I}A^{\Theta}$ form (1)

i.e., $\quad \mathbf{B} = A^{\Theta}$

$\therefore$ From (1), we get $AA^{\Theta} = A^{\Theta}A = I.$ **Hence proved.**

Example 1:

Prove that the matrix $\frac{1}{\sqrt{3}}\begin{bmatrix} 1 & 1+i \\ 1-i & -1 \end{bmatrix}$ *is unitary.*

Solution:

$$\text{Let } A = \frac{1}{\sqrt{3}}\begin{bmatrix} 1 & 1+i \\ 1-i & -1 \end{bmatrix}$$

$$\text{Then } A^{\Theta} = \frac{1}{\sqrt{3}}\begin{bmatrix} 1 & 1+i \\ 1-i & -1 \end{bmatrix}$$

$$\therefore A^{\Theta}A = \frac{1}{\sqrt{3}}\begin{bmatrix} 1 & 1+i \\ 1-i & -1 \end{bmatrix} \times \frac{1}{\sqrt{3}}\begin{bmatrix} 1 & 1+i \\ 1-i & -1 \end{bmatrix}$$

$$= \frac{1}{3}\begin{bmatrix} 1.1 + (1+i).(1-i) & 1.(1+i) + (1+i)(-1) \\ (1-i).1 + (-1)(1-i) & (1-i)(1+i) + (-1)(-1) \end{bmatrix}$$

$$= \frac{1}{3}\begin{bmatrix} 1 + i - i^2 & 0 \\ 0 & 1 - i^2 + 1 \end{bmatrix} = \frac{1}{3}\begin{bmatrix} 3 & 0 \\ 0 & 3 \end{bmatrix} = \begin{bmatrix} 1 & 0 \\ 0 & 1 \end{bmatrix} = I.$$

Example 2:

Show that the matrix $A = \begin{bmatrix} \cos\alpha & \sin\alpha \\ -\sin\alpha & \cos\alpha \end{bmatrix}$ *is orthogonal.*

Solution:

$$A' = \begin{bmatrix} \cos\alpha & -\sin\alpha \\ \sin\alpha & \cos\alpha \end{bmatrix}$$

$$\therefore A'A = \begin{bmatrix} \cos\alpha & -\sin\alpha \\ \sin\alpha & \cos\alpha \end{bmatrix}\begin{bmatrix} \cos\alpha & \sin\alpha \\ -\sin\alpha & \cos\alpha \end{bmatrix}$$

$$= \begin{bmatrix} \cos^2\alpha + \sin^2\alpha & \cos\alpha\sin\alpha - \sin\alpha\cos\alpha \\ \sin\alpha\cos\alpha - \cos\alpha\sin\alpha & \sin^2\alpha\cos^2\alpha \end{bmatrix}$$

$$= \begin{bmatrix} 1 & 0 \\ 0 & 1 \end{bmatrix} = I. \text{ Hence A orthogonal.}$$

1.25 PARTITIONING OF MATRICES

Submatrix

Definition: *A matrix obtained by striking off some of the rows and columns of another matrix A is defined as a* **sub-matrix** *of A.*

For example if $A = \begin{bmatrix} 2 & 3 & 1 \\ 3 & 5 & 7 \end{bmatrix}$, then [2], [3], [5] etc.

$\begin{bmatrix} 2 & 3 \\ 3 & 5 \end{bmatrix}, \begin{bmatrix} 3 & 1 \\ 5 & 7 \end{bmatrix}$ etc. are all sub-martices of A.

It is sometimes found useful to subdivide a matrix into submatrices by drawing lines parallel to its rows and columns and to consider these sub-matrices as the elements of the original matrix.

Consider the matrix,

$$A = \left[\begin{array}{ccc:cc} x_1 & y_1 & z_1 & \alpha_1 & \beta_1 \\ x_2 & y_2 & z_2 & \alpha_2 & \beta_2 \\ x_3 & y_3 & z_3 & \alpha_3 & \beta_3 \\ \hdashline p_2 & q_1 & r_1 & a_1 & b_1 \\ p_2 & q_2 & r_2 & a_2 & b_2 \end{array}\right]$$

Let $A_{11} = \begin{bmatrix} x_1 & y_1 & z_1 \\ x_2 & y_2 & z_2 \\ x_3 & y_3 & z_3 \end{bmatrix}$; $A_{12} = \begin{bmatrix} \alpha_1 & \beta_1 \\ \alpha_2 & \beta_2 \\ \alpha_3 & \beta_3 \end{bmatrix}$;

$A_{21} = \begin{bmatrix} p_1 & q_1 & r_1 \\ p_2 & q_2 & r_2 \end{bmatrix}$; $A_{22} = \begin{bmatrix} a_1 & b_1 \\ a_2 & b_2 \end{bmatrix}$

Then we may write $A = \begin{bmatrix} A_{11} & A_{12} \\ A_{21} & A_{22} \end{bmatrix}$.

The matrix A is then said to have been *partitioned* and the dotted lies indicate the partitions. Here it is obvious that a matrix can be partitioned in several ways. The elements A_{11}, A_{12}, A_{21} and A_{22} are themeselves matrices and are the sub-matrices of A.

Identically Partitioned Matrices

Two matrices of the same size know as identically partitioned matrices if when expressed as matrices of matrices (*i.e.,* when partitioned) they are of the same order and the corresponding sub-matrices (or elements) are also of the same size. Such matrices are said to be *additively coherent.*

For exmaple:

$$\begin{pmatrix} 1 & 2 & 3 & : & 7 & 8 \\ 4 & 5 & 6 & : & 9 & 5 \\ \cdots & \cdots & \cdots & \cdots & \cdots & \cdots \\ 2 & 3 & 4 & : & 2 & 3 \\ 5 & 6 & 7 & : & 4 & 5 \\ 4 & 5 & 8 & : & 6 & 7 \end{pmatrix} \text{ and } \begin{pmatrix} 1 & 2 & 4 & : & 3 & 0 \\ 2 & 0 & 5 & : & 4 & 6 \\ \cdots & \cdots & \cdots & \cdots & \cdots & \cdots \\ 1 & 0 & 2 & : & 1 & 2 \\ 2 & 5 & 4 & : & 3 & 4 \\ 2 & 6 & 2 & : & 5 & 6 \end{pmatrix}$$

Two matrices A and B, which are conformble to the procuct AB, are called *multiplicately coherent* if A and B are parittioned in such a way that the columns of A are partitioned in the same way as the rows of B are partitioned. Here the rows of A and columns of B can be partitioned in any way.

$$\text{Let } A = \begin{bmatrix} 1 & 0 & 0 & 0 \\ 2 & 0 & 0 & 0 \\ 2 & 3 & 1 & 1 \end{bmatrix} \text{ and } B = \begin{bmatrix} 2 & 3 & 5 \\ 3 & 7 & 1 \\ 4 & 0 & 2 \\ 2 & 5 & 1 \end{bmatrix}$$

Here A is a 3×4 matrix and B is a 4×3 matrix, so these are conformable to the product AB (*i.e.,* the product AB exists), Now if we write

$$A = \begin{bmatrix} 1 & 0 & 0 & : & 0 \\ 2 & 1 & 0 & : & 0 \\ \cdots & \cdots & \cdots & \cdots & \cdots \\ 2 & 3 & 1 & : & 1 \end{bmatrix} \text{ and } B = \begin{bmatrix} 2 & : & 3 & 5 \\ 3 & : & 7 & 1 \\ 4 & : & 0 & 2 \\ \cdots & \cdots & \cdots & \cdots \\ 2 & : & 5 & 1 \end{bmatrix}$$

then the partitioning of the columns of A is in the same way as the partitioning of the rows of **B**.(Here we note that after third column in A the partitioning has been done and in **B** the partitioning has been done after thired row). Thus, according to definition given above the matrices A and B are called multiplicative coherent.

SOLVED EXAMPLES

Example 1:

If the product of two non-zero square matrices is a zero matrix, then pove that both of them are singular matrices.

Solution:

Let **A** and **B** be two non-zero $n \times n$ martices.

Given that $\mathbf{AB} = \mathbf{O}$, who **O** is the $n \times n$ null matrix

Let us suppose that **B** is a non-singular matrix then $\mathbf{B^{-1}}$ exists.

Then $\mathbf{AB = O \Rightarrow (AB)\ B^{-1}\ OB^{-1}}$,

post multiplying both sides by $\mathbf{B^{-1}}$

$$\Rightarrow \mathbf{A\ (BB^{-1}) = O},$$

by associative law of multiplication. **(Note)**

$$\Rightarrow \mathbf{AI = O}, \quad \because \quad \mathbf{BB^{-1} = I}$$

$$\Rightarrow \mathbf{A = O},$$

which is against hypothesis as **A** is a non-zero matrix.

Hence B is not a non-singular matrix *i.e.,* B is a singular matrix.

Similarly we can prove that **A** is also a singular matrix.

Example 2:

Show that

$$\begin{bmatrix} 1 & 2 & 3 \\ 2 & 5 & 7 \\ -2 & -4 & -5 \end{bmatrix} \textit{ is the inverse of } \begin{bmatrix} 3 & -2 & -1 \\ -4 & 1 & -1 \\ 2 & 0 & 1 \end{bmatrix}$$

Solution:

$$\begin{bmatrix} 1 & 2 & 3 \\ 2 & 5 & 7 \\ -2 & -4 & -5 \end{bmatrix} \times \begin{bmatrix} 3 & -2 & -1 \\ -4 & 1 & -1 \\ 2 & 0 & 1 \end{bmatrix}$$

$$= \begin{bmatrix} 1.3 + 2(-4) + 3.2 & 1(-2) + 2.1 + 3.0 & 1(-1) + 2(-1) + 3.1 \\ 2.3 + 5(-4) + 7.2 & 2(-2) + 5.1 + 7.0 & 2(-1) + 5(-1) + 7.1 \\ -2.3 - 4(-4) - 5.2 & -2(-2) - 4.1 - 5.0 & -2(-1) - 4.1(-1) - 5.1 \end{bmatrix}$$

$$= \begin{bmatrix} 1 & 0 & 0 \\ 0 & 1 & 0 \\ 0 & 0 & 1 \end{bmatrix}, \text{ which is an unit matrix.}$$

Hence $\begin{bmatrix} 1 & 2 & 3 \\ 2 & 5 & 7 \\ -2 & -4 & -5 \end{bmatrix}$ is the iverse of $\begin{bmatrix} 3 & -2 & 1 \\ -4 & 1 & -1 \\ 2 & 0 & 1 \end{bmatrix}$

Example 3:

If A is a non-singular matrix, then prove that AB = AC ⇒ B = C, where B and C are square matrices of the same order.

Solution:

Since **A** is a non-singular matrix, so $\mathbf{A^{-1}}$ exists.

Now $\mathbf{AB = AC \Rightarrow A^{-1}\,(AB) = A^{-1}\,(AC)}$,

premeultiplying both sides by $\mathbf{A^{-1}}$

$\Rightarrow \mathbf{(A^{-1}\,A)\,B = (A^{-1}\,A)\,C}$,

by associative law of multiplication

$\Rightarrow \mathbf{IB = IC}$, $\quad\because\quad \mathbf{A^{-1}\,A = I}$

$\Rightarrow \mathbf{B = C}$, $\quad\because\quad \mathbf{IB = B}$ etc. **Hence proved.**

Example 4:

Find a if

$$[a \quad 4 \quad 1] \times \begin{bmatrix} 2 & 1 & 0 \\ 1 & 0 & 2 \\ 0 & 2 & 4 \end{bmatrix} \times \begin{bmatrix} a \\ 4 \\ 1 \end{bmatrix} = O, \text{ where } O\ 1 \times 1 \text{ null matrix.}$$

Solution:

$$[a \quad 4 \quad 1] \times \begin{bmatrix} 2 & 1 & 0 \\ 1 & 0 & 2 \\ 0 & 2 & 4 \end{bmatrix}$$

$= [2a + 4 + 0 \quad a + 0 + 2 \quad 0 + 8 + 4]$ **(Note)**

$= [2a + 4 \quad a + 2 \quad 12]$

$$\therefore [a \quad 4 \quad 1] \times \begin{bmatrix} 2 & 1 & 0 \\ 1 & 0 & 2 \\ 0 & 2 & 4 \end{bmatrix} \times \begin{bmatrix} a \\ 4 \\ -1 \end{bmatrix}$$

$$= [2a + 4 \quad a + 2 \quad 12] \times \begin{bmatrix} a \\ 4 \\ -1 \end{bmatrix}$$

$= [\{(2a + 4) \times a\} + (a + 2).4 + 12\,(-1)]$ **(Note)**

$= [2a^2 + 4a + 4a + 8 - 12] = [2a^2 + 8a - 4] = [0]$; given

$\therefore 2a^2 + 8a - 4 = 0 \quad \Rightarrow \quad a^2 + 4a - 2 = 0$

$\Rightarrow \qquad a = \frac{1}{2} [-4 \pm \sqrt{(4 + 8)}] = -2 \pm \sqrt{6}$ **Ans.**

Example 5:

If $A = \begin{bmatrix} 2 & 3 & 4 \\ 1 & 2 & 3 \\ -1 & 1 & 2 \end{bmatrix}$ *and* $B = \begin{bmatrix} 1 & 3 & 0 \\ -1 & 2 & 1 \\ 0 & 0 & 2 \end{bmatrix}$ *find AB and BA.*

Solution:

$$AB = \begin{bmatrix} 2 & 3 & 4 \\ 1 & 2 & 3 \\ -1 & 1 & 2 \end{bmatrix} \times \begin{bmatrix} 1 & 3 & 0 \\ -1 & 2 & 1 \\ 0 & 0 & 2 \end{bmatrix}$$

$$= \begin{bmatrix} 2.1 + 3.(-1) + 4.0 & 2.3 + 3.2 + 4.0 & 2.0 + 3.1 + 4.2 \\ 1.1 + 2.(-1) + 3.0 & 1.3 + 2.2 + 3.0 & 1.0 + 2.1 + 3.2 \\ -1.1 - 1.1 + 2.0 & -1.3 + 1.2 + 2.0 & -1.0 + 1.1 + 2.2 \end{bmatrix}$$

$$= \begin{bmatrix} 2 - 3 + 0 & 6 + 6 + 0 & 0 + 3 + 8 \\ 1 - 2 + 0 & 3 + 4 + 0 & 0 + 2 + 6 \\ -1 - 1 + 0 & -3 + 2 + 0 & 0 + 1 + 4 \end{bmatrix} = \begin{bmatrix} -1 & 12 & 11 \\ -1 & 7 & 8 \\ -2 & - & 5 \end{bmatrix}$$

$$\text{And } BA = \begin{bmatrix} 1 & 3 & 0 \\ -1 & 2 & 1 \\ 0 & 0 & 2 \end{bmatrix} \times \begin{bmatrix} 2 & 3 & 4 \\ 1 & 2 & 3 \\ 1 & 1 & 2 \end{bmatrix}$$

$$= \begin{bmatrix} 1.2 + 3.1 + 0.(-1) & 1.3 + 3.2 + 0.1 & 1.4 + 3.3 + 0.2 \\ (-1).2 + 2.1 + (-1) & (-1).3 + 2.2 + 1.1 & (-1).4 + 2.3 + 1.2 \\ 0.2 + 0.1 + 2.(-1) & 0.3 + 0.2 + 2.1 & 0.4 + 0.3 + 2.2 \end{bmatrix}$$

$$= \begin{bmatrix} 2 + 3 + 0 & 3 + 6 + 0 & 4 + 9 + 0 \\ -2 + 2 - 1 & -3 + 4 + 1 & -4 + 6 + 2 \\ 0 + 0 - 2 & 0 + 0 + 2 & 0 + 0 + 4 \end{bmatrix} = \begin{bmatrix} 5 & 9 & 13 \\ -1 & 2 & 4 \\ -2 & 2 & 4 \end{bmatrix}$$

Example 6:

If $A = \begin{bmatrix} i & 0 \\ 0 & i \end{bmatrix}$, *prove that* $A^n = I_2, A, -I_2 - A$ *according as* $n = 4p$, $4p + 1$, $4p + 2$ *and* $4p + 3$ *respectively.*

Solution:

$$\text{Given } A = \begin{bmatrix} i & 0 \\ 0 & i \end{bmatrix} \quad ...(1)$$

$$\therefore A^2 = A.A = \begin{bmatrix} i & 0 \\ 0 & i \end{bmatrix} \times \begin{bmatrix} i & 0 \\ 0 & i \end{bmatrix}$$

$$= \begin{bmatrix} i.i + 0.0 & i.0 + 0.i \\ 0.i + i.0 & 0.0 + i.i \end{bmatrix} = \begin{bmatrix} i^2 & 0 \\ 0 & i^2 \end{bmatrix} \quad ...(2)$$

$$A^3 = A^2.\ A = \begin{bmatrix} i^2 & 0 \\ 0 & i^2 \end{bmatrix} \times \begin{bmatrix} i & 0 \\ 0 & i \end{bmatrix}$$

$$= \begin{bmatrix} i^2.i + 0.0 & i^2.0 + 0.i \\ 0.i + i^2.0 & 0.0 + i^2.i \end{bmatrix} = \begin{bmatrix} i^3 & 0 \\ 0 & i^3 \end{bmatrix} \quad ...(3)$$

From (2) and (3) we get $A^2 = \begin{bmatrix} i^2 & 0 \\ 0 & i^2 \end{bmatrix}, A^3 = \begin{bmatrix} i^3 & 0 \\ 0 & i^3 \end{bmatrix}$

Let us assume that $A^n = \begin{bmatrix} i^n & 0 \\ 0 & i^n \end{bmatrix}$ ...(4)

and also assume that (4) is true when n = k.

i.e $A^k = \begin{bmatrix} i^k & 0 \\ 0 & i^k \end{bmatrix}$ ...(5)

$$\therefore A^{k+1} = A^k.A = \begin{bmatrix} i^k & 0 \\ 0 & i^k \end{bmatrix} \times \begin{bmatrix} i & 0 \\ 0 & i \end{bmatrix}$$

$$= \begin{bmatrix} i^k.i + 0.0 & i^k.0 + 0.i \\ 0.i + i^k.0 & 0.0 + i^k.i \end{bmatrix} = \begin{bmatrix} i^{k+1} & 0 \\ 0 & i^{k+1} \end{bmatrix}$$

$\therefore$ (4) is true for n = k + 1 provided (5) is true.

Also we have shown in (2) and (3) that (4) is true for n = 2 and 3. So it is true for 3 + 1 *i.e.,* 4 and so on.

Hence (4) is true for all positive integrate values of n.

Also if n = 4p then from (4) we get

$$A^n = \begin{bmatrix} i^{4p} & 0 \\ 0 & i^{4p} \end{bmatrix} = \begin{bmatrix} 1 & 0 \\ 0 & 1 \end{bmatrix}, \text{ Since } i^{4p} = (i^4)^p = (1)^p = 1, \text{ where } i = \sqrt{(-1)}$$

i.e., $A^n = I_2$. **Hence proved.**

If n = 4p + 1, then $i^n = i^{4p+1} = (i)^{4p}.\ i = 1.\ i = i$

$\therefore$ From (4), we get $A^n = \begin{bmatrix} i^n & 0 \\ 0 & i^n \end{bmatrix} = \begin{bmatrix} i & 0 \\ 0 & i \end{bmatrix} = A$. **Hence proved.**

If n = 4p + 2, then $i^n = i^{4p+2} = i^{4p} = \times\ i^2$

$= (1)\ (-1)$, since $i^{4p} = 1$, $i^2 = -1$

$= -1$.

$\therefore$ From (4), we get

$$A^n = \begin{bmatrix} i^n & 0 \\ 0 & i^n \end{bmatrix} = \begin{bmatrix} -1 & 0 \\ 0 & -1 \end{bmatrix} = -\begin{bmatrix} 1 & 0 \\ 0 & 1 \end{bmatrix}$$

$\Rightarrow$ $A^n = -I_2$. **Hence proved.**

If n = 4p + 3, then $i^n = i^{4p+3} = (i^{4p+2}).\ i = (-1)\ i$, as above

$\therefore$ From (4), we get

$$A^n = \begin{bmatrix} i^n & 0 \\ 0 & i^n \end{bmatrix} = \begin{bmatrix} -i & 0 \\ 0 & -i \end{bmatrix} = -\begin{bmatrix} i & 0 \\ 0 & i \end{bmatrix}$$

$\Rightarrow$ $A^n = -A$, from (1). **Hence proved.**

Example 7:

If AB = BA then prove that $(AB)^n = A^nB^m$.

Solution:

We shall prove this by mathematical induction.

If n = 1, then $(AB)^n = A^nB^n \Rightarrow (AB)^1 = AB$, which is true.

If n = 2, then

$(AB)^n = (AB)^2 = (AB)\ (AB)$

$= (ABA)\ B$, by associative law

$= (AAB)\ B$, $\because$ BA = AB, given

$= A^2 B^2$

Hence $(AB)^n = A^n B^n$ is true for n = 2.

Now suppose that it is true for n = m *i.e.,* $(AB)^m = A^m B^m$

$\Rightarrow \quad (AB)^m (AB) = (A^m B^m)(AB)$

$\Rightarrow \quad (AB)^{m+1} = A^m (B^m A) B$, by associative law

$\Rightarrow \quad (AB)^{m+1} = A^m (B^{m-1} BA) B, \quad \because \quad B^m = B^{m-1} B$

$= A^m (B^{m-1} AB) B, \quad \because \quad BA = AB$, given

$= A^m (B^{m-2} BAB) B, \quad \because \quad B^{m-1} = B^{m-2} B$

$= A^m (B^{m-2} ABB) B, \quad \because \quad BA = AB$, given

$= A^m (B^{m-2} AB^2) B$

$= A^m (B^{m-2} B^2) B = (A^m A)(B^{m-2} B^2 B)$

$= A^{m+1} B^{m+1}$

i.e., If $(AB)^n A^n B^n$ is true for n = m, it is true for n = m + 1.

Also we have proved that it is true for n = 1 and 2.

Hence, by mathematical induction it is true for all + ve integral values of n.

Example 8:

If A and B are two matrices such that AB and A + B are both defined, then prove that A and B are square matrices.

Solution:

Let A be an m × n matrix.

Since A + B is defined *i.e.,* A and B are conformable to addition, so B must also be an m × n matrix.

Again AB is defined *i.e.,* A and B are comfortable to multiplication and hence the number of columns in A must be equal to the number of rows in B *i.e.,* n = m.

Hence A and B are m × m matrices *i.e.,* square matrices.

Example 9:

If $p(x) = \begin{bmatrix} \cos x & \sin x \\ -\sin x & \cos x \end{bmatrix}$, *then show that* $P(x) \cdot P(y) = P(x+y)$ $= P(y) P(x)$.

Solution:

$$P(x) \cdot P(y)$$

$$= \begin{bmatrix} \cos x & \sin x \\ -\sin x & \cos x \end{bmatrix} + \begin{bmatrix} \cos y & \sin y \\ -\sin y & \cos y \end{bmatrix}$$

$$= \begin{bmatrix} \cos x \cos y - \cos x \sin y & \cos x \sin y + \sin x \cos y \\ -\sin x \cos y - \cos x \sin y & -\sin x \sin y + \cos x \cos y \end{bmatrix}$$

$$= \begin{bmatrix} \cos(x+y) & \sin(x+y) \\ -\sin(x+y) & \cos(x+y) \end{bmatrix} = P(x+y)$$

Similarly we can prove (to be proved in the exam.) that

$P(y) . P(x) = P(y) . P(x)$

Hence $P(x) \cdot P(y) = P(x+y) = P(y) \cdot P(x)$ **Hence proved.**

Example 10:

Evaluate $\begin{bmatrix} \cos\theta + \sin\theta & \sqrt{2}\sin\theta \\ -\sqrt{2}\sin\theta & \cos\theta - \sin\theta \end{bmatrix}^n$

Solution:

$$\text{Let } A = \begin{bmatrix} \cos\theta + \sin\theta & \sqrt{2}\sin\theta \\ -\sqrt{2}\sin\theta & \cos\theta - \sin\theta \end{bmatrix} \quad ...(1)$$

Then $A^2 = A. A$

$$= \begin{bmatrix} \cos\theta + \sin\theta & \sqrt{2}\sin\theta \\ -\sqrt{2}\sin\theta & \cos\theta - \sin\theta \end{bmatrix} \begin{bmatrix} \cos\theta + \sin\theta & \sqrt{2}\sin\theta \\ -\sqrt{2}\sin\theta & \cos\theta - \sin\theta \end{bmatrix}$$

$$= \begin{bmatrix} (\cos\theta + \sin\theta)^2 - 2\sin^2\theta & (\cos + \sin\theta)\sqrt{2}\sin\theta + \sqrt{2}\sin\theta(\cos\theta - \sin\theta \\ -\sqrt{2}\sin\theta(\cos\theta + \sin\theta) - \sqrt{2}\sin\theta(\cos\theta - \sin\theta) & -\sqrt{2}\sin\theta\sqrt{2}\sin\theta + (\cos\theta - \sin\theta)^2 \end{bmatrix}$$

$$= \begin{bmatrix} (\cos^2\theta - \sin^2\theta) + 2\sin\theta\cos\theta & 2\sqrt{2}\sin\theta\cos\theta \\ -2\sqrt{2}\sin\theta\cos\theta & (\cos^2\theta - \sin^2\theta) - 2\cos\theta\sin\theta \end{bmatrix}$$

$$\Rightarrow \quad A^2 = \begin{bmatrix} \cos 2\theta + \sin 2\theta & \sqrt{2}\sin 2\theta \\ -\sqrt{2}\sin 2\theta & (\cos 2\theta - \sin 2\theta \end{bmatrix} \quad ...(2)$$

Looking at (1) and (2), let us assume that

$$A^n = \begin{bmatrix} \cos n\theta + \sin n\theta & \sqrt{2}\sin n\theta \\ -\sqrt{2}\sin n\theta & (\cos n\theta - \sin n\theta \end{bmatrix} \quad ...(3)$$

Let (3) be true for an = k

i.e.,
$$A^k = \begin{bmatrix} \cos k\theta + \sin k\theta & \sqrt{2}\sin k\theta \\ -\sqrt{2}\sin k\theta & (\cos k\theta - \sin k\theta \end{bmatrix} \quad ...(4)$$

$$\therefore \quad A^{k+1} = A^k \cdot A$$

$$= \begin{bmatrix} \cos k\theta + \sin k\theta & \sqrt{2}\sin k\theta \\ -\sqrt{2}\sin k\theta & \cos k\theta - \sin k\theta \end{bmatrix} \times \begin{bmatrix} \cos\theta + \sin\theta & \sqrt{2}\sin\theta \\ -\sqrt{2}\sin\theta & \cos\theta - \sin\theta \end{bmatrix}$$

$$= \begin{bmatrix} (\cos k\theta + \sin k\theta)(\cos\theta + \sin\theta) & (\cos k\theta + \sin k\theta)(\sqrt{2}\sin\theta) \\ +(\sqrt{2}\sin k\theta)(-\sqrt{2}\sin\theta) & +(\sqrt{2}\sin k\theta)(\cos\theta - \sin\theta) \\ -\sqrt{2}\sin k\theta(\cos\theta + \sin\theta) & (-\sqrt{2}\sin k\theta)(\sqrt{2}\sin\theta) \\ +(\cos k\theta - \sin k\theta)(-\sqrt{2}\sin\theta) & +(\cos k\theta - \sin k\theta(\cos\theta - \sin\theta) \end{bmatrix}$$

$$= \begin{bmatrix} \cos k\theta\cos\theta + \cos k\theta\sin\theta + \sin k\theta\cos\theta & \sqrt{2}(\sin k\theta\cos\theta \\ +\sin k\theta\sin\theta - 2\sin k\theta\sin\theta & +\cos k\theta\sin\theta) \\ & -2\sin k\theta + \sin\theta + \cos k\theta\cos\theta \\ -\sqrt{2}(\sin k\theta\cos\theta + \cos k\theta\sin\theta) & -\cos k\theta\sin\theta - \sin k\theta\cos\theta \\ & +\sin k\theta\sin\theta \end{bmatrix}$$

$$= \begin{bmatrix} \cos(k\theta + \theta) + \sin(k\theta + \theta) & \sqrt{2}\sin(k\theta + \theta) \\ -\sqrt{2}\sin(k\theta + \theta) & \cos(k\theta + \theta) - \sin(k\theta + \theta) \end{bmatrix}$$

$$= \begin{bmatrix} \cos(k + 1)\theta + \sin(k + 1)\theta & \sqrt{2}\sin(k + 1)\theta \\ -\sqrt{2}\sin(k + 1)\theta & \cos(k + 1)\theta - \sin(k + 1)\theta \end{bmatrix}$$

(3) is true for n = k + 1 provided (4) is true.

Also we have shown in (2) that (3) is true for n = 2.

Hence it is true for n = 2 + 1 *i.e.,* 3 and so on.

Hence (3) is true for all positive integral values of n.

Hence $A^n = \begin{bmatrix} \cos n\theta + \sin n\theta & \sqrt{2}\sin n\theta \\ -\sqrt{2}\sin n\theta & \cos n\theta - \sin n\theta \end{bmatrix}$ **Ans.**

Example 11:

If $A = \begin{bmatrix} 1 & -3 & 2 \\ 2 & 1 & -3 \\ 4 & -3 & -1 \end{bmatrix}$; $B = \begin{bmatrix} 1 & 4 & 1 & 0 \\ 2 & 1 & 1 & 1 \\ 1 & -2 & 1 & 2 \end{bmatrix}$ *and* $C = \begin{bmatrix} 2 & 1 & -1 & -2 \\ 3 & -2 & -1 & -1 \\ 2 & -5 & -1 & 0 \end{bmatrix}$,

show that $AB = AC$

Solution:

$$AB = \begin{bmatrix} 1 & -3 & 2 \\ 2 & 1 & -3 \\ 4 & -3 & -1 \end{bmatrix} \times \begin{bmatrix} 1 & 4 & 1 & 0 \\ 2 & 1 & 1 & 1 \\ 1 & -2 & 1 & 0 \end{bmatrix}$$

$$= \begin{bmatrix} 1.1 - 3.2 + 2.1 & 1.4 - 3.1 + 2(-2) & 1.1 - 3.1 + 2.1 & 1.0 - 3.1 + 2.2 \\ 2.1 + 1.2 - 3.1 & 2.4 + 1.1 - 3(-2) & 2.1 + 1.1 - 3.1 & 2.0 + 1.1 - 3.2 \\ 4.1 - 3.2 - 1.1 & 4.4 - 3.1 - 1(-2) & 4.1 - 3.2 - 1.1 & 4.0 - 3.1 - 1.2 \end{bmatrix}$$

$$= \begin{bmatrix} -3 & -3 & 0 & 1 \\ 1 & 15 & 0 & -5 \\ -3 & 15 & 0 & -5 \end{bmatrix}$$ **Ans.**

$$AC = \begin{bmatrix} 1 & -3 & 2 \\ 2 & 1 & -3 \\ 4 & -3 & -1 \end{bmatrix} \times \begin{bmatrix} 2 & 1 & -1 & -2 \\ 3 & -2 & -1 & -1 \\ 2 & -5 & -1 & 0 \end{bmatrix}$$

$$= \begin{bmatrix} 1.2 - 3.3 + 2.2 & 1.1 - 3(-2) + 2(-5) & 1(-1) - 3(-1) + 2(-1) & 1(-2) - 3(-1) + 2.0 \\ 2.2 + 1.3 - 3.2 & 2.1 + 1(-2) - 3(-5) & 2(-1) + 1(-1) - 3(-1) & 2(-2) + 1(-1) - 3.0 \\ 4.2 - 3.3 - 1.2 & 4.1 - 3(-2) - 1(-5) & 4(-1) - 3(-1) - 1(-1) & 4(-2) - 3(-1) - 1.0 \end{bmatrix}$$

$$= \begin{bmatrix} -3 & -3 & 0 & 1 \\ 1 & 15 & 0 & -5 \\ -3 & 15 & 0 & -5 \end{bmatrix}$$

∴ From (1) and (2), we get AB = AC.

[**Note:** Students should see that AB = AC does not necessarily simply that B = C.

Example 12:

If $A = \begin{bmatrix} 1 & 2 & 3 \\ 3 & -2 & 1 \\ 4 & 2 & 1 \end{bmatrix}$, *find the matrix X such that* $A + X + I = 0$, *where I and O are unit and zero* 3×3 *matrices respectively.*

Solution:

Given that $A + X + I = O \Rightarrow X = O - A - I$

$$= \begin{bmatrix} 0 & 0 & 0 \\ 0 & 0 & 0 \\ 0 & 0 & 0 \end{bmatrix} - \begin{bmatrix} 1 & 2 & 3 \\ 3 & -2 & 1 \\ 4 & 2 & 1 \end{bmatrix} - \begin{bmatrix} 1 & 0 & 0 \\ 0 & 1 & 0 \\ 0 & 0 & 1 \end{bmatrix}$$

subtituting values of A, I and O

$$= \begin{bmatrix} 0-1-1 & 0-2-0 & 0-3-0 \\ 0-3-0 & 0+2-1 & 0-1-0 \\ 0-4-0 & 0-2-0 & 0-1-1 \end{bmatrix} = \begin{bmatrix} -2 & -2 & -3 \\ -3 & 1 & -1 \\ -4 & -2 & -2 \end{bmatrix}$$

Example 13:

Given $A = \begin{bmatrix} 1 & 2 & -3 \\ 5 & 0 & 2 \\ 1 & -1 & 1 \end{bmatrix}$ *and* $B = \begin{bmatrix} 3 & -1 & 2 \\ 4 & 2 & 5 \\ 2 & 0 & 3 \end{bmatrix}$ *find the matrix C such that* $A + C = B$.

Solution:

$A + C = B$ or $C = B - A$

$$\Rightarrow \qquad C = \begin{bmatrix} 3 & -1 & 2 \\ 4 & 2 & 5 \\ 2 & 0 & 3 \end{bmatrix} - \begin{bmatrix} 1 & 2 & -3 \\ 5 & 0 & 2 \\ 1 & -1 & 1 \end{bmatrix}$$

$$= \begin{bmatrix} 3-1 & -1-2 & 2+3 \\ 4-5 & 2-0 & 5-2 \\ 2-1 & 0-(0-1) & 3-1 \end{bmatrix} = \begin{bmatrix} 2 & -3 & 5 \\ -1 & 2 & 3 \\ 1 & 1 & 2 \end{bmatrix}$$

Ans.

Example 14:

If $A = \begin{bmatrix} 1 & 2 \\ 3 & 4 \\ 5 & 6 \end{bmatrix}$ *and* $B = \begin{bmatrix} -3 & -2 \\ 1 & -5 \\ 4 & 3 \end{bmatrix}$ *find* $D = \begin{bmatrix} p & q \\ r & s \\ t & u \end{bmatrix}$, *such that* $A + B - D = 0$

Solution:

$$A + B - D = O \text{ or } D = A + B$$

$$\Rightarrow \qquad D = \begin{bmatrix} 1 & 2 \\ 3 & 4 \\ 5 & 6 \end{bmatrix} + \begin{bmatrix} -3 & -2 \\ 1 & -5 \\ 4 & 3 \end{bmatrix}$$

$$= \begin{bmatrix} 1-3 & 2-2 \\ 3+1 & 4-5 \\ 5+4 & 6+3 \end{bmatrix} = \begin{bmatrix} -2 & 0 \\ 4 & -1 \\ 9 & 9 \end{bmatrix} = \begin{bmatrix} p & q \\ r & s \\ t & u \end{bmatrix} \text{ given}$$

$\therefore$ We have $p = -2$, $q = 0$, $r = 4$, $s = -1$, $t = 9$, $u = 9$ which gives D.

Ans.

Example 15:

If $A = \begin{bmatrix} 1 & -1 \\ 2 & -1 \end{bmatrix}$, $B = \begin{bmatrix} a & 1 \\ b & -1 \end{bmatrix}$ *and* $(A + B)^2 = A^2 + B^2$, *find a and b.*

Solution:

Here we have

$$A^2 = \begin{bmatrix} 1 & -1 \\ 2 & - \end{bmatrix} \times \begin{bmatrix} 1 & -1 \\ 2 & -1 \end{bmatrix}$$

$$= \begin{bmatrix} 1-2 & -1+1 \\ 2-2 & -2+1 \end{bmatrix} = \begin{bmatrix} -1 & 0 \\ 0 & -1 \end{bmatrix}$$

$$B^2 = \begin{bmatrix} a & 1 \\ b & -1 \end{bmatrix} \times \begin{bmatrix} a & 1 \\ b & -1 \end{bmatrix} = \begin{bmatrix} a^2+b & a-1 \\ ab-b & b+1 \end{bmatrix}$$

$$\therefore A^2 + B^2 = \begin{bmatrix} -1 & 0 \\ 0 & -1 \end{bmatrix} + \begin{bmatrix} a^2+b & a-1 \\ ab-b & b+1 \end{bmatrix}$$

$$= \begin{bmatrix} -1+a^2+b & 0+a-1 \\ 0+ab-b & -1+b+1 \end{bmatrix} = \begin{bmatrix} a^2+b-1 & a-1 \\ ab-b & b \end{bmatrix} \quad ...(1)$$

$$\text{Also } A + B = \begin{bmatrix} 1 & -1 \\ 2 & -1 \end{bmatrix} + \begin{bmatrix} a & 1 \\ b & -1 \end{bmatrix}$$

$$= \begin{bmatrix} 1+a & -1+1 \\ 2+b & -1-1 \end{bmatrix} = \begin{bmatrix} 1+a & 0 \\ 2+b & -2 \end{bmatrix}$$

$$\therefore (A + B)^2 = \begin{bmatrix} 1+a & 0 \\ 2+b & -2 \end{bmatrix} \times \begin{bmatrix} 1+a & 0 \\ 2+b & -2 \end{bmatrix}$$

$$= \begin{bmatrix} (1+a)^2 + 0 & 0+0 \\ (2+b)(1+a) - 2(2+b) & 0+4 \end{bmatrix}$$

$$= \begin{bmatrix} (1+a)^2 & 0 \\ (2+b)(1+a) & 4 \end{bmatrix} \quad ...(2)$$

Now it is given that $(A + B)^2 = A^2 + B^2$.

$\Rightarrow \quad \begin{bmatrix} (1+a)^2 & 0 \\ (2+b)(1+a) & 4 \end{bmatrix} = \begin{bmatrix} a^2+b-1 & a-1 \\ ab-b & b \end{bmatrix}$, from (1) and (2)

$\Rightarrow \quad 0 = a - 1$ and $4 = b$, comparing the elements of second column on both sides

$\Rightarrow \quad a = 1$ and $b = 4$. **Ans.**

Example 16:

If A, B are two n × n matrices and if

$$C = A + B, AB = BA, B^2 = O,$$

then show that for every integer m,

$$C^{m\ 1} = A^m\ [A + (m + 1)\ B].$$

Solution:

We hall prove that $C^{m+1} = A^m\ [A + (m + 1)\ B]$, ...(1)

by mathematical induction.

For m = 1, from (1) we get $C^2 = A\ [A + 2B]$ ...(2)

Also C = A + B, given

$\therefore \quad C^2 = (A + B)^2 = (A + B)\ (A + B)$

$= A^2 + BA + AB + B^2,$

$= A^2 + 2\ AB$, since $AB = BA$, $B^2 = O$ (given)

$\Rightarrow \quad C^2 = A\ (A + 2B)$, which is the same as (2).

Hence (1) is true for m = 1.

Let us now assume that (1) holds when m = k

i.e., $\quad C^{k+1} = A^k\ [A + (k + 1)\ B]$ ...(3)

Now $C^{k+1} = C^{k+1}\ C$, by

$= A^k\ [A + (k + 1)\ B].\ (A + B)$, from (3) and C = A + B (given)

$\Rightarrow \quad C^k = A^k[A\ (A + B) + (k + 1)\ B\ (A + B)]$

$= A^k[A^2 + AB + (k + 1)\ BA + (k + 1)\ B^2]$

$= A^k[A^2 + AB + (k + 1)\ AB]$, $\quad \because BA = AB,\ B^2 = O$

$= A^k[A^2 + (1 + k + 1)\ AB]$,

$= A^k.A\ [A + \{(k + 1) + 1\}\ B]$

$\Rightarrow \quad C^{k+2} = A^{k+1}\ [A + \{(k + 1) + 1\}\ B]$.

Hence (1) is true for m = k + 1 provided (3) is true *i.e.,* for m = k. Also we have shown that (1) true for m = 1, so it is true for m = 1 + *i.e.,* m = 2 and so on. Hence (1) is true for all positive integral values of m.

Hence proved.

Example 17:

Show that

$$\begin{bmatrix} \cos\theta & -\sin\theta \\ \sin\theta & \cos\theta \end{bmatrix} = \begin{bmatrix} 1 & -\tan\frac{1}{2}\theta \\ \tan\frac{1}{2}\theta & 1 \end{bmatrix} \begin{bmatrix} 1 & \tan\frac{1}{2}\theta \\ -\tan\frac{1}{2}\theta & 1 \end{bmatrix}^{-1}$$

Solution:

We have

$$\begin{bmatrix} \cos\theta & -\sin\theta \\ \sin\theta & \cos\theta \end{bmatrix} \times \begin{bmatrix} 1 & \tan\frac{1}{2}\theta \\ -\tan\frac{1}{2}\theta & 1 \end{bmatrix}$$

$$= \begin{bmatrix} \cos\theta + \sin\theta\tan\frac{1}{2}\theta & \cos\theta\tan\frac{1}{2}\theta - \sin\theta \\ \sin\theta - \cos\theta\tan\frac{1}{2}\theta & \sin\theta\tan\frac{1}{2}\theta + \cos\theta \end{bmatrix}$$

$$= \begin{bmatrix} \dfrac{\cos\theta\cos\frac{1}{2}\theta + \sin\theta\sin\frac{1}{2}\theta}{\cos\frac{1}{2}\theta} & \dfrac{\cos\theta\sin\frac{1}{2}\theta - \sin\theta\cos\frac{1}{2}\theta}{\cos\frac{1}{2}\theta} \\ \dfrac{\sin\theta\cos\frac{1}{2}\theta - \cos\theta\sin\frac{1}{2}\theta}{\cos\frac{1}{2}\theta} & \dfrac{\cos\theta\sin\frac{1}{2}\theta + \cos\theta\cos\frac{1}{2}\theta}{\cos\frac{1}{2}\theta} \end{bmatrix}$$

$$= \frac{1}{\cos\frac{1}{2}\theta}\begin{bmatrix} \cos\theta\cos\frac{1}{2}\theta + \sin\theta\sin\frac{1}{2}\theta & \cos\theta\sin\frac{1}{2}\theta - \sin\theta\cos\frac{1}{2}\theta \\ \sin\theta\cos\frac{1}{2}\theta - \cos\theta\sin\frac{1}{2}\theta & \cos\theta\cos\frac{1}{2}\theta + \sin\theta\sin\frac{1}{2}\theta \end{bmatrix}$$

$$= \left(\sec\frac{1}{2}\theta\right)\begin{bmatrix} \cos\left(\theta - \frac{1}{2}\theta\right) & -\sin\left(\theta - \frac{1}{2}\theta\right) \\ \sin\left(\theta - \frac{1}{2}\theta\right) & \cos\left(\theta - \frac{1}{2}\theta\right) \end{bmatrix}$$

$$= \left(\sec\frac{1}{2}\theta\right)\begin{bmatrix} \cos\frac{1}{2}\theta & -\sin\frac{1}{2}\theta \\ \sin\frac{1}{2}\theta & \cos\frac{1}{2}\theta \end{bmatrix}$$

$$= \begin{bmatrix} \cos\frac{1}{2}\theta\sec\frac{1}{2}\theta & -\sin\frac{1}{2}\theta\sec\frac{1}{2}\theta \\ \sin\frac{1}{2}\theta\sec\frac{1}{2}\theta & \cos\frac{1}{2}\theta\sec\frac{1}{2}\theta \end{bmatrix}$$

$$= \begin{bmatrix} 1 & -\tan\frac{1}{2}\theta \\ \tan\frac{1}{2}\theta & 1 \end{bmatrix}$$

$$\Rightarrow \begin{bmatrix} \cos\theta & -\sin\theta \\ \sin\theta & \cos\theta \end{bmatrix} = \begin{bmatrix} 1 & -\tan\frac{1}{2}\theta \\ \tan\frac{1}{2}\theta & 1 \end{bmatrix} \begin{bmatrix} 1 & \tan\frac{1}{2}\theta \\ -\tan\frac{1}{2}\theta & 1 \end{bmatrix}^{-1}$$

Hence proved

Example 18:

If A and B be n-rowed square matrices, then show that

1. $(A + B)^2 = A^2 + AB + BA + B^2$;
2. $(A + B)(A - B) = A^2 - AB + BA - B^2$;
3. $(A - B)(A + B) = A^2 + AB - BA - B^2$; *and*
4. $(A - B)^2 = A^2 - AB - BA + B^2$.

Solution:

As A and B are n-rowed square matrices therefore A + B and A – B are also n-rowed square matrices and as such distributive law is true.

1. $(A + B)^2 = (A + B) \times (A + B)$

$= (A + B) A + (A + B) B$, by distributive law

$= AA + BA + AB + BB$, by distributive law

$= A^2 + BA + AB + B^2$.

2. $(A + B)(A - B)$

$= (A + B) A + (A + B)(-B)$, by distributive law

$= AA + BA + A(-B) + B(-B)$, by distributive law

$= A^2 + BA - AB - B^2$.

3. $(A - B)(A + B) = (A - B) A + (A - B) B$, by distributive law

$= AA + BA + AB - BB$, by distributive law

$= A^2 - BA + AB - B^2$.

4. $(A - B)^2 = (A - B).(A - B)$

$= AA + A(-B) + (-B) A + (-B)(-B)$, by distributive law

$= A^2 - AB - BA + B^2$. **Hence proved.**

Example 19:

Show that if A, B, C are matrices, such that A (BC) is defined, then (AB) C is also defined and A (BC) = (AB) C.

Solution:

Since A (BC) is defined so the matrices A, B, C are conformable to multiplication and we can take $A = [a_{ij}]$, $B = [b_{jk}]$ and $C = [c_{kl}]$, where A, B, C are $m \times n$, $n \times p$, $p \times q$ matrices.

Then $AB = [a_{ij}] [b_{jk}]$ is a $m \times p$ matrix

i.e., (i, k)th element of the product $AB = \sum_{j=1}^{n} a_{ij} b_{jk}$ **(Note)**

Similarly (j, l)th element of the product $BC = \sum_{k=1}^{p} b_{jk} c_{kl}$ **(Note)**

Also (AB) C is the product of an $m \times p$ and a $p \times q$ matrices and so is conformable to multiplication, hence defined.

$\therefore$ (i, l)th element in the product of (AB) and C

= sum of products of corresponding elements in the ith row of AB and lth column of C with k common

$$= \sum_{k=1}^{p}\left[\left(\sum_{j=1}^{n} a_{ij} b_{jk}\right) c_{kl}\right] \quad \textbf{(Note)}$$

$$= \sum_{k=1}^{p}\sum_{j=1}^{n} a_{ij} b_{jk} c_{kl} \quad ...(1)$$

Again (i, l)th element in the product of A and (BC)

= sum of products of cor corresponding elements in the ith row of A and lth column of (BC)

$$= \sum_{j=1}^{n} a_{ij} \sum_{k=1}^{p} b_{jk} c_{kl} \quad \textbf{(Note)}$$

$$= \sum_{k=1}^{p} \sum_{j=1}^{n} a_{ij} b_{jk} c_{kl} \quad ...(2)$$

$\therefore$ From (1) and (2) we conclude that (AB) C = A (BC).

Example 20:

If $A = \begin{bmatrix} 2 & 0 & 0 \\ 0 & 2 & 0 \\ 0 & 0 & 2 \end{bmatrix}$ *and* $B = \begin{bmatrix} x_1 & y_1 & z_1 \\ x_2 & y_2 & z_2 \\ x_3 & y_3 & z_3 \end{bmatrix}$ *then prove that AB = 2B.*

Solution:

$$AB = \begin{bmatrix} 2 & 0 & 0 \\ 0 & 2 & 0 \\ 0 & 0 & 2 \end{bmatrix} \times \begin{bmatrix} x_1 & y_1 & z_1 \\ x_2 & y_2 & z_2 \\ x_3 & y_3 & z_3 \end{bmatrix}$$

$$= \begin{bmatrix} 2x_1 + 0 + 0 & 2y_1 + 0 + 0 & 2z_1 + 0 + 0 \\ 0 + 2x_2 + 0 & 0 + 2y_2 + 0 & 0 + 2z_2 + 0 \\ 0 + 0 + 2x_3 & 0 + 0 + 2y_3 & 0 + 0 + 2z_3 \end{bmatrix}$$

$$= \begin{bmatrix} 2x_1 & 2y_1 & 2z_1 \\ 2x_2 & 2y_2 & 2z_2 \\ 2x_3 & 2y_3 & 2z_3 \end{bmatrix} = 2 \begin{bmatrix} x_1 & y_1 & z_1 \\ x_2 & y_2 & z_2 \\ x_3 & y_3 & z_3 \end{bmatrix}$$

= 2B. **Hence Proved.**

EXERCISES

1. Show that the matrices $\begin{bmatrix} 0 & 1 \\ 1 & 0 \end{bmatrix}$ *and* $\begin{bmatrix} 1 & 0 \\ 0 & -1 \end{bmatrix}$ anti-commute.

2. Show that the matrices $\begin{bmatrix} 1 & 2 \\ 2 & 1 \end{bmatrix}$ *and* $\begin{bmatrix} 5 & 7 \\ 7 & 5 \end{bmatrix}$ commute.

3. If $A = \begin{bmatrix} 1 & 0 & 0 & 0 \\ 1 & -1 & 0 & 0 \\ 1 & -2 & 1 & 0 \\ 1 & -3 & 3 & 1 \end{bmatrix}$, whow that $A^2 = I$, where I is the unite matrix.

4. If A and B are idempotent, then A + B will be idempotent if AB = BA = O, where O is the null matrix.

   [**Hint:** $(A + B)^2 = A^2 + AB + BA + B^2 = A + O + O + B$]

5. Show that the matrix $\begin{bmatrix} ab & b^2 \\ -a^2 & -ab \end{bmatrix}$ is nilpotent.

6. Show that $\begin{bmatrix} 1 & 1 & 3 \\ 5 & 2 & 6 \\ -2 & -1 & -3 \end{bmatrix}$ is a nilpotent matrix of order 3.

7. If A = $\begin{bmatrix} 1 & -2 & 3 \\ -4 & 2 & 5 \end{bmatrix}$; B = $\begin{bmatrix} 1 & 2 \\ -1 & 0 \\ 2 & 4 \end{bmatrix}$, then show that (AB)" = B'A'.

8. If A = $\begin{bmatrix} 1 & -1 & 0 \\ 2 & 1 & 3 \\ 4 & 1 & 8 \end{bmatrix}$ *and* $B = \begin{bmatrix} 4 & 1 & 0 \\ 2 & -3 & 1 \\ 1 & 1 & -1 \end{bmatrix}$ then verify that (AB)' =B'A'

9. If A and B are symmetric (or skew-symmetric) matrices, then so is A + B.

10. If A and B are symmetric matrices, then prove that AB + BA is symmertic and AB – BA is Skew Symmettic,

11. show that all positive integal powers of s symmetric matrix are symmertic.

12. If A is any matrix, then show that A' A is a symmetric matrix.

13. If A is symmetric matrix, then show that AA' = A'A and A^2 is symmetic.

14. Show that the matrix A = $\frac{1}{\sqrt{2}}\begin{bmatrix} 1 & i \\ -i & -1 \end{bmatrix}$ is unitary.

15. For any two orthogonal matrices A and B, show that BA is an orthogonal matrix.

16. For any two unitary matrices A and B, show that BA is an unitary matrix.

Ans. XY = $\begin{bmatrix} -2 & -4 \\ 3 & 2 \end{bmatrix}$

17. If A = $\begin{bmatrix} 1 & 2 \\ 3 & 4 \\ 5 & 6 \end{bmatrix}$ and B = $\begin{bmatrix} 9 & 8 & 7 \\ 6 & 5 & 4 \end{bmatrix}$, then find AB and BA.

Ans. AB = $\begin{bmatrix} 21 & 18 & 15 \\ 51 & 44 & 37 \\ 81 & 70 & 59 \end{bmatrix}$ and BA = $\begin{bmatrix} 68 & 92 \\ 41 & 56 \end{bmatrix}$

18. Find AB when $A = \begin{bmatrix} 2 & -1 & 0 \\ 0 & 2 & 1 \\ 1 & 0 & 1 \end{bmatrix}$ and $B = \begin{bmatrix} -2 & 1 & -1 \\ 1 & 2 & -2 \\ 2 & -1 & -4 \end{bmatrix}$

Ans. $\begin{bmatrix} -5 & 0 & 0 \\ 4 & 3 & -8 \\ 0 & 0 & -5 \end{bmatrix}$

19. Show that the matrix $A = \begin{bmatrix} 1 & 2 \\ 3 & 1 \end{bmatrix}$ satisfies the equation $A^2 - 2A - 5I = O$, where O is the 2×2 null matrix.

20. Evaluate $A^2 - 3A - 13I$, where I is the 2×2 unit matrix and $A = \begin{bmatrix} 2 & 5 \\ 3 & 1 \end{bmatrix}$

Ans. $\begin{bmatrix} 0 & 0 \\ 0 & 0 \end{bmatrix} = O$

21. Show that matrix $A = \begin{bmatrix} 1 & 0 & 0 \\ 2 & 1 & 0 \\ 3 & 2 & 1 \end{bmatrix}$

satisfies the equation $A^3 - 3A^2 + 3A - I = O$, where I is the unit matrix and O the null matrix of order 3.

22. If $A = \begin{bmatrix} 2 & 3 \\ 4 & -1 \end{bmatrix}$, $B = \begin{bmatrix} 3 & -2 \\ 2 & 1 \end{bmatrix}$, $C = \begin{bmatrix} 1 & 2 \\ 3 & 4 \end{bmatrix}$ verify (1) (AB) C = A (BC); (2) (A + B) C = AC + BC.

23. If $A = \begin{bmatrix} 1 & 2 \\ 3 & 4 \end{bmatrix}$, $B = \begin{bmatrix} 2 & 1 \\ 4 & 2 \end{bmatrix}$, $C = \begin{bmatrix} 5 & 1 \\ 7 & 4 \end{bmatrix}$,

show that A (B + C) = AB + AC.

24. If $A_\alpha = \begin{bmatrix} \cos\alpha & \sin\alpha \\ -\sin\alpha & \cos\alpha \end{bmatrix}$, then show that $A_\alpha^n = \begin{bmatrix} \cos n\alpha & \sin n\alpha \\ -\sin n\alpha & \cos n\alpha \end{bmatrix}$,

where n is any positive integer.

Also prove that A_α and A_β commute and $A_\alpha . A_\beta = A_{\alpha+\beta}$.

25. Show that $\begin{bmatrix} 1 & 1 & 1 \\ 0 & 1 & 1 \\ 0 & 0 & 1 \end{bmatrix}^n = \begin{bmatrix} 1 & n & \frac{1}{2}n(n+1) \\ 0 & 1 & n \\ 0 & 0 & 1 \end{bmatrix}$ for all natural numbers n

26. A fruit seller has in stock 20 dozen mangoes, 16 dozen apples and 32 dozen bananas. Suppose the selling prices are Rs. 0.35, Rs. 0.75 and Rs. 0.08 per mango, apple and banana respectively. Find the total amount the fruit seller will get by selling his whole stock.

Ans. Rs. 258.72

27. If $A = \begin{bmatrix} 2 & 3 & 11 \\ 0 & 2 & -1 \end{bmatrix}$ and $B = \begin{bmatrix} 1 & -2 \\ 3 & 0 \\ -1 & 2 \end{bmatrix}$ then find Ab and BA.

Ans. $AB = \begin{bmatrix} 10 & -2 \\ 7 & -2 \end{bmatrix}$, $BA = \begin{bmatrix} 2 & -1 & 3 \\ 6 & 9 & 3 \\ -2 & 1 & -3 \end{bmatrix}$

28. If $A = \begin{bmatrix} 1 & 2 & 1 \\ 4 & 0 & 2 \end{bmatrix}$ and

$B = \begin{bmatrix} 3 & -4 \\ 1 & 5 \\ -2 & 2 \end{bmatrix}$ find AB and show that $AB \neq BA$.

29. Find AB and BA if

$A = \begin{bmatrix} 3 & 4 & -2 \\ -2 & -1 & -1 \\ -1 & -3 & -1 \end{bmatrix}$ and $B = \begin{bmatrix} -1 & -1 & -1 \\ 2 & 2 & 2 \\ 1 & 1 & 1 \end{bmatrix}$

Ans. $AB = \begin{bmatrix} 7 & 7 & 7 \\ -1 & -1 & -1 \\ -6 & -6 & -6 \end{bmatrix}$, $BA = \begin{bmatrix} 0 & 0 & 0 \\ 0 & 0 & 0 \\ 0 & 0 & 0 \end{bmatrix}$

30. If $A = \begin{bmatrix} 2 & -3 & -5 \\ -1 & 4 & 5 \\ 1 & -3 & -4 \end{bmatrix}$ and $B = \begin{bmatrix} 2 & -2 & -4 \\ -1 & 3 & 4 \\ 1 & -2 & -3 \end{bmatrix}$

verify that AB = A and BA = B.

31. Show that

$\begin{bmatrix} 1 & 0 & 0 & 0 \\ 2 & 1 & 0 & 0 \\ 4 & 2 & 1 & 0 \\ -2 & 3 & 1 & 1 \end{bmatrix}$ *is the inverse oi* $\begin{bmatrix} 1 & 0 & 0 & 0 \\ -2 & 1 & 0 & 0 \\ 0 & -2 & 1 & 0 \\ 8 & -1 & -1 & 1 \end{bmatrix}$

32. If A be any square matrix, then show that

$$A + A^{\Theta} \text{ is Hermitian.}$$

33. If A and B are symmetric and they commute, then $A^{-1}B$ and $A^{1}B^{1}$ are symmetic

34. Show that every square matrix can be expressed is one and only one way as P + iQ, where P and Q are Hermitian.

35. If B is any square matrix, show that B' AB is symmetric provided B' AB is defined.

36. Find the matrices A and B, when

$$A + B = \begin{bmatrix} 1 & 0 & 2 \\ 2 & 2 & 2 \\ 1 & 1 & 2 \end{bmatrix} \text{ and } B = \begin{bmatrix} 1 & 4 & 4 \\ 4 & 2 & 0 \\ -1 & -1 & 2 \end{bmatrix}$$

Ans. $A = \begin{bmatrix} 1 & 2 & 3 \\ 3 & 2 & 1 \\ 0 & 0 & 2 \end{bmatrix}$ $B = \begin{bmatrix} 0 & -2 & -1 \\ -1 & 0 & 1 \\ 1 & 1 & 0 \end{bmatrix}$

37. If X, Y are two matrices given by the equations $X + Y = \begin{bmatrix} 1 & -2 \\ 3 & 4 \end{bmatrix}$

and $X - Y = \begin{bmatrix} 3 & 2 \\ -1 & 0 \end{bmatrix}$, find X, Y. **Ans.** $X = \begin{bmatrix} 2 & 0 \\ 1 & 2 \end{bmatrix}, Y = \begin{bmatrix} -1 & -2 \\ 2 & 2 \end{bmatrix}$

38. If $A = \begin{bmatrix} 1 & 2 & 3 \\ 0 & 5 & 7 \\ 6 & 8 & 9 \end{bmatrix}, B = \begin{bmatrix} 2 & 0 & 3 \\ 3 & 0 & 5 \\ 5 & 7 & 0 \end{bmatrix}$ evaluate 2A – 3B.

Ans. $\begin{bmatrix} -4 & 4 & -3 \\ -9 & 10 & -1 \\ -3 & -5 & 18 \end{bmatrix}$

39. Multiply [4 5 6] and $\begin{bmatrix} 2 \\ 3 \\ -1 \end{bmatrix}$ **Ans.** [17]

40. Multiply [1 2 3] and $\begin{bmatrix} 4 & -6 & 9 & 6 \\ 0 & -7 & 10 & 7 \\ 5 & 8 & -11 & -8 \end{bmatrix}$ **Ans.** [19 4 –4 –4]

41. If $A = \begin{bmatrix} 1 & -1 \\ -1 & 1 \end{bmatrix}$ and $B = \begin{bmatrix} 1 & 1 \\ 1 & 1 \end{bmatrix}$, wow that AB is a null matrix.

42. Show that $\begin{bmatrix} 0 & 0 & 1 \\ 0 & 1 & 0 \\ 1 & 0 & 0 \end{bmatrix} \times \begin{bmatrix} 0 & 1 & 0 \\ 0 & 0 & 1 \\ 1 & 0 & 0 \end{bmatrix} = \begin{bmatrix} 1 & 0 & 0 \\ 0 & 0 & 1 \\ 0 & 1 & 0 \end{bmatrix}$

43. If $A = \begin{bmatrix} 1 & 1 & -1 \\ -2 & 3 & -4 \\ 3 & -2 & 3 \end{bmatrix}$, $B = \begin{bmatrix} -1 & -2 & -1 \\ 6 & 12 & 6 \\ 5 & 10 & 5 \end{bmatrix}$ then prove that AB = O but BA ≠ O.

44. Show that

$$\begin{bmatrix} 4 & 2 & -1 & 2 \\ 3 & -7 & 1 & -8 \\ 2 & 4 & -3 & 1 \end{bmatrix} \times \begin{bmatrix} 2 & 3 \\ -3 & 0 \\ 1 & 5 \\ 3 & 1 \end{bmatrix} = \begin{bmatrix} 7 & 9 \\ 4 & 6 \\ -8 & -8 \end{bmatrix}$$

45. If $A = \begin{bmatrix} 1 & -2 & 3 \\ 2 & 3 & -1 \\ -3 & 1 & 2 \end{bmatrix}$ and $B = \begin{bmatrix} 1 & 0 & 2 \\ 0 & 1 & 2 \\ 1 & 2 & 0 \end{bmatrix}$ then prove that AB ≠ BA.

46. Show that $\begin{bmatrix} -1 & 1 & 1 \\ 1 & -1 & 1 \\ 1 & 1 & -1 \end{bmatrix} \times \begin{bmatrix} 0 & \frac{1}{2} & \frac{1}{2} \\ \frac{1}{2} & 0 & \frac{1}{2} \\ \frac{1}{2} & \frac{1}{2} & 0 \end{bmatrix} = \begin{bmatrix} 1 & 0 & 0 \\ 0 & 1 & 0 \\ 0 & 0 & 1 \end{bmatrix}$

47. Form the products AB and BA, when

$A = [1 \quad 2 \quad 3 \quad 4]$ and $B = \begin{bmatrix} 5 \\ 4 \\ 3 \\ 2 \end{bmatrix}$ **Ans.** [30]

48. If $A = \begin{bmatrix} 1 & 2 \\ 4 & -3 \end{bmatrix}$, evaluate A^2. **Ans.** $\begin{bmatrix} 9 & -4 \\ -8 & 17 \end{bmatrix}$

2

Transpose Matrix

2.1 INTRODUCTION OF TRANSPOSE MATRIX

Definition: *The matrix of order $n \times m$ obtained by interchanging the rows and columns of a matrix A of order $m \times n$ is called the transpose matrix of A or transpose of the matrix A and is denoted by A' or A^t.*

Or

If $A = [a_{ij}]$ be a matrix of order $m \times n$, then the matrix $B = [b_{ij}]$ of order $n \times m$, such that $b_{ij} = a_{ij}$ is known as *transposed matrix* of A or *the transpose of the matrix* A and is denoted by **A'** or $\mathbf{A}^t$.

$$\text{If } A = \begin{bmatrix} 1 & 3 & 5 \\ 2 & 4 & 6 \end{bmatrix} \text{ then } A' = \begin{bmatrix} 1 & 2 \\ 3 & 4 \\ 5 & 6 \end{bmatrix}$$

Theorem:

The transpose of the transpose of a matrix is the matrix itself i.e., **(A') = A.**

Proof:

Let $\mathbf{A} = [a_{ij}]$ be an $m \times n$ matrix. The **A'** *i.e.,* the transpose of **A** is an $n \times m$ matrix and **(A')** *i.e.,* The transpose of **A'** (or the transpose of **A**) is an $m \times n$ matrix.

Therefore, the matrices **A** and **(A')** are both $m \times n$ matrices and hence comparable. ...(1)

Also, the element in the ith row and jth column of **(A')'**

= the element in the jth row and ith column of **A'**

= the element in the jth row and ith column of **A**

i.e.,, the corresponding elements of **(A')** and **A** are equal ...(2)

∴ From (1) and (2), we conclude that **(A') = A.** **Hence proved.**

2.2 SOME IMPORTANT THEOREMS ON TRANSPOSED MATRICES

Theorem 1:

The transpose of the sum of two matrices is the sum of their transpose i.e., **(A + B)' = A' + B.**

Proof:

Let A = $[a_{ij}]$ and B = $[b_{ij}]$.

Then **A + B** = $[a_{ij} + b_{ij}] = [c_{ij}]$, say

then $c_{ij} = a_{ij} + b_{ij}$

∴ **(A + B)'** $[d_{ji}]$, where $d_{ji} = c_{ij}$ for all $1 \le i \le m$, $1 \le j \le n$

i.e., $d_{ji} = a_{ij} + b_{ij}$, for all $1 \le i \le m$, $1 \le j \le n$

⇒ **(A + B)'** = $[c_{ij}] = [a_{ij} + b_{ij}]$

Also **A'** = $[f_{ji}]$, where $f_{ji} = a_{ij}$ for all $1 \le i \le m$, $1 \le j \le n$

and **B'** = $[f_{ji}]$, where $g_{ji} = b_{ij}$ for all $1 \le i \le m$, $1 \le j \le n$

∴ **A' + B'** = $[f_{ji}] + [g_{ji}] = [f_{ji} + g_{ji}]$

$= [a_{ij} + b_{ij}]$ ...(2)

∴ From (1) and (2) we get **(A + B)' = A' + B'ss**

Theorem 2:

The transpose of the product of two matrices is the product in reverse order of their transpose i.e. **(AB)' = B'A'.**

Proof:

Let A = $[a_{ij}]$ and B = $[b_{ij}]$ be the two matrices of orderes m × n and n × p respectively.

Let **C = AB** = $[a_{ij}] \times [b_{ij}] = [c_{ij}]$, say,

where **C** is a matrix of order m × p.

∴ The element in the ith row and jth column of **AB** is $c_{ij} = \sum_{k=1}^{n} a_{jk} b_{kj}$.

This is also the element in the ith row and jth column of **(AB)'**. ...(1)

The elements in the jth rwo of **B'** are $b_{1j}, b_{2j}, b_{3j}, \ldots, b_{nj}$ and the elements in the ith column of **A'** are $a_{i1}, a_{i2}, a_{i3}, \ldots, a_{in}$. Then the elements in the jth row and ith column of **B' A'** is

$$\sum_{k=1}^{n} b_{kj}\, a_{ik} = \sum_{k=1}^{n} a_{ik}\, b_{kj} = c_{ij} \qquad \ldots(2)$$

Hence from (1) and (2) we conclude the **(AB)' = B'A'**.

Theorem 3:

if A is any m × n matrix, then $(k\mathbf{A})' = k\mathbf{A}'$*, where k is any number*

Proof:

Let $A = [a_{ij}]$ be any m × n matrix. Then k**A** is also an m × n matrix and therefore (k**A**)' *i.e.,* the transpose of the matrix k**A** is an n × m matrix.

Also A', the transpose of the matrix A, is an n × m matrix and kA' is also an n × m matrix.

Thus, we find that the matrices (kA)' and kA' are both n × m matrices and hence comparable. ...(1)

Again the element in ith row and jth column of (kA)'

= the element in jth row and ith column of kA

= k times the element in jth row and ith column of A **(Note)**

= k times the element in ith row and jth column of A' **(Note)**

= $k\, a_{ji}$ **(Note)**

= the element in ith row and jth column of kA'

i.e., the corresponding elements of (kA)' and kA' are equal ...(2)

∴ From (1) and (2), we conclude that (kA)' = kA'. **Hence proved.**

Example 1:

If $A = \begin{bmatrix} 2 & 3 \\ 0 & 1 \end{bmatrix}$, $B = \begin{bmatrix} 3 & 4 \\ 2 & 1 \end{bmatrix}$ *then verify that [AB]' = B'A'.*

Solution:

$$AB = \begin{bmatrix} 2 & 3 \\ 0 & 1 \end{bmatrix} \times \begin{bmatrix} 3 & 4 \\ 2 & 1 \end{bmatrix}$$

$$= \begin{bmatrix} 2.3 + 3.2 & 2.4 + 3.1 \\ 0.3 + 1.2 & 0.4 + 1.1 \end{bmatrix} = \begin{bmatrix} 12 & 11 \\ 2 & 1 \end{bmatrix}$$

$\therefore$ [AB]' = transposed matrix of AB

$\Rightarrow$ $[AB]' = \begin{bmatrix} 12 & 2 \\ 11 & 1 \end{bmatrix}$, by defintion ...(1)

Again $B' = \begin{bmatrix} 3 & 2 \\ 4 & 1 \end{bmatrix}$ and $A' = \begin{bmatrix} 2 & 0 \\ 3 & 1 \end{bmatrix}$

$$\therefore \; B'A' = \begin{bmatrix} 3 & 2 \\ 4 & 1 \end{bmatrix} \times \begin{bmatrix} 2 & 0 \\ 3 & 1 \end{bmatrix}$$

$$= \begin{bmatrix} 3.2 + 2.3 & 3.0 + 2.1 \\ 4.2 + 1.3 & 4.0 + 1.1 \end{bmatrix} = \begin{bmatrix} 12 & 2 \\ 11 & 1 \end{bmatrix}$$

= [AB]', from (1). **Hence proved.**

Example 2:

If $A = \begin{bmatrix} \cos\alpha & -\sin\alpha \\ -\sin\alpha & \cos\alpha \end{bmatrix}$, *verify that* $AA' = I_2 = A'A$.

Solution:

Here $A' = \begin{bmatrix} \cos\alpha & -\sin\alpha \\ \sin\alpha & \cos\alpha \end{bmatrix}$

$$\therefore \; AA' = \begin{bmatrix} \cos\alpha & \sin\alpha \\ -\sin\alpha & \cos\alpha \end{bmatrix} \times \begin{bmatrix} \cos\alpha & -\sin\alpha \\ \sin\alpha & \cos\alpha \end{bmatrix}$$

$$= \begin{bmatrix} \cos^2\alpha + \sin^2\alpha & -\cos\alpha\sin\alpha + \sin\alpha\cos\alpha \\ -\sin\alpha\cos\alpha + \cos\alpha\sin\alpha & \sin^2\alpha + \cos^2\alpha \end{bmatrix}$$

$$= \begin{bmatrix} 1 & 0 \\ 0 & 1 \end{bmatrix} = I_2.$$

Similarly we can prove that

$$AA' = \begin{bmatrix} \cos\alpha & -\sin\alpha \\ \sin\alpha & \cos\alpha \end{bmatrix} \times \begin{bmatrix} \cos\alpha & \sin\alpha \\ -\sin\alpha & \cos\alpha \end{bmatrix}$$

$$= \begin{bmatrix} \cos^2\alpha + \sin^2\alpha & \cos\alpha\sin\alpha - \sin\alpha\cos\alpha \\ \sin\alpha\cos\alpha - \cos\alpha\sin\alpha & \sin^2\alpha + \cos^2\alpha \end{bmatrix}$$

$$= \begin{bmatrix} 1 & 0 \\ 0 & 1 \end{bmatrix} = \mathbf{I_2}.$$

Hence $\mathbf{AA' = I_2 = A'A}$.

Example 3:

Verify that $(B)^t (A)^t = (AB)^t$, when

$$A = \begin{bmatrix} 1 & 1 & 2 \\ 2 & 1 & 0 \end{bmatrix} \text{ and } B = \begin{bmatrix} 1 & 2 \\ 2 & 0 \\ -1 & 1 \end{bmatrix}$$

Solution:

$$AB = \begin{bmatrix} 1 & 1 & 2 \\ 2 & 1 & 0 \end{bmatrix} \times \begin{bmatrix} 1 & 2 \\ 2 & 0 \\ -1 & 1 \end{bmatrix}$$

$$= \begin{bmatrix} 1.1 + 1.2 + 2.(-1) & 1.2 + 1.0 + 2.1 \\ 2.1 + 1.2 + 0.(-1) & 2.2 + 1.0 + 0.1 \end{bmatrix} = \begin{bmatrix} 1 & 4 \\ 4 & 4 \end{bmatrix}$$

$\therefore$ $\mathbf{(AB)^t}$ = **transposed matrix of AB**

$$= \begin{bmatrix} 1 & 4 \\ 4 & 4 \end{bmatrix}, \text{ by definition} \qquad ...(1)$$

Again $\mathbf{(B)^t} = \begin{bmatrix} 1 & 2 & -1 \\ 2 & 0 & 1 \end{bmatrix}$ and $\mathbf{(A)^t} = \begin{bmatrix} 1 & 2 \\ 1 & 1 \\ 2 & 0 \end{bmatrix}$

$$\therefore \mathbf{B^t A^t} = \begin{bmatrix} 1 & 2 & -1 \\ 2 & 0 & 1 \end{bmatrix} \times \begin{bmatrix} 1 & 2 \\ 1 & 1 \\ 2 & 0 \end{bmatrix}$$

$$= \begin{bmatrix} 1.1 + 2.1 - 1.2 & 1.2 + 2.1 - 1.0 \\ 2.1 + 0.1 + 1.2 & 2.2 + 0.1 + 1.0 \end{bmatrix}$$

$$= \begin{bmatrix} 1 & 4 \\ 4 & 4 \end{bmatrix} = \mathbf{(AB)^t}, \text{ from (1).}$$

Hence proved.

Example 4:

Write down the transpose of the matrix.

$$A = \begin{bmatrix} 1 & 2 & 4 \\ 6 & 8 & 1 \end{bmatrix}$$

Solution:

Let A' be the required transpose of the matrix A. then A' = matrix obtained by interchanging the rows and columns of the matrix A

$$= \begin{bmatrix} 1 & 6 \\ 2 & 8 \\ 4 & 1 \end{bmatrix}$$

Ans.

Example 4:

Verify that $(AB)' = B' A'$, where

$$A = \begin{bmatrix} 1 & 1 & 1 \\ 2 & 2 & 3 \\ 2 & 4 & 9 \end{bmatrix} \text{ and } B = \begin{bmatrix} 0 & 1 & -1 \\ -3 & 2 & 4 \\ 1 & 1 & 0 \end{bmatrix}$$

Solution:

$$AB = \begin{bmatrix} 1 & 1 & 1 \\ 2 & 2 & 3 \\ 2 & 4 & 9 \end{bmatrix} \times \begin{bmatrix} 0 & 1 & -1 \\ -3 & 2 & 4 \\ 1 & 1 & 0 \end{bmatrix}$$

$$= \begin{bmatrix} 1.0 + 1.(-3) + 1.1 & 1.1 + 1.2 + 1.1 & 1.(-1) + 1.4 + 1.0 \\ 2.0 + 2.(-3) + 3.1 & 2.1 + 2.2 + 3.1 & 2.(-1) + 2.4 + 3.0 \\ 2.0 + 4.(-3) + 9.1 & 2.1 + 4.2 + 9.1 & 2.(-1) + 4.4 + 9.0 \end{bmatrix}$$

$$= \begin{bmatrix} -2 & 4 & 3 \\ -3 & 9 & 6 \\ -3 & 19 & 14 \end{bmatrix}$$

$\therefore$ $(\mathbf{AB})^t$ = Transposed matrix of **(AB)**

$\Rightarrow$ $(\mathbf{AB})^t = \begin{bmatrix} -2 & -3 & -3 \\ 3 & 9 & 19 \\ 3 & 6 & 14 \end{bmatrix}$, by defintion

Again $A^t = \begin{bmatrix} 1 & 2 & 2 \\ 1 & 2 & 4 \\ 1 & 3 & 9 \end{bmatrix}$, $B^t = \begin{bmatrix} 0 & -3 & 1 \\ 1 & 2 & 1 \\ -1 & 4 & 0 \end{bmatrix}$

$$\therefore \quad B^tA^t = \begin{bmatrix} 0 & -3 & 1 \\ 1 & 2 & 1 \\ -1 & 4 & 0 \end{bmatrix} \times \begin{bmatrix} 1 & 2 & 2 \\ 1 & 2 & 4 \\ 1 & 3 & 9 \end{bmatrix}$$

$$= \begin{bmatrix} 0.1 - 3.1 + 1.1 & 0.2 - 3.2 + 1.3 & 0.2 - 3.4 + 1.9 \\ 1.1 + 2.1 + 1.1 & 1.2 + 2.2 + 1.3 & 1.2 + 2.4 + 1.9 \\ -1.1 + 4.1 + 0.1 & -1.2 + 4.2 + 0.3 & -1.2 + 4.4 + 0.9 \end{bmatrix}$$

$$= \begin{bmatrix} -2 & -3 & -3 \\ 4 & 9 & 19 \\ 3 & 6 & 14 \end{bmatrix} = (\mathbf{AB})^t, \text{ from (1)}$$

Hence proved.

2.3 COMPLEX CONJUGATE (OR CONJUGATE) OF A MATRIX

Definition: The matrix obtained from any given matrix A of order $m \times n$ with complex elements a_{ij} by replaclng its elements by the corresponding conjugate complex numbers is called the complex conjugate or conjugate of **A** and is denoted by $\overline{A}$ and read as '**A** conjugate'.

$\Rightarrow$ if $A = [a_{ij}]$ and $\overline{a}_{ij}$ is the complex conjugate of the element a_{ij} then $\overline{A} = [\overline{a}_{ij}]$, for all $1 \le i \le m$, $1 \le j \le n$.

For exmaple: If $\mathbf{A} = \begin{bmatrix} 1+i & 2+3i \\ 2 & 3i \end{bmatrix}$,

then $\overline{\mathbf{A}} = \begin{bmatrix} \overline{1+i} & \overline{2+3i} \\ \overline{2} & \overline{3i} \end{bmatrix}$

Rea Matrix

Definition: **A** matrix **A** is called real provided it satisfles the relation

$$\mathbf{A} = \overline{\mathbf{A}}.$$

Imaginary Matrix

Definition: *A m,atrix A is called imaginary provided it satisfies the relation*

$$\mathbf{A} = -\overline{\mathbf{A}}.$$

2.4 THEOREMS ON COMPLEX CONJUGATE OF A MATRIX

Theorem 1:

If $\mathbf{A} = [a_{ij}]$ *be any* $m \times n$ *matrix and* $\mathbf{B} = [b_{ij}]$ *be any* $n \times p$ *matrix i.e., if* **A** *and* **B** *are conformable to the product* **AB**, *then* $\overline{\mathbf{A}\mathbf{B}} = \overline{\mathbf{A}} + \overline{\mathbf{B}}$.

Proof:

Since **A** and **B** are conformable to the product **AB,** so **AB** = $[a_{ij}] \times [b_{jk}]$ = $[c_{ij}]$, where $c_{ik} = a_{ij}\, b_{jk}$, for all $1 \le i \le m$, $1 \le k \le p$ and there is summation on j where j = 1, 2, 3,...,n.

Also $\overline{\mathbf{A}} = [\bar{a}_{ij}]$, for all $1 \le i \le m$, $1 \le j \le n$

and $\overline{\mathbf{B}} = [\bar{b}_{jk}]$, for all $1 \le j \le n$, $1 \le k \le p$

$\overline{\mathbf{A}}\,\overline{\mathbf{B}}$ is defined and we have

$\overline{\mathbf{A}}\,\overline{\mathbf{B}}\,[\bar{a}_{ij}] \times [\bar{b}_{jk}] = [d_{ik}]$ ...(1)

where $d_{ik} = \bar{a}_{ij}\,\bar{b}_{jk}$ for all $1 \le i \le m$, $1 \le k \le p$ and j = 1, 2,...,n.

Again $\overline{\mathbf{AB}}$ = complex conjugate of **AB** *i.e.,* $[c_{ik}]$

$= [\bar{c}_{ik}]$, where $c_{ik} = a_{ij}\, b_{jk}$

$= [\overline{a_{ij}\, b_{jk}}] = [\bar{a}_{ij}\,\bar{b}_{jk}]$ $\because \overline{z_1 z_2} = \bar{z}_1 \bar{z}_2$, for any complex numbers z_1 and z_2

$= [d_{ik}]$, since $d_{ik} = \bar{a}_{ij}\,\bar{b}_{jk}$ for all $1 \le i \le m$,

$1 \le k \le p$ and j = 1, 2,..., n ...(2)

$\therefore$ From (1) and (2), we conclude that $\overline{\mathbf{AB}} = \overline{\mathbf{A}}\,\overline{\mathbf{B}}$.

Theorem 2:

If **A** = $[a_{ij}]$ *be any m × n matrix with complex elements* a_{ij}*, then* $\overline{\lambda \mathbf{A}} = \lambda \overline{\mathbf{A}}$.

Proof:

By defintion, we know

$\overline{\mathbf{A}}\,[\bar{a}_{ij}]$, for all $1 \le i \le m$, $1 \le j \le n$ and $\bar{a}_{ij}$ is the complex conjugate of a_{ij}.

Also $\lambda A = [\lambda a_{ij}]$, for all $1 \le i \le m$, $1 \le j \le n$

$\therefore\ \overline{\lambda \mathbf{A}} = [\overline{\lambda\, a_{ij}}] = [\overline{\lambda\, a_{ij}}]$ for all $1 \le i \le m$, $1 \le j \le n$ ...(1)

and we know that $\overline{z_1 z_2} = \bar{z}_1 \bar{z}_2$, where z_2, z_2 are any two complex numbers.

Again $\bar{\lambda}\,\bar{A} = [b_{ij}]$, where b_{ij}

$= \overline{\lambda\, a_{ij}}$ for all $1 \le i \le m,\ 1 \le j \le n$

$= [\bar{\lambda}\,\bar{a}_{ij}]$, for all $1 \le i \le m,\ 1 \le j \le n.$ ...(2)

∴ From (1) and (2) we conclude that the corresponding elements of $\overline{\lambda A}$ and $\bar{\lambda}\,\bar{A}$ are equal. Also it is evident that $\overline{\lambda A}$ and $\bar{\lambda}\,\bar{A}$ are matrices of the same order. Hence, we conclude that

$$\overline{\lambda A} = \bar{\lambda}\,\bar{A}.$$

Theorem 3:

If $A = [a_{ij}]$ be any $m \times n$ matrix with complex elements a_{ij}, then the complex conjugate of $\bar{A}$ is the matrix A itself.

Proof:

We know that $\bar{A} = [\bar{a}_{ij}]$, for all $1 \le i \le m,\ 1 \le j \le n$ and $\bar{a}_{ij}$ is the complex conjugate of a_{ij}.

i.e., the element in the ith row and jth column of complex conjugate of a *i.e.*, $\bar{A}$.

= the complex conjugate of the element in ith row and jth column of A

∴ The element in ith row and jth column of the complex conjugate of $\bar{A}$ *i.e.*, $\bar{\bar{A}}$

= the complex conjugate of the element in ith row and jth column of $\bar{A}$

= the complex conjuate of $\bar{a}_{ij}$ **(Note)**

= a_{ij} *i.e.*, the element in ith row and jth column of **A** **(Note)**

i.e., the corresponding elements of **A** and the complex conjugate of $\bar{A}$ are equal. ...(1)

Also it is evident that **A**, $\bar{A}$ and its complex conjugate are $m \times n$ matrices and hence comparable. ...(2)

∴ From (1) and (2), we conclude that the complex conjugate of $\bar{A}$ is equal to **A** or $\bar{\bar{A}} = A$.

Theorem 4:

If **A** *and* **B** *are two matrices conformable to addition, then* $\overline{\mathbf{A}+\mathbf{B}} = \overline{\mathbf{A}} + \overline{\mathbf{B}}$.

Proof:

Let A = $[a_{ij}]$ and B = $[b_{ij}]$ be any two matrices of order m × n. Then as these matrices are given as conformable to addition, so we have

$$\mathbf{A} + \mathbf{B} = [a_{ij} + b_{ij}], \text{ for all } 1 \le i \le m,\ 1 \le j \le n. \qquad ...(1)$$

Also $\overline{\mathbf{A}} = [\bar{a}_{ij}]$ and $\overline{\mathbf{B}} = [\bar{b}_{ij}]$, by definition.

$$\therefore\ \overline{\mathbf{A}} + \overline{\mathbf{B}} = \left[\bar{a}_{ij} + \bar{b}_{ij}\right] = \left[\overline{a_{ij} + b_{ij}}\right], \qquad ...(2)$$

for all $1 \le i \le m,\ 1 \le j \le n$.

and also a $\bar{z}_j + \bar{z}_2 = \overline{z + z_2}$, where z_1, z_2

are any two complex numbers.

Again from (1), we have

$\overline{\mathbf{A}+\mathbf{B}}$ = complex conjugate of $[a_{ij} + b_{ij}]$

= complex conjugate of $[c_{ij}]$, were $c_{ij} = a_{ij} + b_{ij}$

= $[\bar{c}_{ij}] = [\overline{a_{ij} + b_{ij}}]$, for all $1 \le i \le m,\ 1 \le j \le n$

i.e., $\overline{\mathbf{A}+\mathbf{B}} = [\overline{a_{ij} + b_{ij}}]$, for all $1 \le i \le m,\ 1 \le j \le n$ ...(3)

∴ From (2) and (3) we conclude that the corresponding elements of $\overline{\mathbf{A}} + \overline{\mathbf{B}}$ and $\overline{\mathbf{A}+\mathbf{B}}$ are matrices of order m × **n** as **A** and **B** are given as conformable to addition. Hence we conclude that $\overline{\mathbf{A}+\mathbf{B}} = \overline{\mathbf{A}} + \overline{\mathbf{B}}$.

2.5 TRANSPOSED CONJUGATE OF A MATRIX

Definition: *The transpose of conjugate of a matrix A i.e.,* $(\overline{\mathbf{A}})'$ *is defined as transposed conjugate or tranjugate of A and is denoted by* A^{Θ}

i.e.,, $A^{\Theta} = (\overline{\mathbf{A}})$.

For example: If A = $\begin{bmatrix} 1+i & 2+3i \\ 2 & 3i \end{bmatrix}$

$$\text{then} \quad \overline{A} = \begin{bmatrix} 1-i & 2-3i \\ 2 & -3i \end{bmatrix}$$

$$\therefore A^{\Theta} = \text{transpose of } \overline{A} = (\overline{A})'$$

$$= \begin{bmatrix} 1-i & 2 \\ 2-3i & -3i \end{bmatrix}$$

2.6 THEOREMS ON TRANSPOSED CONJUGATE OF A MATRIX

Theorem 1:

For an matrix **A**, $(A^{\Theta})^{\Theta} = A$.

Proof:

Let $A^{\Theta} = B$ *i.e.*, $B = (\overline{A})'$

Then B' = tanspose of **B**

$= \text{transpose of } (\overline{A})'$

$= \overline{A}$, since we know $(A')' = A$

$\therefore \quad (\overline{B})$ = complex conjugate of **B'**

= complex conjugate of $\overline{A}$

= **A**, since we know $\overline{\overline{A}} = A$

i.e., $\quad B^{\Theta} = A$, since $B^{\Theta} = (\overline{B}) = (\overline{B})'$

i.e., $\quad (A^{\Theta})^{\Theta} = A$, since $A^{\Theta} = B$. **Hence proved.**

2.7 TRIANGULAR ROW OPERATIONS

Let us consider the matrices

$$A = \begin{bmatrix} 1 & 2 & 3 \\ 4 & 5 & 6 \\ 7 & 8 & 9 \end{bmatrix}, B = \begin{bmatrix} 4 & 5 & 6 \\ 1 & 2 & 3 \\ 7 & 8 & 9 \end{bmatrix},$$

$$C = \begin{bmatrix} 3 & 6 & 9 \\ 4 & 5 & 6 \\ 7 & 8 & 9 \end{bmatrix}, D = \begin{bmatrix} 1 & 2 & 3 \\ 6 & 9 & 12 \\ 7 & 8 & 9 \end{bmatrix}$$

Here we observe that the matrices B, C, D are related to the matrix A in as much as:

(a) B can be obtained from A by interchanging first and second rows of A;

(b) C can be obtained from A by multiplying the first row of A by 3 and

(c) D can be obtained from A by adding two times the first row to the second row of A.

Such operations on the rows of a matrix are known as elementary row operations. Formal definition is given below:

Definition: *Let A_i denote the ith row of the matrix $A - [a_{ij}]$, then the elementary row operations on the matrix A are defined as:*

1. the interchanging of any two rows A_i and A_j (*i.e.*, ith and jth rows). The symbols R_{ij} or $R_i \leftrightarrow R_j$ are generally employed for this operation.
2. the multiplication of every element of A_i by a non-zero scalar c *i.e.*, replacing the ith rwo A_i by cA_i. The symbols R_i (c) pr $R_i \rightarrow cR_j$ are employed for this operation.
3. the addition to the elements of row A_i of c (a scalar) times the corresponding elements of the row A_k *i.e.*, replacing the row A_i by $A_i + cA_k$.

The symbols R_{ik} (c) or $R_i \rightarrow R_j + cR_k$ are used for this operation.

Example:

$$\text{Let } A = \begin{bmatrix} 1 & 2 & 3 \\ 3 & 4 & 5 \\ 5 & 6 & 7 \end{bmatrix}$$

The effect of the elementary rwo operation $A_2 - A_1$ or R_{21} (-1) is to product the matrix.

$$B = \begin{bmatrix} 1 & 2 & 3 \\ 3-1 & 4-2 & 5-3 \\ 5 & 6 & 7 \end{bmatrix} = \begin{bmatrix} 1 & 2 & 3 \\ 2 & 2 & 2 \\ 5 & 6 & 7 \end{bmatrix}$$

Again the effect of elementary row operation $B_2 + B_1$ or R_{21} (1) is to produce the matrix

$$\begin{bmatrix} 1 & 3 & 3 \\ 2+1 & 2+2 & 2+3 \\ 5 & 6 & 7 \end{bmatrix} = \begin{bmatrix} 1 & 2 & 3 \\ 3 & 4 & 5 \\ 5 & 6 & 7 \end{bmatrix} \text{ i.e. the matrix A}$$

Thus the above two operations are the inverse elementary row operations.

2.8 ROW EQUIVALENT MATRICES

Definition: *If an $m \times n$ matrix B can be obtained from and $m \times n$ matrix A by a finite number of elementary row operations, then B is called the row equivalent to A and is written as*

$$B \overset{\text{row}}{\sim} A.$$

Note: Equivalent matrices have the same order.

Example:

$$\begin{bmatrix} 1 & 3 & 4 & 7 \\ 2 & -3 & 5 & 6 \\ 1 & 0 & 3 & 2 \end{bmatrix} \overset{\text{row}}{\sim} \begin{bmatrix} 2 & -3 & 5 & 6 \\ 1 & 3 & 4 & 7 \\ 1 & 0 & 3 & 2 \end{bmatrix}$$

(interchanging first and second rows).

2.9 ELEMENTARY ROW MATRIX

Definition: *The matrix obtained by the application of one elementary row operation to the identity matrix I_n is called an elementary row matrix.*

Example:

Examples of elementary matrices obtained from I_3, where

$$I_3 = \begin{bmatrix} 1 & 0 & 0 \\ 0 & 1 & 0 \\ 0 & 0 & 1 \end{bmatrix}$$

1. $$I_3 \sim \begin{bmatrix} 0 & 1 & 0 \\ 1 & 0 & 0 \\ 0 & 0 & 1 \end{bmatrix} = E_a \text{ (say)},$$

obtained by interchanging first two rows.

2. $$I_3 \sim \begin{bmatrix} 1 & 0 & 0 \\ 0 & c & 0 \\ 0 & 0 & 1 \end{bmatrix} = E_b \text{ (say)},$$

obtained by multiplying the elements of second row by c.

3. $$I_3 \sim \begin{bmatrix} 1 & 2 & 0 \\ 0 & 1 & 0 \\ 0 & 0 & 1 \end{bmatrix} = E_c \text{ (say)},$$

obtained by adding two times the elements of second row to the corresponding elements of first row *i.e.,* replacing A_1 by $A_1 + 2A2$ *i.e.,,* R_{12} (2).

Theorem:

Each elementary row operation on an m × n matrix can be effected by premultiplying it by the corresponding elementary matrix.

Example:

Let $$A = \begin{bmatrix} a_{11} & a_{12} & a_{13} & a_{14} \\ a_{21} & a_{22} & a_{23} & a_{24} \\ a_{31} & a_{32} & a_{33} & a_{34} \end{bmatrix}$$

1. Interchanging the first and third rows, we have

$$A \sim \begin{bmatrix} a_{31} & a_{32} & a_{33} & a_{34} \\ a_{21} & a_{22} & a_{23} & a_{24} \\ a_{11} & a_{12} & a_{13} & a_{14} \end{bmatrix}$$

The corresponding elementary matrix (obtained by interchanging first and third rows of I_3) is given by

$$E_{13} = \begin{bmatrix} 0 & 0 & 1 \\ 0 & 1 & 0 \\ 1 & 0 & 0 \end{bmatrix}$$

[Here students should not that as we are to premultiply A therefore the number of columns of E_{13} should be 3, the number of rows of A].

$$\text{Now } E_{13}A = \begin{bmatrix} 0 & 0 & 1 \\ 0 & 1 & 0 \\ 1 & 0 & 0 \end{bmatrix} \times \begin{bmatrix} a_{11} & a_{12} & a_{13} & a_{14} \\ a_{21} & a_{22} & a_{23} & a_{24} \\ a_{31} & a_{32} & a_{33} & a_{34} \end{bmatrix}$$

$$= \begin{bmatrix} a_{31} & a_{32} & a_{33} & a_{34} \\ a_{21} & a_{22} & a_{23} & a_{24} \\ a_{11} & a_{12} & a_{13} & a_{14} \end{bmatrix}$$

This shows that **B** can be obtained from **A** by pre-multiplying it by E_{13}, the corresponding elementary matrix.

2. Let $A = \begin{bmatrix} 1 & 2 & 3 \\ 4 & 5 & 6 \\ 7 & 8 & 9 \end{bmatrix}$

Multiplying the elements of second row by 2, we get

$$A \sim \begin{bmatrix} 1 & 2 & 3 \\ 8 & 10 & 12 \\ 7 & 8 & 9 \end{bmatrix} = B \text{ (say)}$$

The corresponding elementary matrix (obtained by multiplying the elements of second row of I_3 by)2 is given by

$$E_2(2) = \begin{bmatrix} 1 & 0 & 0 \\ 0 & 2 & 0 \\ 0 & 0 & 1 \end{bmatrix}$$

$$\begin{bmatrix} 1 & 0 & 0 \\ 0 & 2 & 0 \\ 0 & 0 & 1 \end{bmatrix} \times \begin{bmatrix} 1 & 2 & 3 \\ 4 & 5 & 6 \\ 7 & 8 & 9 \end{bmatrix}$$

$$= \begin{bmatrix} 1.1 + 0.4 + 0.7 & 1.2 + 0.5 + 0.8 & 1.3 + 0.6 + 0.9 \\ 0.1 + 2.4 + 0.7 & 0.2 + 2.5 + 0.8 & 0.3 + 2.6 + 0.9 \\ 0.1 + 0.4 + 1.7 & 0.2 + 0.5 + 1.8 & 0.3 + 0.6 + 1.9 \end{bmatrix}$$

$$= \begin{bmatrix} 1 & 2 & 3 \\ 8 & 10 & 12 \\ 7 & 8 & 9 \end{bmatrix} = B$$

i.e.,, B can be obtained from A by pre-multiplying it by E_3 (2).

3. Let $A = \begin{bmatrix} 1 & -2 & 3 \\ -3 & 4 & 5 \\ 5 & 6 & -7 \end{bmatrix}$

Replacing A_1 by $A_1 + 2A_2$ i.e, adding two times the elements of second row to the corresponding elements of first row, we get

$$A \sim \begin{bmatrix} -5 & 2 & 13 \\ -3 & 4 & 5 \\ 5 & 6 & -7 \end{bmatrix} = B \text{ (say)}$$

The corresponding elementary matrix (obtained by adding two times the elements of second row of I_3 to the corresponding elements of the first row) is given by

$$E_{12}(2) = \begin{bmatrix} 1 & 2 & 0 \\ 0 & 1 & 0 \\ 0 & 0 & 1 \end{bmatrix}$$

Then $E_{12}(2) \times A = \begin{bmatrix} 1 & 2 & 0 \\ 0 & 1 & 0 \\ 0 & 0 & 1 \end{bmatrix} \times \begin{bmatrix} 1 & -2 & 3 \\ -3 & 4 & 5 \\ 5 & 6 & -7 \end{bmatrix}$

$$= \begin{bmatrix} 1.1 + 2(-3) + 0.5 & 1.(-2) + 2.4 + 0.6 & 1.3 + 2.5 + 0.(-7) \\ 0.1 + 1(-3) + 0.5 & 0.(-2) + 1.4 + 0.6 & 0.3 + 1.5 + 0.(-7) \\ 0.1 + 0(-3) + 1.5 & 0.(-2) + 0.4 + 1.6 & 0.3 + 0.5 + 1.(-7) \end{bmatrix}$$

$$= \begin{bmatrix} -5 & 6 & 13 \\ -3 & 4 & 5 \\ 5 & 6 & -7 \end{bmatrix} = B$$

i.e.,, **B** can be obtained from A by pre-multiplying it by $E_{12}(2)$.

2.10 TYPES OF ELEMENTARY ROW MATRICES AND THEIR SYMBOLS

1. E_{ij} denotes the elementary matrix obtained by interchanging the ith and jth rows (or columns) of an identity (or unit) matrix.
2. $E_i(c)$ denotes the elementary matrix obtained by multiplying the ith row (or column) of the identity matrix by c.
3. $E_{ik}(c)$ denotes the elementary matrix obtained by adding to the elements of the ith row of the identity matrix c times the corresponding elements of the kth row.
4. $E'_{ik}(c)$ denotes the transpose of $E_{ik}(c)$ and can be obtained by adding to the elements of the ith column of the identity matrix c times the corresponding elements of the kth column.

Theorem 1:

The elementary matrix E_{ij} $E_i(c)$, $E_{jk}(1)$ are non-singular.

Proof:

1. The elementary matrix $\mathbf{E}_{ij}$ is obtained by interchanging the ith and jth rows of **I**. We shall get back **I** if we now apply the same row

operation upon E_{ij} which can also be effected by pro-multiplying E_{ij} by E_{ij}

$\therefore \quad \mathbf{E}_{ij}\,\mathbf{E}_{ij} = \mathbf{I}.$

i.e., $\mathbf{E}_{ij}$ is its own inverse *i.e.,,* $\mathbf{E}_{ij}$ is non-singular.

2. The elementary matrix $\mathbf{E}_i$ (c) is obtained by multiplying the ith row of the identity matrix by c (where $c \neq 0$). We shall get back **I** if we now multiply the elements of ith row of $\mathbf{E}_i$ (c) by 1/c which can also be effected by pre-multiplying $\mathbf{E}_i$ (c) with the corresponding elementary matrix which is obtained from **I** by multiplying its ith row by 1/c, which is therefore the inverse of $\mathbf{E}_i$ (c).

For example, let $I = \begin{bmatrix} 1 & 0 & 0 \\ 0 & 1 & 0 \\ 0 & 0 & 1 \end{bmatrix}$ and $E_3\,(c) = \begin{bmatrix} 1 & 0 & 0 \\ 0 & 1 & 0 \\ 0 & 0 & c \end{bmatrix}$

Then $\{E_3\,(c)\}^{-1} = \begin{bmatrix} 1 & 0 & 0 \\ 0 & 1 & 0 \\ 0 & 0 & 1/c \end{bmatrix}$, where $\{E_3\,(c)\}^{-1}$ is the inverse of E_3 (c)

3. The elementary matrix E_{jk} (1) obtained from **I** by replacing its jth row by (jth row + kth row).

We shall get back **I** if we now replace the jth row of E_{ji} (1) by (jth row –kth row).

Hence the inverse of $\mathbf{E}_{jk}$ (1) is the elementary matrix obtained from **I** by replacing its jth row by (jth row –kth row)

Let $\mathbf{I} = \begin{bmatrix} 1 & 0 & 0 \\ 0 & 1 & 0 \\ 0 & 0 & 1 \end{bmatrix}$ and $E_{13}\,(1) = \begin{bmatrix} 1 & 0 & 1 \\ 0 & 1 & 0 \\ 0 & 0 & 1 \end{bmatrix}$ obtained from **I** by replacing its lst row by (lst row + 3rd row).

Then $\{E_{13}\,(1)\}^{-1} = \begin{bmatrix} 1 & 0 & -1 \\ 0 & 0 & 1 \\ 0 & 0 & 1 \end{bmatrix}$, obtained from I by replacing its first row by (lst row – 3rd row).

Theorem 2:

If the matrix **B** *is row equivalent to the matrix* **A** *the* **B** = **SA**, *where* **S** *is non-singular.*

We know that

if $\mathbf{B} \overset{\text{row}}{\sim} \mathbf{A}$ then $\mathbf{B} = \mathbf{SA}$ where $\mathbf{S}$ is the product of the elementary matrices and we have proved that elementary matrices are non-singular and hence their product is also non-singular.

This proves the above theorem.

Theorem 3:

If **a** *square matrix* **A** *of order n is row equivalent to the identity matrix* $\mathbf{I}_n$ *then* **A** *is non-singular.*

Proof:

We know that

$\mathbf{A} \overset{\text{row}}{\sim} \mathbf{I}_n$ then $\mathbf{A} = \mathbf{S}.\mathbf{I}_n$, where $\mathbf{S}$ is non-singular.

Now S I_n being the product of two non-singular matrices is non-singular. Therefore A is non-singular.

Note. The converse of this theorem is also true.

Theorem 4:

If **a** *sequence of row operations applied to a square matrix* **A** *reduces it to the identity matrix I, then the same sequence of row operations applied to the identity matrix gives the inverse of* **A** *(i.e.* $\mathbf{A}^{-1}$*).*

Proof:

We know that $\mathbf{SA} = \mathbf{I}$, where $\mathbf{S}$ is the product of the elementary matrices.

i.e., $(\mathbf{E}_k \ldots \mathbf{E3. E2. E1})\, \mathbf{A} = \mathbf{I}$, where $\mathbf{E}_i$ denotes the elementary matrices.

$\Rightarrow$ $(\mathbf{E}_k \ldots \mathbf{E}_3.\mathbf{E}_2.\mathbf{E}_1)\, \mathbf{AA}^{-1} = \mathbf{IA}^{-1}$

$\Rightarrow$ $(\mathbf{E}_k \ldots \mathbf{E}_3.\mathbf{E}_2.\mathbf{E}_1)\, \mathbf{I} = \mathbf{A}^{-1}$, **since** $\mathbf{A}^{-1} = \mathbf{I}$ **and** $\mathbf{IA}^{-1} = \mathbf{A}^{-1}$

Hence the theorem.

In the following examples we shall show the successive matrices row equivalent to **A** and **I** in the left hand and right hand columns respectively. When ultimately A is reduced to **I** in the left hand column, I is reduced to A^{-1} in the right hand column.

Also R_1, R_2, R_3,... etc, stand for first row, second row, third row etc.

Example 1:

Find the inverse of the matrix $A = \begin{bmatrix} 1 & 2 & 2 \\ -1 & 3 & 0 \\ 0 & -2 & 1 \end{bmatrix}$

Solution:

$$\overset{A}{\begin{bmatrix} 1 & 2 & -2 \\ -1 & 3 & 0 \\ 0 & -2 & 1 \end{bmatrix}} \Bigg| \sim \overset{I}{\begin{bmatrix} 1 & 0 & 0 \\ 0 & 1 & 0 \\ 0 & 0 & 1 \end{bmatrix}}$$

$$\sim \begin{bmatrix} 1 & -2 & 0 \\ -1 & 3 & 0 \\ 0 & -2 & 1 \end{bmatrix} \Bigg| \sim \begin{bmatrix} 1 & 0 & 2 \\ 0 & 1 & 0 \\ 0 & 0 & 1 \end{bmatrix}$$

(Replacing R_1 by $R_1 + 2R_3$)

$$\sim \begin{bmatrix} 1 & -2 & 0 \\ -1 & 1 & 0 \\ 0 & -2 & 1 \end{bmatrix} \Bigg| \sim \begin{bmatrix} 1 & 0 & 2 \\ 1 & 1 & 2 \\ 0 & 0 & 1 \end{bmatrix}$$

(Replacing R_2 by $R_2 + R_1$)

$$\sim \begin{bmatrix} 1 & 0 & 0 \\ 0 & 1 & 0 \\ 0 & 0 & 1 \end{bmatrix} \Bigg| \sim \begin{bmatrix} 3 & 2 & 6 \\ 1 & 1 & 2 \\ 2 & 2 & 5 \end{bmatrix}$$

(Replacing R_1 by $R_1 + 2R_2$ and R_3 by $R_3 + 2R_2$)

$= I \qquad | \qquad = A^{-1}$

$$\therefore A^{-1} = \begin{bmatrix} 3 & 2 & 6 \\ 1 & 1 & 2 \\ 2 & 2 & 5 \end{bmatrix}$$ **Ans.**

Example 2:

$A = \begin{bmatrix} 1 & 2 & 1 \\ 3 & 2 & 3 \\ 1 & 1 & 2 \end{bmatrix}$, *evaluate* A^{-1}.

Solution:

$$\overset{A}{\begin{bmatrix} 1 & 2 & 1 \\ 3 & 2 & 3 \\ 1 & 1 & 2 \end{bmatrix}} \Bigg| \sim \overset{I}{\begin{bmatrix} 1 & 0 & 0 \\ 0 & 1 & 0 \\ 0 & 0 & 1 \end{bmatrix}}$$

$$\sim \begin{bmatrix} 1 & 2 & 1 \\ 0 & -4 & 0 \\ 0 & -1 & 1 \end{bmatrix} \Bigg| \sim \begin{bmatrix} 1 & 0 & 0 \\ -3 & 1 & 0 \\ -1 & 0 & 1 \end{bmatrix}$$

(Replacing R_2 by $R_2 - 3R_1$ and R_4 by $R_3 - R_1$)

$$\sim \begin{bmatrix} 1 & 3 & 0 \\ 0 & 1 & 0 \\ 0 & -1 & 1 \end{bmatrix} \Bigg| \sim \begin{bmatrix} 2 & 0 & -1 \\ \frac{3}{4} & -\frac{1}{4} & 0 \\ -1 & 0 & 1 \end{bmatrix}$$

(Replacing R_1 by $R_1 - R_3$ and R_2 by $- 1/4R_2$)

$$\sim \begin{bmatrix} 1 & 0 & 0 \\ 0 & 1 & 0 \\ 0 & 0 & 1 \end{bmatrix} \Bigg| \sim \begin{bmatrix} -\frac{1}{4} & \frac{3}{4} & -1 \\ \frac{3}{4} & -\frac{1}{4} & 0 \\ -\frac{1}{4} & \frac{1}{4} & 1 \end{bmatrix}$$

(Replacing R_1 by $R_1 - 2R_2$ and R_3 by $R_3 + R_2$)

$= I \quad | \quad = A^{-1}$

$$\therefore A^{-1} \begin{bmatrix} -\frac{1}{4} & \frac{3}{4} & -1 \\ \frac{3}{4} & -\frac{1}{4} & 0 \\ -\frac{1}{4} & \frac{1}{4} & 1 \end{bmatrix}$$

Example 3:

Find the inverse of the matrix $A = \begin{bmatrix} 1 & -3 & 2 \\ 2 & 0 & 0 \\ 1 & 4 & 1 \end{bmatrix}$

Solution:

$$\begin{matrix} A & & I \\ \begin{bmatrix} 1 & -3 & 2 \\ 2 & 0 & 0 \\ 1 & 4 & 1 \end{bmatrix} & \sim & \begin{bmatrix} 0 & 0 & 0 \\ 0 & 1 & 0 \\ 0 & 0 & 1 \end{bmatrix} \end{matrix}$$

$$\sim \begin{bmatrix} 1 & -3 & 2 \\ 1 & 0 & 0 \\ 1 & 4 & 1 \end{bmatrix} \sim \begin{bmatrix} 1 & 0 & 0 \\ 0 & \frac{1}{2} & 0 \\ 0 & 0 & 1 \end{bmatrix}$$

(Replacing R_2 by $\frac{1}{2}$ R_2)

$$\sim \begin{bmatrix} 1 & 0 & 0 \\ 1 & 0 & 2 \\ 1 & 4 & 1 \end{bmatrix} \sim \begin{bmatrix} 0 & \frac{1}{2} & 0 \\ 1 & 0 & 0 \\ 0 & 0 & 1 \end{bmatrix}$$

(Interchanging R_1 and R_2)

$$\sim \begin{bmatrix} 1 & 0 & 0 \\ 0 & -3 & 2 \\ 0 & 4 & 1 \end{bmatrix} \sim \begin{bmatrix} 0 & \frac{1}{2} & 0 \\ 1 & -\frac{1}{2} & 0 \\ 0 & -\frac{1}{2} & 1 \end{bmatrix}$$

(Replacing R_2 by $R_2 - R_1$ and R_3 by $R_3 - R_1$)

$$\sim \begin{bmatrix} 1 & 0 & 0 \\ 0 & -11 & 0 \\ 0 & 4 & 1 \end{bmatrix} \sim \begin{bmatrix} 0 & \frac{1}{2} & 0 \\ 1 & \frac{1}{2} & -2 \\ 0 & -\frac{1}{2} & 1 \end{bmatrix}$$

(Replacing R_2 by $R_2 - 2\ R_2$)

$$\sim \begin{bmatrix} 1 & 0 & 0 \\ 0 & 1 & 0 \\ 0 & 4 & 1 \end{bmatrix} \sim \begin{bmatrix} 0 & \frac{1}{2} & 0 \\ -\frac{1}{11} & -\frac{1}{22} & \frac{2}{11} \\ 0 & -\frac{1}{2} & 1 \end{bmatrix}$$

(Replacing R_2 by $-\frac{1}{11} R_2$)

$$\sim \begin{bmatrix} 1 & 0 & 0 \\ 0 & 1 & 0 \\ 0 & 0 & 1 \end{bmatrix} \Bigg| \sim \begin{bmatrix} 0 & \frac{1}{2} & 0 \\ -\frac{1}{11} & -\frac{1}{22} & \frac{2}{11} \\ \frac{4}{11} & -\frac{1}{2} & 1 \end{bmatrix}.$$

(Replacing R_3 by $R_3 - 4 R_2$)

$= I \qquad | \qquad = A^{-2}$

$$\therefore A^{-1} = \begin{bmatrix} 0 & \frac{1}{2} & 0 \\ -\frac{1}{11} & -\frac{1}{22} & \frac{2}{11} \\ \frac{4}{11} & -\frac{7}{22} & \frac{2}{11} \end{bmatrix}$$ **Ans.**

Example 4:

Find the inverse of the matrix $A = \begin{bmatrix} 1 & 2 & 3 \\ 3 & 1 & 2 \\ 0 & 1 & 2 \end{bmatrix}$

Solution:

$$\begin{matrix} A & & I \end{matrix}$$

$$\sim \begin{bmatrix} 1 & 2 & 1 \\ 3 & 1 & 2 \\ 0 & 1 & 2 \end{bmatrix} \Bigg| \sim \begin{bmatrix} 1 & 0 & 0 \\ 0 & 1 & 0 \\ 0 & 0 & 1 \end{bmatrix}$$

$$\sim \begin{bmatrix} 1 & 2 & 1 \\ 3 & -5 & -1 \\ 0 & 1 & 2 \end{bmatrix} \Bigg| \sim \begin{bmatrix} 1 & 0 & 0 \\ -3 & 1 & 0 \\ 0 & 0 & 1 \end{bmatrix}$$

(Replacing R_2 by $R_2 - 3 R_1$)

$$\sim \begin{bmatrix} 1 & 2 & 1 \\ 0 & 1 & 11 \\ 0 & 1 & 2 \end{bmatrix} \Bigg| \sim \begin{bmatrix} 1 & 0 & 0 \\ -3 & 1 & 6 \\ 0 & 0 & 1 \end{bmatrix}$$

(Replacing R_2 by $R_2 + 6 R_3$)

$$\sim\begin{bmatrix}1 & 0 & -21\\0 & 1 & 11\\0 & 1 & 2\end{bmatrix}\Bigg\|\sim\begin{bmatrix}7 & -2 & -12\\-3 & 1 & 6\\0 & 0 & 1\end{bmatrix}$$

(Replacing R_1 by $R_1 - 2R_2$)

$$\sim\begin{bmatrix}1 & 0 & -21\\0 & 1 & 11\\0 & 0 & -9\end{bmatrix}\Bigg\|\sim\begin{bmatrix}7 & -2 & -12\\-3 & 1 & 6\\3 & -1 & -5\end{bmatrix}$$

(Replacing R_3 by $R_3 - R_2$)

$$\sim\begin{bmatrix}1 & 0 & -21\\0 & 1 & 11\\0 & 0 & 1\end{bmatrix}\Bigg\|\sim\begin{bmatrix}7 & -2 & -12\\-3 & 1 & 6\\-\frac{1}{3} & \frac{1}{9} & \frac{5}{9}\end{bmatrix}$$

(Replacing R_3 by $-\frac{1}{9}R_3$). **(Note)**

$$\sim\begin{bmatrix}1 & 0 & 0\\c & 1 & 11\\0 & 0 & 1\end{bmatrix}\Bigg\|\sim\begin{bmatrix}0 & \frac{1}{2} & -\frac{1}{3}\\-3 & 1 & 6\\-\frac{1}{3} & \frac{1}{9} & \frac{5}{9}\end{bmatrix}$$

(Replacing R_1 by $R_1 + 21R_3$)

$$\sim\begin{bmatrix}1 & 0 & 0\\0 & 1 & 0\\0 & 0 & 1\end{bmatrix}\Bigg\|\sim\begin{bmatrix}0 & \frac{1}{2} & -\frac{1}{3}\\\frac{2}{3} & -\frac{2}{9} & -\frac{1}{9}\\-\frac{1}{3} & \frac{1}{9} & \frac{5}{9}\end{bmatrix}$$

(Replacing R_2 by $R_2 - 11R_3$)

$= I \quad | \quad = A^{-1}$

Therefore $A^{-1} = \begin{bmatrix}0 & \frac{1}{2} & -\frac{1}{3}\\\frac{2}{3} & -\frac{2}{9} & -\frac{1}{9}\\-\frac{1}{3} & \frac{1}{9} & \frac{5}{9}\end{bmatrix}$ **Ans.**

2.11 ELEMENTARY COLUMN OPERATION AND COLUMN EQUIVALENT MATRICES

Here the letter R in the symbols are to be replaced by C e.g. C_{ij}, C_i (c), C_{ik} (c), where C_{ij} stands for the interchange of ith and jth columns etc. or $C_i \leftrightarrow C_j$; $C_i \to c\ C_i$, $C_i \to C_i + c\ C_k$.

Theorem 1:

If there the two m × n matrices A and B, then $B \overset{col}{\sim} A$ *if B = AT, where T is an n × n non-singular matrix.*

Proof:

If $B \overset{col}{\sim} A$, then $B' \overset{row}{\sim} A'$, where **B'** and **A'** are the transposed matrices of **B** and **A** respectively.

Therefore **B'** = **SA'**, where **S** is an n × n non-singular matrix.

Consequently **B** = **AS'**, where **S'** is the transposed matrix of **S**.

= **AT**, where **T** = S', an n × n-singular matrix.

Hence the theorem.

Theorem 2:

Each elementary column operation on an m × n matrix A can be effected by post multiplying A by then n × n matrix obtained from the n × n identity matrix I_n *by the same elementary column operation.*

Proof:

If $B \overset{col}{\sim} A$,

then $B' \overset{row}{\sim} A'$ where **B'** and **A'** are the transposed matrices of **B** and **A**.

Since if **B** is obtained from **A** by elementary column operation, then **B'** can be obtained from **A'** by an elementary row operation.

Hence **B'** = **EA'**, where **E** is the elementary matrix obtained from $\mathbf{I_n}$ by an elementary row operation.

Therefore **B** = **A E'**, where E', the transposed matrix of E, can be obtained from I_n by the same elementary column operation.

Hence the theorem

2.12 EQUIVALENT MATRICES (GENERAL DEFINITION).

Definition: *Two n × n matrices A and* ***B*** *are called equivalent if one can be obtained from the other by a finite number of row and column operations (or elementary operation) and written as* ***B*** *~ A.*

2.13 TRIANGULAR MATRIX

Definition: *A matrix $[a_{ij}]$ is called a triangular matrix if*

$a_{ij} = 0$ *for* $i > j$.

$$\begin{bmatrix} 2 & 3 & 1 & 4 \\ 0 & 1 & 2 & 3 \\ 0 & 0 & 5 & 7 \end{bmatrix} \text{ or } \begin{bmatrix} 2 & 3 & 4 & 5 \\ 0 & 1 & 3 & 4 \\ 0 & 0 & 2 & 5 \\ 0 & 0 & 0 & 7 \end{bmatrix}$$

Note 1: The elements a_{ij} for which $i \leq j$ are not necessarily zero.

Note 2: Triangular matrix need not be square. If it is square, then it is called upper triangular matrix.

Theorem:

Every matrix can be reduced to triangle form by elementary row operations.

Proof:

We shall prove this theorem by Mathematical Induction.

Assume that this theorem holds for all matrices containing n – 1 rows and let $A = [a_{ij}]$ be an $n \times m$ matrix given below:

$$A = \begin{bmatrix} a_{11} & a_{12} & a_{13} & \dots & \dots & \dots & a_{1m} \\ a_{21} & a_{22} & a_{23} & \dots & \dots & \dots & a_{2m} \\ \dots & \dots & \dots & \dots & \dots & \dots & \dots \\ \dots & \dots & \dots & \dots & \dots & \dots & \dots \\ a_{n1} & a_{n2} & a_{n3} & \dots & \dots & \dots & a_{nm} \end{bmatrix}$$

Now the following cases arise:

Case I: If $a_{11} \neq 0$, then replacing R_1 by $(1/a_{11}) R_1$ (*i.e.*, by applying elementary row operation) the matrix A reduces to an $n \times m$ matrix.

$$B = [b_{ij}] = \begin{bmatrix} b_{11} & b_{12} & \dots & \dots & b_{1m} \\ b_{21} & b_{22} & \dots & \dots & b_{2m} \\ \dots & \dots & \dots & \dots & \dots \\ b_{n1} & b_{n2} & \dots & \dots & b_{nm} \end{bmatrix}$$

where $b_{11} = 1$.

Now apply elementary row-operation $R_k - R_{ji} R_1$ to R_k where $k = 1, 2, \dots, n$ *i.e.*, subtract b_{ki} times R_1 from R_k, where k takes value from 1 to n. This reduces the matrix B to matrix $C = [c_{ij}]$ where $c_{k1} = 0$ whenever $k > 1$ and we have

$$C = \begin{bmatrix} 1 & c_{12} & c_{13} & \cdots & c_{1m} \\ 0 & c_{22} & c_{23} & \cdots & c_{2m} \\ \cdots & \cdots & \cdots & \cdots & \cdots \\ 0 & c_{n2} & c_{n3} & \cdots & c_{nm} \end{bmatrix} \quad ...(1)$$

Now by our assumption that the theorem which we are going to prove holds for matrices containing (n – 1) rows we find that (n – 1) rowed matrix

$$\begin{bmatrix} 0 & c_{22} & c_{23} & \cdots & c_{2m} \\ 0 & c_{32} & c_{33} & \cdots & c_{3m} \\ \cdots & \cdots & \cdots & \cdots & \cdots \\ 0 & c_{n2} & c_{n3} & \cdots & c_{nm} \end{bmatrix}$$

can always be reduced to triangular form by elementary row operations and hence from (1) the matrix C will reduce to triangular form when the same elementary row operations are applied to c.

Case II: If $a_{11} = 0$ but $a_{k1} \neq 0$ for some value of k then interchanging R_1 and R_k the matrix **A** reduces to the matrix $\mathbf{D} = [d_{ij}]$ where $d_{11} \neq 0$.

Then the matrix D can always be reduced to the triangular from as in case **I** above.

Case III: If $a_{k1} = 0$ for all values of k, then we have

$$A = \begin{bmatrix} 0 & a_{12} & a_{13} & \cdots & a_{1m} \\ 0 & a_{22} & a_{23} & \cdots & a_{2m} \\ \cdots & \cdots & \cdots & \cdots & \cdots \\ 0 & a_{n2} & a_{n3} & \cdots & a_{nm} \end{bmatrix}$$

By hypothesis (inductive) then (n – 1) rowed matrix

$$\begin{bmatrix} 0 & a_{22} & a_{23} & \cdots & a_{2m} \\ \cdots & \cdots & \cdots & \cdots & \cdots \\ 0 & a_{n2} & a_{n3} & \cdots & a_{nm} \end{bmatrix}$$

as in case I above can be reduced to triangular form by elementary row operations and the same elementary operations when applied on A will reduce A to triangular form.

Hence the matrix A can always be reduced to triangular form and the proof is complete by mathematical induction.

Example 1:

Reduce $\begin{bmatrix} -1 & 2 & 1 & 8 \\ 2 & 1 & -1 & 0 \\ 3 & 2 & 1 & 7 \end{bmatrix}$ *to triangular form.*

Solution:

Let $A = \begin{bmatrix} -1 & 2 & 1 & 8 \\ 2 & 1 & -1 & 0 \\ 3 & 2 & 1 & 7 \end{bmatrix}$

$\sim \begin{bmatrix} 1 & -2 & -1 & -8 \\ 2 & 1 & -1 & 0 \\ 3 & 2 & 1 & 7 \end{bmatrix}$, replacing R_1 by $-R_1$

$\sim \begin{bmatrix} 1 & -2 & -1 & -8 \\ 0 & 5 & 1 & 16 \\ 0 & 8 & 4 & 31 \end{bmatrix}$, replacing R_1 by $R_2 - 2R_1$ and R_3 by $R_3 - 3R_1$

$\sim \begin{bmatrix} 1 & -2 & -1 & -8 \\ 0 & 5 & 1 & 16 \\ 0 & 40 & 20 & 155 \end{bmatrix}$, replacing R_3 by $5R_3$

$\sim \begin{bmatrix} 1 & -2 & -1 & -8 \\ 0 & 5 & 1 & 16 \\ 0 & 0 & 12 & 27 \end{bmatrix}$, replacing R_3 by $R_3 - 8R_2$

This is the required triangular form.

Example 2:

Reduce the matrix $\begin{bmatrix} 3 & 1 & 4 \\ 1 & 2 & -5 \\ 0 & 1 & 2 \end{bmatrix}$ *to triangular form.*

Solution:

Let $A = \begin{bmatrix} 3 & 1 & 4 \\ 1 & 2 & -5 \\ 0 & 1 & 2 \end{bmatrix}$

$$\sim \begin{bmatrix} 1 & \frac{1}{3} & \frac{4}{3} \\ 1 & 2 & -5 \\ 0 & 1 & 2 \end{bmatrix}, \text{ replacing } R_1 \text{ by } \frac{1}{3} R_1$$

$$\sim \begin{bmatrix} 1 & \frac{1}{3} & \frac{4}{8} \\ 1 & \frac{5}{8} & -\frac{19}{8} \\ 0 & 1 & 2 \end{bmatrix}, \text{ replacing } R_1 \text{ by } \frac{1}{3} R_1$$

$$\sim \begin{bmatrix} 0 & \frac{1}{3} & \frac{4}{8} \\ 0 & \frac{5}{8} & -\frac{19}{8} \\ 0 & 0 & \frac{29}{5} \end{bmatrix}, \text{ replacing } R_3 \text{ by } R_3 - \frac{3}{5} R_2$$

This is the required triangular form.

Aliter. $A = \begin{bmatrix} 3 & 1 & 4 \\ 1 & 2 & -5 \\ 0 & 1 & 2 \end{bmatrix}$

$$\sim \begin{bmatrix} 1 & 2 & -5 \\ 3 & 1 & 4 \\ 0 & 1 & 2 \end{bmatrix}, \text{ int erchanging } R_1 \text{ and } R_2$$

$$\sim \begin{bmatrix} 1 & 2 & -5 \\ 0 & -5 & 19 \\ 0 & 1 & 2 \end{bmatrix}, \text{ replacing } R_2 \text{ by } R_2 - 3R_1$$

$$\sim \begin{bmatrix} 1 & 2 & -5 \\ 0 & -5 & 19 \\ 0 & 5 & 10 \end{bmatrix}, \text{ replacing } R_3 \text{ by } 5R_3$$

$$\sim \begin{bmatrix} 1 & 2 & -5 \\ 0 & -5 & 19 \\ 0 & 0 & 29 \end{bmatrix}, \text{ replacing } R_3 \text{ by } R_3 + R_2$$

This is also a triangular matrix

SOLVED EXAMPLES

Example 1:

Apply successively the row transformations (or operation R_{23}, $R_3(-2)$

and R_{13} (4) to the matrix $\begin{bmatrix} 3 & 1 & 2 & 1 \\ 2 & 0 & 3 & 2 \\ 1 & 2 & 3 & 4 \\ 3 & 1 & 4 & 1 \end{bmatrix}$

Solution:

1. Applying R_{23} operation to the given matrix we have

$$\begin{bmatrix} 3 & 1 & 2 & 1 \\ 1 & 2 & 3 & 4 \\ 2 & 0 & 3 & 2 \\ 3 & 1 & 4 & 1 \end{bmatrix}$$ [Here we have interchanged second and third rows]

2. Applying $R_3(-2)$ operation to the given matrix we have

$$\begin{bmatrix} 3 & 1 & 2 & 1 \\ 2 & 0 & 3 & 2 \\ -2 & -4 & -6 & -8 \\ 2 & 1 & 4 & 1 \end{bmatrix}$$ [Here we have replaced third row R_3 by $-2R_3$]

3. Applying R_{13} (4) operation to the given matrix we have

$$\begin{bmatrix} 7 & 9 & 14 & 17 \\ 2 & 0 & 3 & 2 \\ 1 & 2 & 3 & 4 \\ 3 & 1 & 4 & 1 \end{bmatrix}$$ [Here we have replaced the first row R_1 by $R_1 + 4R_3$]

Example 2:

Find the inverse of the matrix $A = \begin{bmatrix} i & -1 & 2i \\ 2 & 0 & 2 \\ -1 & 0 & 1 \end{bmatrix}$

Solution:

$$\overset{A}{\begin{bmatrix} i & -1 & 2i \\ 2 & 0 & 2 \\ -1 & 0 & 1 \end{bmatrix}} \Bigg| \sim \overset{I}{\begin{bmatrix} 1 & 0 & 0 \\ 0 & 1 & 0 \\ 0 & 0 & 1 \end{bmatrix}}$$

$$\sim \begin{bmatrix} i & -1 & 2i \\ 1 & 0 & 1 \\ -1 & 0 & 1 \end{bmatrix} \Bigg| \sim \begin{bmatrix} 1 & 0 & 0 \\ 0 & 1/2 & 0 \\ 0 & 0 & 1 \end{bmatrix}$$

(Replacing R_2 by $\frac{1}{2}$ R_2)

$$\sim \begin{bmatrix} i & -1 & 2i \\ 1 & 0 & 1 \\ 0 & 0 & 2 \end{bmatrix} \Bigg| \sim \begin{bmatrix} 1 & 0 & 0 \\ 0 & 1/2 & 0 \\ 0 & 1/2 & 1 \end{bmatrix}$$

(Replacing R_3 by $R_3 + R_2$)

$$\sim \begin{bmatrix} i & -1 & 2i \\ 1 & 0 & 1 \\ 0 & 0 & 1 \end{bmatrix} \Bigg| \sim \begin{bmatrix} 1 & 0 & 0 \\ 0 & 1/4 & 0 \\ 0 & 1/4 & 1/2 \end{bmatrix}$$

(Replacing R_3 by $\frac{1}{2}$ R_3)

$$\sim \begin{bmatrix} i & -1 & 2i \\ 1 & 0 & 0 \\ 0 & 0 & 1 \end{bmatrix} \Bigg| \sim \begin{bmatrix} 1 & 0 & 0 \\ 0 & 1/2 & -1/2 \\ 0 & 1/4 & 1/2 \end{bmatrix}$$

(Replacing R_2 by $R_2 - R_3$)

$$\sim \begin{bmatrix} i & -1 & 0 \\ 1 & 0 & 0 \\ 0 & 0 & 1 \end{bmatrix} \Bigg| \sim \begin{bmatrix} 1 & -\frac{3}{4}i & -\frac{1}{2}i \\ 0 & 1/4 & -1/2 \\ 0 & 1/4 & 1/2 \end{bmatrix}$$

(Replacing R_1 by $R_1 - i\ R_2 - 2i\ R_2$) **(Note)**

$$\sim \begin{bmatrix} 0 & 1 & 0 \\ 1 & 0 & 0 \\ 0 & 0 & 1 \end{bmatrix} \Bigg| \sim \begin{bmatrix} -1 & \frac{3}{4}i & \frac{1}{2}i \\ 0 & 1/4 & -1/2 \\ 0 & 1/4 & 1/2 \end{bmatrix}$$

(Replacing R_1 by $-R_1$)

$$\sim \begin{bmatrix} 1 & 0 & 0 \\ 0 & 1 & 0 \\ 0 & 0 & 1 \end{bmatrix} = I \Bigg| \sim \begin{bmatrix} 0 & 1/4 & -1/2 \\ -1 & \frac{3}{4} & \frac{1}{2}i \\ 0 & 1/4 & 1/2 \end{bmatrix} = A^{-1}$$

(Interchanging R_1 and R_2)

$$\text{Therefore } A^{-1} = \begin{bmatrix} 0 & \frac{1}{4} & -\frac{1}{2} \\ -1 & \frac{3}{4}i & \frac{1}{2}i \\ 0 & \frac{1}{4} & \frac{1}{2} \end{bmatrix}$$

Ans.

Example 3:

Compute the following elementary matrices of order 4 E_{23}, E_2 (4), E_{34} (-2), E'_{34} (–2).

Solution:

$$I_4 = \begin{bmatrix} 1 & 0 & 0 & 0 \\ 0 & 1 & 0 & 0 \\ 0 & 0 & 1 & 0 \\ 0 & 0 & 0 & 1 \end{bmatrix}$$

1. $$E_{23} = \begin{bmatrix} 1 & 0 & 0 & 0 \\ 0 & 0 & 1 & 0 \\ 0 & 1 & 0 & 0 \\ 0 & 0 & 0 & 1 \end{bmatrix}$$ [Interchanging R_2 and R_3 or C_2 and C_3. (Note]

2. $$E_2(4) = \begin{bmatrix} 1 & 0 & 0 & 0 \\ 0 & 4 & 0 & 0 \\ 0 & 0 & 1 & 0 \\ 0 & 0 & 0 & 1 \end{bmatrix}$$ [Replacing R_3 by $4R_2$ or C_2 by $4C_2$]

3. $E_{34}(-2) = \begin{bmatrix} 1 & 0 & 0 & 0 \\ 0 & 1 & 0 & 0 \\ 0 & 0 & 1 & -2 \\ 0 & 0 & 0 & 1 \end{bmatrix}$ [Replacing R_3 by $R_3 - 2R_4$]

4. $E'_{34}(-2) = \begin{bmatrix} 1 & 0 & 0 & 0 \\ 0 & 1 & 0 & 0 \\ 0 & 0 & 1 & 0 \\ 0 & 0 & -2 & 1 \end{bmatrix}$ [Replacing C_3 by $C_3 - 2C_4$]

Here students should note that E'_{34} (-2) is nothing but the transpose of E_{34} (–2).

Example 4:

Find the inverse of the matrix $A = \begin{bmatrix} 1 & 2 & -1 \\ -1 & 1 & 2 \\ 2 & -1 & 1 \end{bmatrix}$

Solution:

$$\overset{A}{\begin{bmatrix} 1 & 2 & -1 \\ -1 & 1 & 2 \\ 2 & -1 & 1 \end{bmatrix}} \Bigg\| \sim \overset{I}{\begin{bmatrix} 1 & 0 & 0 \\ 0 & 1 & 0 \\ 0 & 0 & 1 \end{bmatrix}}$$

$$\sim \begin{bmatrix} 1 & 2 & -1 \\ 0 & 3 & 1 \\ 0 & -5 & 3 \end{bmatrix} \Bigg\| \sim \begin{bmatrix} 1 & 0 & 0 \\ 1 & 1 & 0 \\ -2 & 0 & 1 \end{bmatrix}$$

(Replacing R_2 by $R_2 + R_1$ and R_3 by $R_3 - 2R_1$)

$$\sim \begin{bmatrix} 1 & 5 & 0 \\ 0 & 3 & 1 \\ 0 & -11 & 1 \end{bmatrix} \Bigg\| \sim \begin{bmatrix} 2 & 1 & 0 \\ 1 & 1 & 0 \\ -4 & -2 & 1 \end{bmatrix}$$

(Replacing R_1 by $R_1 + R_2$ and R_3 by $R_3 - 2R_2$)

$$\sim \begin{bmatrix} 1 & 5 & 0 \\ 0 & 14 & 0 \\ 0 & -11 & 1 \end{bmatrix} \Bigg\| \sim \begin{bmatrix} 2 & 1 & 0 \\ 5 & 3 & -1 \\ -4 & -2 & 1 \end{bmatrix}$$

(Replacing R_2 by $R_2 - R_3$),

$$\sim \begin{bmatrix} 1 & 5 & 0 \\ 0 & 1 & 0 \\ 0 & -11 & 1 \end{bmatrix} \sim \begin{bmatrix} 2 & 1 & 0 \\ \frac{5}{14} & \frac{3}{14} & -\frac{1}{14} \\ -4 & -2 & 1 \end{bmatrix}$$

(Replacing R_2 by $\frac{1}{14}$ R_2)

$$\sim \begin{bmatrix} 1 & 0 & 0 \\ 0 & 1 & 0 \\ 0 & 0 & 1 \end{bmatrix} = I \quad \begin{bmatrix} \frac{3}{14} & -\frac{1}{14} & \frac{5}{14} \\ \frac{5}{14} & \frac{3}{14} & -\frac{1}{14} \\ -\frac{1}{14} & \frac{5}{14} & \frac{3}{14} \end{bmatrix} = A^{-1}$$

(Replacing R_1 by $R_1 - 5R_2$ and R_3 by $R_3 + 11R_2$)

Therefore $A^{-1} = \begin{bmatrix} \frac{3}{14} & -\frac{1}{14} & \frac{5}{14} \\ \frac{5}{14} & \frac{3}{14} & -\frac{1}{14} \\ -\frac{1}{14} & \frac{5}{14} & \frac{3}{14} \end{bmatrix} \Rightarrow \frac{1}{14} \begin{bmatrix} 3 & -1 & 5 \\ 5 & 3 & -1 \\ -1 & 5 & 3 \end{bmatrix}$ **Ans.**

Example 5:

Evaluate the inverses of the following elementary matrices of order four: $E_3(-2)$, $E_{23}(4)$.

Solution:

The identity matrix of order four is given by

$$I_4 = \begin{bmatrix} 1 & 0 & 0 & 0 \\ 0 & 1 & 0 & 0 \\ 0 & 0 & 1 & 0 \\ 0 & 0 & 0 & 1 \end{bmatrix}$$

1. Then $E_3(-2) = \begin{bmatrix} 1 & 0 & 0 & 0 \\ 0 & 1 & 0 & 0 \\ 0 & 0 & -2 & 0 \\ 0 & 0 & 0 & 1 \end{bmatrix}$, replacing R_3 by $-2R_3$

$\therefore$ Inverse of $E_3(-2)$ *i.e.*, $\{E_3(-2)\}^{-1} = \begin{bmatrix} 1 & 0 & 0 & 0 \\ 0 & 1 & 0 & 0 \\ 0 & 0 & -\frac{1}{2} & 0 \\ 0 & 0 & 0 & 1 \end{bmatrix}$ **Ans.**

(obtained by replacing R_3 of I_4 by $-\frac{1}{2}R_3$) **(Note)**

2. $E_{23}(4) = \begin{bmatrix} 1 & 0 & 0 & 0 \\ 0 & 1 & 4 & 0 \\ 0 & 0 & 1 & 0 \\ 0 & 0 & 0 & 1 \end{bmatrix}$, replacing R_2 of I_4 by $R_2 + 4R_3$

Then the inverse of $E_{23}(4)$ *i.e.*, $\{E_{23}(4)\}^{-1}$is given by

$\begin{bmatrix} 1 & 0 & 0 & 0 \\ 0 & 1 & -4 & 0 \\ 0 & 0 & 1 & 0 \\ 0 & 0 & 0 & 1 \end{bmatrix}$, replacing R_2 of I_4 by $R_2 - 4R_3$ **(Note)**

Example 6:

Apple successively the column operations C_{13} : C_2 (-4) and C_{23} (–2) to the matrix $\begin{bmatrix} 1 & -1 & 2 & 3 & 4 \\ 2 & 1 & -2 & 1 & 3 \\ 3 & 2 & 1 & -2 & 5 \\ 4 & 5 & 6 & 7 & 8 \end{bmatrix}$

Solution:

1. Applying C_{13} operation to the given matrix, we have

$\begin{bmatrix} 2 & -1 & 1 & 3 & 4 \\ -2 & 1 & 2 & 1 & 3 \\ 1 & 2 & 3 & -2 & 5 \\ 6 & 5 & 4 & 7 & 8 \end{bmatrix}$ [Here we have int erchanged C_1 and C_3 i.e. first and third columns].

2. Applying C_2 (–4) operation to the given matrix, we have

$$\begin{bmatrix} 1 & 4 & 2 & 3 & 4 \\ 2 & -4 & -2 & 1 & 3 \\ 3 & -8 & 1 & -2 & 5 \\ 4 & -20 & 6 & 7 & 8 \end{bmatrix}$$ [Here we have replaced second columns C_2 by $-4C_2$]

3. Applying $C_{23}(-2)$ operation to the given matrix, we have

$$\begin{bmatrix} 1 & -5 & 2 & 3 & 4 \\ 2 & 5 & -2 & 1 & 3 \\ 3 & 0 & 1 & -2 & 5 \\ 4 & -7 & 6 & 7 & 8 \end{bmatrix}$$ [Here we have replaced second columns C_2 by $C_2 - 2C_3$]

2.14 ORDER OF A MINOR

Definition: *If any r rows and any r columns from an m × n matrix A are retained and remaining (m – r) rows and (n – r) columns removed, then the determinant of the remaining r × r sub-matrix of A is called minor of A of order r.*

In the matrix $$\begin{bmatrix} a_{11} & a_{12} & a_{13} & a_{14} \\ a_{21} & a_{22} & a_{23} & a_{24} \\ a_{31} & a_{32} & a_{33} & a_{34} \\ a_{41} & a_{42} & a_{43} & a_{44} \\ a_{51} & a_{52} & a_{53} & a_{54} \end{bmatrix}$$

element a_{11}, a_{13}, a_{31} etc. are minors of order unity;

$$\begin{vmatrix} a_{11} & a_{12} \\ a_{21} & a_{22} \end{vmatrix}, \begin{vmatrix} a_{11} & a_{13} \\ a_{21} & a_{23} \end{vmatrix}, \begin{vmatrix} a_{33} & a_{34} \\ a_{53} & a_{54} \end{vmatrix} \text{ etc.}$$

are minors of order 2;

$$\begin{vmatrix} a_{11} & a_{12} & a_{13} \\ a_{21} & a_{22} & a_{23} \\ a_{31} & a_{32} & a_{33} \end{vmatrix}, \begin{vmatrix} a_{21} & a_{23} & a_{24} \\ a_{41} & a_{43} & a_{44} \\ a_{51} & a_{53} & a_{54} \end{vmatrix} \text{ etc.}$$

$$\begin{vmatrix} a_{11} & a_{12} & a_{13} & a_{14} \\ a_{21} & a_{22} & a_{23} & a_{24} \\ a_{41} & a_{42} & a_{43} & a_{44} \\ a_{51} & a_{52} & a_{53} & a_{54} \end{vmatrix}, \begin{vmatrix} a_{21} & a_{22} & a_{23} & a_{24} \\ a_{31} & a_{32} & a_{33} & a_{34} \\ a_{41} & a_{42} & a_{43} & a_{44} \\ a_{51} & a_{52} & a_{53} & a_{54} \end{vmatrix} \text{ etc.}$$

are minors of order 4.

EXERCISES

1. *Apply the column operation $C_3(4)$ and $C_{12}(-3)$ to the matrix*

$$\begin{bmatrix} 0 & 1 & 2 & 3 & 4 \\ 1 & 2 & 3 & 4 & 0 \\ 3 & 4 & 0 & 1 & 2 \\ 2 & 0 & 1 & 3 & 4 \end{bmatrix}$$

Ans. $\begin{bmatrix} 0 & 1 & 8 & 3 & 4 \\ 1 & 2 & 12 & 1 & 2 \\ 3 & 4 & 0 & 1 & 2 \\ 2 & 0 & 4 & 3 & 4 \end{bmatrix}$ and $\begin{bmatrix} 3 & 1 & 2 & 3 & 4 \\ -5 & 2 & 3 & 4 & 0 \\ -9 & 4 & 0 & 1 & 2 \\ 2 & 0 & 1 & 3 & 4 \end{bmatrix}$

2. *Compute E_{23}, $E_2(-2)$ and $E_{34}(-1)$ for the identity matrix of order 4.*

Ans. $\begin{bmatrix} 1 & 0 & 0 & 0 \\ 0 & 0 & 1 & 0 \\ 0 & 1 & 0 & 0 \\ 0 & 0 & 0 & 1 \end{bmatrix}$; $\begin{bmatrix} 1 & 0 & 0 & 0 \\ 0 & -2 & 0 & 0 \\ 0 & 0 & 1 & 0 \\ 0 & 1 & 0 & 1 \end{bmatrix}$; $\begin{bmatrix} 1 & 0 & 0 & 0 \\ 0 & 1 & 0 & 0 \\ 0 & 0 & 1 & -1 \\ 0 & 0 & 0 & 1 \end{bmatrix}$

3. *Evaluate the inverses of the following elementary matrices of order 4: E_{14}, $E_4(3)$, $E_{23}(2)$.*

Ans. E_{14}, $\begin{bmatrix} 1 & 0 & 0 & 0 \\ 0 & 1 & 0 & 0 \\ 0 & 0 & 1 & 0 \\ 0 & 0 & 0 & 1/3 \end{bmatrix}$ and $\begin{bmatrix} 1 & 0 & 0 & 0 \\ 0 & 1 & -2 & 0 \\ 0 & 0 & 1 & 0 \\ 0 & 0 & 0 & 1 \end{bmatrix}$

4. *Find the inverse of the matrix*

$$A = \begin{bmatrix} 1 & 1 & 1 & 1 \\ 1 & 2 & 3 & -4 \\ 2 & 3 & 5 & -5 \\ 3 & -4 & -5 & 8 \end{bmatrix}$$

Ans. $\frac{1}{18}\begin{bmatrix} 2 & 16 & 6 & 4 \\ 22 & 41 & -30 & -1 \\ -10 & -44 & 30 & -2 \\ 4 & -13 & 6 & -1 \end{bmatrix}$

5. *Find the inverse of the matrix* $\begin{bmatrix} 1 & 3 & 3 & 2 & 1 \\ 1 & 4 & 3 & 3 & -1 \\ 1 & 3 & 4 & 1 & 1 \\ 1 & 1 & 1 & 1 & -1 \\ 1 & -2 & -1 & 2 & 2 \end{bmatrix}$

Ans. $\frac{1}{15}\begin{bmatrix} 30 & -20 & -15 & 25 & -5 \\ 30 & -11 & -18 & 7 & -8 \\ -30 & -12 & 21 & -9 & 6 \\ -15 & 12 & 6 & -9 & 6 \\ 15 & -7 & -6 & -1 & -1 \end{bmatrix}$

6. *Has the following matrix an inverse?*

$$\begin{bmatrix} 2 & 1 & 3 & 1 \\ 1 & 2 & -1 & 4 \\ 3 & 3 & 2 & 5 \\ 1 & -1 & 4 & -3 \end{bmatrix}$$

Ans. No.

7. *Find* A^{-1} *if* $A = \begin{bmatrix} 2 & 4 & 3 \\ 0 & 1 & 1 \\ 2 & 2 & -1 \end{bmatrix}$ **Ans.** $\frac{1}{4}\begin{bmatrix} 3 & -10 & -1 \\ -2 & 8 & 2 \\ 2 & -4 & -2 \end{bmatrix}$

8. *Find the reciprocal matrix of* $\begin{bmatrix} 1 & 1 & 1 \\ 2 & 2 & 3 \\ 1 & 4 & 9 \end{bmatrix}$ **Ans.** $\frac{1}{3}\begin{bmatrix} -6 & 5 & -1 \\ 15 & -8 & 1 \\ -6 & 3 & 0 \end{bmatrix}$

9. *Reduce the matrix* $\begin{bmatrix} 1 & -1 & 1 \\ 2 & 3 & 4 \\ 3 & -1 & 4 \end{bmatrix}$ *to the triangular form.*

4. *Apply the row operations* $R_4(-3)$ *and* $R_{21}(4)$ *to the matrix*

$$\begin{bmatrix} 4 & -1 & 2 & 3 \\ -1 & 8 & -3 & -4 \\ 2 & 3 & 4 & -1 \\ -3 & -4 & -1 & 8 \end{bmatrix}$$

Ans. $\begin{bmatrix} 4 & -1 & 2 & 3 \\ -1 & 8 & -3 & -4 \\ 2 & 3 & 4 & -1 \\ 9 & 12 & 3 & -24 \end{bmatrix}$ and $\begin{bmatrix} 4 & -1 & 2 & 3 \\ 15 & 4 & 5 & 8 \\ 2 & 3 & 4 & -1 \\ -3 & -4 & -1 & 8 \end{bmatrix}$

3

Rank of a Matrix

3.1 INTRODUCTION

Definition: *If in an m × n matrix A, at least one of its r × r minors is different from zero while all the minors of order (r + 1) are zero, then r is defined as the rank of the matrix A.*

Or

A number r is defined as the rank of an m × n matrix A provide (i) A has at least one minor of order r which does not vanish and (ii) there is no minor of order (r + 1) which is nor equal to zero.

Note 1: The rank of a matrix remains unaltered by the application of elementary row or column operations *i.e.*, all equivalent matrices have the same rank.

Note 2: From the definition of the rank of a matrix we conclude that :

(a) If a matrix A does not posses any minor of order (r+1) then p(A)< r.

(b) If at least one minor of order r of the matrix A is not equal to zero, then p (A) < r.

Note 3: If every minor of order p of a matrix A is zero, the every minor of order higher than p is definitely zero.

Note 4: The rank of a matrix A is also denoted by p (A).

Note 5: The rank of a zero matrix by definition is 0 *i.e.*, p (O) = 0.

Consider the matrix $A = \begin{bmatrix} 1 & 2 & 3 \\ 2 & 3 & 4 \\ 3 & 5 & 7 \end{bmatrix}$

This matrix A has only one three rowed minor *i.e.*, minor of order 3, *viz.* $\begin{vmatrix} 1 & 2 & 3 \\ 2 & 3 & 4 \\ 3 & 5 & 7 \end{vmatrix}$ and its value can easily be calculated to zero, by expanding with respect to first row.

The matrix A has 9 minors of order 2 (or two-rowed minors) and one of them is $\begin{vmatrix} 3 & 4 \\ 5 & 7 \end{vmatrix}$ which has the value

$$(3 \times 7) - (5 \times 4) = 21 - 20 = 1 \neq 0.$$

This fact that A is a matrix whose every minor of order 3 is zero and there is at least one minor of order 2 which is not equal to zero is also expressed as 'the rank of the matrix A is 2'.

Example 1:

Find the rank of the matrix $A = \begin{bmatrix} 6 & 1 & 3 & 8 \\ 4 & 2 & 6 & -1 \\ 10 & 3 & 9 & 7 \\ 16 & 4 & 12 & 15 \end{bmatrix}$

Solution:

The det. of order 4 formed by this matrix

$$= \begin{vmatrix} 6 & 1 & 3 & 8 \\ 4 & 2 & 6 & -1 \\ 10 & 3 & 9 & 7 \\ 16 & 4 & 12 & 15 \end{vmatrix}$$

$$= \begin{vmatrix} 6 & 1 & 3 & 8 \\ 4 & 2 & 6 & -1 \\ 6 & 1 & 3 & 8 \\ 6 & 1 & 3 & 5 \end{vmatrix},$$ replacing R_3 and R_4 by $R_3 - R_2$ and $R_4 - R_3$ respectively

= 0, as its three rows are identical

A minor of order 3

$$= \begin{vmatrix} 6 & 1 & 3 \\ 4 & 2 & 6 \\ 10 & 3 & 9 \end{vmatrix} = \begin{vmatrix} 6 & 1 & 3 \\ 4 & 2 & 6 \\ 6 & 1 & 2 \end{vmatrix}, \text{ repacing } R_3 \text{ by } R_3 - R_2$$

= 0, two rows being identical.

In a similar way we can prove that all the minors of order 3 are zero.

Now a minor of order 2 = $\begin{vmatrix} 6 & 1 \\ 4 & 2 \end{vmatrix} = 12 - 4 = 8 \neq 0$

Hence, the rank of the given matrix = 2. **Ans.**

Example 2:

Find the rank of the matrix $\begin{bmatrix} 3 & 4 & 5 & 6 & 7 \\ 4 & 5 & 6 & 7 & 8 \\ 5 & 6 & 7 & 8 & 9 \\ 15 & 16 & 17 & 18 & 19 \end{bmatrix}$

Solution:

One minor of order 3 of A

$$= \begin{vmatrix} 5 & 7 & 8 \\ 6 & 8 & 9 \\ 6 & 18 & 19 \end{vmatrix} = \begin{vmatrix} 5 & 7 & 8 \\ 1 & 1 & 1 \\ 11 & 11 & 11 \end{vmatrix}, \text{ replacing } R_2 \text{ and } R_3 \text{ by } R_2 - R_1 \text{ and } R_3 - R_1 \text{ respectively.}$$

$$= \begin{vmatrix} 5 & 7 & 8 \\ 1 & 1 & 1 \\ 0 & 0 & 0 \end{vmatrix}, \text{ replacing } R_3 \text{ by } R_3 - 11R_2$$

= 0.

In a similar way we can prove that all the minors of order 3 of A are zero.

This shows that all minors of order 4 of A are automatically zero.

Now one minor of order 2 of A

$$= \begin{vmatrix} 7 & 8 \\ 8 & 9 \end{vmatrix} = (7 \times 9) - (8 \times 8) = 63 - 64 == -1 \neq 0.$$

Hence the rank of A is 2. **Ans.**

Example 3:

Find the rank of $A = \begin{bmatrix} 1 & 1 & 1 \\ b+c & c+a & a+b \\ bc & ca & ab \end{bmatrix}$

Solution:

$$|A| = \begin{vmatrix} 1 & 1 & 1 \\ b+c & c+a & a+b \\ bc & ca & ab \end{vmatrix}$$

$= -(a-b)(b-c)(c-a)$, on evaluating. ...(i)

Now following cases arises :

Case I: $a = b = c$.

If $a = b = c$, then $A = \begin{vmatrix} 1 & 1 & 1 \\ 2a & 2a & 2a \\ a^2 & a^2 & a^2 \end{vmatrix}$

Therefore all minors of order 2 and 3 of A vanish.

Also A has non-zero minor of order 1, since no element of A is zero.

Hence the rank of A in this case is 1. **Ans.**

Case II: *Two of numbers a, b, c are equal but are different from the third.*

Let $a = b \neq c$.

Then $|A| = \begin{vmatrix} 1 & 1 & 1 \\ a+c & c+a & 2a \\ ac & ca & a^2 \end{vmatrix} = 0$, as C_1, C_2 are identical.

Also A has a minor order 2 viz. $\begin{vmatrix} 1 & 1 \\ a+c & 2a \end{vmatrix}$

$= 2a - (a + c) = a - c \neq 0$, $\because a \neq c$,

Hence the rank of A in this case is 2.

Similarly we can discuss the cases $b = c \neq a$, $c = a \neq b$. **Ans.**

Case III: a, b, c are all different.

In this case $|A| \neq 0$, as is evident from (i) above.

i.e., A has a non-zero minor of order 3 and there exists no minor of order greater than 3.

Example 4:

Find the rank of the matrix $A = \begin{bmatrix} 1 & a & b & 0 \\ 0 & c & d & 1 \\ 1 & a & b & 0 \\ 0 & c & d & 1 \end{bmatrix}$

Solution:

$$|A| = \begin{vmatrix} 1 & a & b & 0 \\ 0 & c & d & 1 \\ 1 & a & b & 0 \\ 0 & c & d & 1 \end{vmatrix}$$

$= 0$, as R_1, R_3 are identical.

A minor of order 3 of A

$$= \begin{vmatrix} a & b & 0 \\ c & d & 1 \\ a & b & 0 \end{vmatrix} = 0, \text{ as } R_1, R_3 \text{ are identical.}$$

In a similar way we can show that all the minors of order 3 are zero in value.

A minor order 2 of A

$$= \begin{vmatrix} a & b \\ c & d \end{vmatrix} = ad - bc \neq 0$$

Hence the rank of the matrix A is 2. **Ans.**

Example 5:

Find the rank of the matrix

$A = \begin{bmatrix} 1 & 1 & 1 \\ a & b & c \\ a^3 & b^3 & c^3 \end{bmatrix}$, *where a, b, c are all real.*

Solution:

$$|A| = \begin{vmatrix} 1 & 1 & 1 \\ a & b & c \\ a^3 & b^3 & c^3 \end{vmatrix} = \begin{vmatrix} 1 & 0 & 0 \\ a & b-a & c-a \\ a^3 & b^3-a^3 & c^3-a^3 \end{vmatrix}, \text{ replacing } C_2, C_3 \text{ by } C_2 - C_1, C_3 - C_1$$

$$= \begin{vmatrix} b-a & c-a \\ b^3-a^3 & c^3-a^3 \end{vmatrix}, \text{ expanding with respect to } R_1$$

$$= (b-a)(c-a) \begin{vmatrix} 1 & 1 \\ b^2+ab+a^2 & c^2+ca+a^2 \end{vmatrix},$$

taking (b – a), (c – a) common from C_1 and C_2

$$= (b-a)(c-a) \begin{vmatrix} 1 & 1 \\ b^2+ab+a^2 & c^2+ca-b^2-ab \end{vmatrix},$$

replacing C_2 by $C_2 - C_1$

$$= (b-a)(c-a)[(c^2+ca-b^2-ab)-0]$$

$$= (b-a)(c-a)[(c^2-b^2)+a(c-b)] \qquad \textbf{(Note)}$$

$$= (b-a)(c-a)[(c-b)(c+b+a)]$$

$$\Rightarrow \quad |A| = (a-b)(b-c)(c-a)(a+b+c) \qquad \ldots(i)$$

Now following cases arise :

Case I: $a = b = c$.

If a = b = c, then $A = \begin{bmatrix} 1 & 1 & 1 \\ a & a & a \\ a^3 & a^3 & a^3 \end{bmatrix}$

Therefore all minors of order 3 and 2 of A are zero.

Also as no element of A is zero, so A has non zero minors of order 1.

Hence in this case, the rank of A is 1. **Ans.**

Case II: *Two of the numbers a, b, c are equal but are different from the third.*

let $a = b \neq c$

Then = $|A| = \begin{vmatrix} 1 & 1 & 1 \\ a & a & a \\ a^3 & a^3 & a^3 \end{vmatrix} = 0$, as C_1 and C_2 are identical.

Also A has a minor of order 2, viz. $\begin{vmatrix} 1 & 1 \\ a & c \end{vmatrix} = c - a \neq 0. \quad \because a \neq c$

Hence in this case the rank of A is 2. **Ans.**

Similarly we can discuss the cases $b = c \neq a$, $c = a \neq b$.

Case III: *a, b, c are all different but* $a + b + c = 0$.

In this case from (i), it is evident that $|A| = 0$. **(Note)**

Also A has a minor of order 2, viz. $\begin{vmatrix} 1 & 1 \\ a & b \end{vmatrix} = b - a \neq 0, \quad \because a \neq b$

Hence in this case the rank of A is 2. **Ans.**

Case IV: *a, b, c are all different but* $a + b + c \neq 0$.

In this case from (i), it is evident that $|A| \neq 0$. **(Note)**

i.e., A has a non-zero minor of order 3.

Also A has no minor of order greater than 3.

Hence in this case the rank of A is 3. **Ans.**

Example 6:

Under what condition the rank of the following matrix A is 3? Is it possible for the rank to be 1? Why?

$$A = \begin{bmatrix} 2 & 4 & 3 \\ 3 & 1 & 2 \\ 1 & 0 & x \end{bmatrix}$$

Solution:

If the rank of the matrix A is 3, then the minor of order 3 of A should be non-zero

i.e.,, $\begin{vmatrix} 2 & 4 & 3 \\ 3 & 1 & 2 \\ 1 & 0 & x \end{vmatrix} \neq 0$, which is the required condition.

Also the rank of A can not be 1 as at least one minor of order 1 of A *i.e.*, one element of A is zero.

[If we are to find the condition under which the rank of A is 2, then the same is $|A| = 0$ *i.e.*, minor of order 3 of A must be zero.]

i.e., $\begin{vmatrix} 2 & 4 & 2 \\ 3 & 1 & 2 \\ 1 & 0 & x \end{vmatrix} = 0$ *i.e.*, $\begin{vmatrix} 2 & 4 & 2 \\ 1 & -3 & 0 \\ 1 & 0 & x \end{vmatrix} = 0,$

replacing R_2 by $R_2 - R_1$

i.e., $\begin{vmatrix} 0 & 10 & 2 \\ 1 & -3 & 0 \\ 1 & 0 & x \end{vmatrix} = 0$, replacing R_1 by $R_1 - 2R_2$

i.e., $\begin{vmatrix} 10 & 2 \\ -3 & 0 \end{vmatrix} - 0 + x \begin{vmatrix} 0 & 10 \\ 1 & -3 \end{vmatrix} = 0$, expanding with respect to R_2

i.e., $6 - 10x = 0$ *i.e.*, $x = 6/10 = 3/5$. **Ans.**

Example 7:

Prove that the points (x_1, y_1), (x_2, y_2), (x_3, y_3) are collinear if the rank of the matrix $\begin{bmatrix} x_1 & y_1 & 1 \\ x_2 & y_2 & 1 \\ x_3 & y_3 & 1 \end{bmatrix}$ *is less than 3.*

Solution:

If the rank of the given matrix is less than 3, then the minor of order 3 of this matrix must be zero.

i.e., $\begin{vmatrix} x_1 & y_1 & 1 \\ x_2 & y_2 & 1 \\ x_3 & y_3 & 1 \end{vmatrix} = 0$ ...(i)

Now the area of the triangle whose vertices are (x_1, y_1), (x_2, y_2) and

$(x_3, y_3) = \frac{1}{2}\begin{vmatrix} x_1 & y_1 & 1 \\ x_2 & y_2 & 1 \\ x_3 & y_3 & 1 \end{vmatrix}$

$= 0$, from (i)

Since the area of this triangle is zero, so its vertices (x_1, y_1) (x_2, y_2) and (x_3, y_3) are collinear. **Hence proved.**

Example 8:

Are the matrices $A = \begin{bmatrix} 1 & 2 & 3 \\ 2 & 5 & 4 \\ 3 & 7 & 9 \end{bmatrix}$ *and*

$B = \begin{bmatrix} 1 & 0 & -5 & 6 \\ 3 & -2 & 1 & 2 \\ 5 & -2 & -9 & 14 \\ 4 & -2 & -4 & 8 \end{bmatrix}$ *equivalent?*

Example 9:

Find the rank of the matrix $A = \begin{bmatrix} 1 & 3 & 2 \\ 1 & 2 & 3 \\ 1 & 5 & 4 \end{bmatrix}$

Solution:

The determinant of order 3 formed by A

$$= \begin{vmatrix} 1 & 3 & 2 \\ 1 & 2 & 3 \\ 1 & 5 & 4 \end{vmatrix} = \begin{vmatrix} 1 & 3 & 2 \\ 0 & -1 & 1 \\ 0 & 2 & 2 \end{vmatrix},$$ replacing R_2, R_3 by $R_2 - R_1$, $R_3 - R_1$ respectively.

$$= \begin{vmatrix} -1 & 1 \\ 2 & 2 \end{vmatrix} = -2 - 2 = -4 \neq 0.$$

$\therefore$ $\rho(A) \geq 3$ ...(i)

Also the matrix A does not possess any matrix of order 4 *i.e.*, 3 + 1, so

$\rho(A) \leq 3.$ ...(ii)

$\therefore$ From (i) and (ii) we get $\rho(A) = 3$. **Ans.**

Example 10:

Find the rank of the matrix $A = \begin{bmatrix} 0 & 1 & 2 \\ 1 & 2 & 3 \\ 3 & 1 & 1 \end{bmatrix}$

Solution:

The determinant of order 3 formed by A

$$= \begin{vmatrix} 0 & 1 & 2 \\ 1 & 2 & 3 \\ 3 & 1 & 1 \end{vmatrix} = \begin{vmatrix} 0 & 1 & 0 \\ 1 & 2 & -1 \\ 3 & 1 & -1 \end{vmatrix}, \text{ replacing } C_3 \text{ by } C_3 - 2C_3$$

$$= - \begin{vmatrix} 1 & -1 \\ 3 & -1 \end{vmatrix} = -[-1+3] = -2 \neq 0.$$

$\therefore$ $\quad p(A) \geq 3$ ...(i)

Also the matrix A does not possesses any matrix of order 4 *i.e.*, 3 + 1, so p (A) ≤ 3. ...(ii)

$\therefore$ From (i) and (ii) we get p (A) = 3. **Ans.**

Example 11:

Find the rank of the matrix $A = \begin{bmatrix} 1 & 2 & 3 \\ 2 & 5 & 8 \\ 4 & 10 & 18 \end{bmatrix}$

Solution:

The determinant of order 2 formed by A

$$= \begin{vmatrix} 1 & 2 & 3 \\ 2 & 5 & 8 \\ 4 & 10 & 18 \end{vmatrix}$$

$$= \begin{vmatrix} 1 & 0 & 0 \\ 2 & 1 & 2 \\ 4 & 2 & 6 \end{vmatrix}, \text{replacing } C_2, C_3 \text{ by } C_2 - 2C_1, C_3 - 3C_1 \text{ respectively}$$

$$= \begin{vmatrix} 1 & 2 \\ 2 & 6 \end{vmatrix} = (1 \times 6) - (2 \times 2) = 6 - 4 = 2 \neq 0$$

$\therefore$ $\quad p(A) \geq 3$ ...(i)

Also matrix A does not possesses any matrix of order 4 *i.e.*, 3 + 1, so

p (A) ≤ 3. ...(ii)

$\therefore$ From (i) and (ii) we get p (A) = 3. **Ans.**

Example 12:

Find the rank of the matrix $A = \begin{bmatrix} 1 & 2 & 3 \\ 2 & 3 & 4 \\ 4 & 10 & 18 \end{bmatrix}$

Solution:

The determinant of order 3 formed by A

$$= \begin{vmatrix} 1 & 2 & 3 \\ 2 & 3 & 4 \\ 4 & 10 & 18 \end{vmatrix} = \begin{vmatrix} 1 & 0 & 0 \\ 2 & -1 & -2 \\ 4 & 2 & 6 \end{vmatrix}$$, replacing C_2, C_3 by $C_2 - 2C_1, C_3 - 3C_1$ respectively.

$$= \begin{vmatrix} -1 & -2 \\ 2 & 6 \end{vmatrix} = -6 + 4 = -2 \neq 0$$

$\therefore$ $\rho(A) \geq 3$ ...(i)

Also the matrix A does not possesses any matrix of order 4 *i.e.*, 3 + 1, so $\rho(A) \leq 3$. ...(ii)

$\therefore$ From (i) and (ii) we get $\rho(A) = 3$. **Ans.**

Example 13:

Find the rank of the matrix $A = \begin{bmatrix} 1 & -3 & 2 \\ 3 & -9 & 6 \\ -2 & 6 & -4 \end{bmatrix}$

Solution:

The determinant of order 3 formed by this matrix A

$$= \begin{vmatrix} 1 & -3 & 2 \\ 3 & -9 & 6 \\ -2 & 6 & -4 \end{vmatrix} = \begin{vmatrix} 1 & 0 & 0 \\ 3 & 0 & 0 \\ -2 & 0 & 0 \end{vmatrix}$$, replacing C_2, C_3 by $C_2 + 3C_1$ and $C_3 - 2C_1$ respectively

$$= 0$$

Also there exists no minor of order 2 of A which is not equal to zero. (Students can verify for themselves).

Finally all minors of order 1 of the matrix A are non-zero, as no element of the matrix A is zero.

Hence the rank of A is 1. **Ans.**

Example 14:

Find the rank of the matrix $\begin{bmatrix} 1 & 2 & 3 & 1 \\ 2 & 4 & 6 & 2 \\ 1 & 2 & 3 & 2 \end{bmatrix}$

Solution:

In this matrix, a minor of order

$$= \begin{vmatrix} 1 & 2 & 3 \\ 2 & 4 & 6 \\ 1 & 2 & 3 \end{vmatrix} = 0, \because R_1, R_2 \text{ are identical}$$

In a similar way we prove that all the minors of order 3 are zero.

Now a minor order 2 = $\begin{vmatrix} 1 & 2 \\ 2 & 4 \end{vmatrix} = 0.$

But another minor order 2 = $\begin{vmatrix} 3 & 1 \\ 3 & 2 \end{vmatrix} \neq 0$

Hence rank of the given matrix is 2. **Ans.**

Example 15:

Find the rank of the matrix $\begin{bmatrix} 13 & 16 & 19 \\ 14 & 17 & 20 \\ 15 & 18 & 21 \end{bmatrix}$

Solution:

The determinant of order 3 formed by this matrix

$$= \begin{vmatrix} 13 & 16 & 19 \\ 14 & 17 & 20 \\ 15 & 18 & 21 \end{vmatrix}$$

$$= \begin{vmatrix} 13 & 3 & 3 \\ 14 & 3 & 3 \\ 15 & 3 & 3 \end{vmatrix},$$ replacing C_2 and C_3 by $C_2 - C_1$ and $C_3 - C_2$ respectively.

= 0, since two columns viz. C_2 and C_3 are identical.

A minor of order 2 = $\begin{vmatrix} 13 & 16 \\ 14 & 17 \end{vmatrix} \neq 0.$

Hence the rank of the given matrix is 2. **Ans.**

Example 16:

Find the rank of the matrix $A = \begin{bmatrix} 1 & 2 & 3 \\ 2 & 3 & 1 \\ -2 & -3 & -1 \end{bmatrix}$

Solution:

The determinant of order 3 formed by this matrix A

$$= \begin{vmatrix} 1 & 2 & 3 \\ 2 & 3 & 1 \\ -2 & -3 & -1 \end{vmatrix} = \begin{vmatrix} 1 & 2 & 3 \\ 2 & 3 & 1 \\ 0 & 0 & 0 \end{vmatrix}, \text{ replacing } R_3 \text{ by } R_3 + R_2$$

$= 0$

Also there exists a minor of order 2 of A

viz. $\begin{vmatrix} 1 & 2 \\ 2 & 3 \end{vmatrix} = 3 - 4 = -1 \neq 0$

Hence the rank of the given matrix A is 2. **Ans.**

Example 17:

Find the rank of the matrix $A = \begin{bmatrix} 1 & 2 & 3 \\ 2 & 3 & 4 \\ 4 & 5 & 6 \end{bmatrix}$

Solution:

The determinant of order 3 formed by A

$$= \begin{vmatrix} 1 & 2 & 3 \\ 2 & 3 & 4 \\ 4 & 5 & 6 \end{vmatrix} = \begin{vmatrix} 1 & 0 & 0 \\ 2 & -1 & -2 \\ 4 & -3 & -6 \end{vmatrix}, \text{ replacing } C_2, C_3 \text{ by } C_2 - 2C_1, C_3 - 3C_1 \text{ respectively.}$$

$$= \begin{vmatrix} -1 & -2 \\ - & -6 \end{vmatrix} = 6 - 6 = 0$$

Also there exists a minor of order 2 of A

viz. $\begin{vmatrix} 3 & 4 \\ 5 & 6 \end{vmatrix} = 18 - 20 = -2 \neq 0.$

Hence the rank of the given matrix A is 2. **Ans.**

Example 18:

Find the rank of the matrix $\begin{bmatrix} 1 & -1 & -2 & -4 \\ 3 & 1 & 3 & -2 \\ 6 & 3 & 0 & -7 \\ 2 & 3 & -1 & -1 \end{bmatrix}$

Solution:

The determinant of order 4 formed by the given matrix

$$= \begin{vmatrix} 1 & -1 & -2 & -4 \\ 3 & 1 & 3 & -2 \\ 6 & 3 & 0 & -7 \\ 2 & 3 & -1 & -1 \end{vmatrix}$$

$$= \begin{vmatrix} 1 & 0 & 0 & 0 \\ 3 & 4 & 9 & 10 \\ 6 & 9 & 12 & 17 \\ 2 & 5 & 3 & 7 \end{vmatrix}, \text{ replacing } C_2, C_3, C_4 \text{ by } C_2 + C_1, C_3 + 2C_1, C_4 + 4C_1 \text{ respectively.}$$

$$= \begin{vmatrix} 4 & 9 & 10 \\ 9 & 12 & 17 \\ 5 & 3 & 7 \end{vmatrix} = \begin{vmatrix} 4 & 9 & 10 \\ 5 & 3 & 7 \\ 5 & 3 & 7 \end{vmatrix}, \text{ replacing } R_2 \text{ by } R_2 - R_1$$

= 0, as its two row and identical.

A minor of order 3

$$= \begin{vmatrix} 1 & -1 & -2 \\ 3 & 1 & 3 \\ 6 & 3 & 0 \end{vmatrix} = \begin{vmatrix} 1 & 0 & 0 \\ 3 & 4 & 9 \\ 6 & 9 & 12 \end{vmatrix}, \text{ replacing } C_2, C_3 \text{ by } C_2 + C_1, C_3 + 2C_1 \text{ respectively.}$$

$$= \begin{vmatrix} 4 & 9 \\ 9 & 12 \end{vmatrix} = 48 - 81 = -33 \neq 0.$$

Hence the rank of the given matrix is 3. **Ans.**

Example 19:

Find the rank of the matrix $A = \begin{bmatrix} 1 & 1 & 1 & -1 \\ 1 & 2 & 3 & 4 \\ 3 & 4 & 5 & 2 \end{bmatrix}$

Solution:

In this matrix there are four minors of order 3 of a Viz.

$$\begin{vmatrix} 1 & 1 & 1 \\ 1 & 2 & 3 \\ 3 & 4 & 5 \end{vmatrix}; \begin{vmatrix} 1 & 1 & -1 \\ 1 & 2 & 4 \\ 3 & 4 & 2 \end{vmatrix}; \begin{vmatrix} 1 & 1 & -1 \\ 1 & 3 & 4 \\ 3 & 5 & 2 \end{vmatrix}; \begin{vmatrix} 1 & 1 & -1 \\ 2 & 3 & 4 \\ 4 & 5 & 2 \end{vmatrix}$$

and each one of them is zero in value (prove it).

Also therexsts a minor of or 2 of A viz.

$$\begin{vmatrix} 2 & 3 \\ 4 & 5 \end{vmatrix} = 10 - 12 = -2 \neq 0$$

Hence the rank of the given matrix A is 2. **Ans.**

Example 20:

Find the rank of the matrix $A = \begin{bmatrix} 6 & 1 & 3 & 8 \\ 16 & 4 & 12 & 15 \\ 5 & 3 & 3 & 4 \\ 4 & 2 & 6 & -1 \end{bmatrix}$

Solution:

The determinant of order 4 formed by A

$$= \begin{vmatrix} 6 & 1 & 3 & 8 \\ 16 & 4 & 12 & 15 \\ 5 & 3 & 3 & 4 \\ 4 & 2 & 6 & -1 \end{vmatrix} = \begin{vmatrix} 0 & 1 & 0 & 0 \\ -8 & 4 & 0 & -17 \\ -13 & 3 & -6 & -20 \\ -8 & 2 & 0 & -17 \end{vmatrix},$$

replacing C_1, C_3, C_4 by $C_1 - 6C_2$, $C_3 - 3C_2$ and $C_4 - 8C_2$ respectively.

$$= \begin{vmatrix} -8 & 0 & -17 \\ -13 & -6 & -20 \\ -8 & 0 & -17 \end{vmatrix} = 0, \text{ as } R_1\ R_2 \text{ are identical.}$$

Also one minor of order 3 viz.

$$\begin{vmatrix} 1 & 3 & 8 \\ 3 & 3 & 4 \\ 2 & 6 & -1 \end{vmatrix} = \begin{vmatrix} 1 & 0 & 0 \\ 3 & -6 & -20 \\ 2 & 0 & -17 \end{vmatrix} = \begin{vmatrix} -6 & -20 \\ 0 & -17 \end{vmatrix} \neq 0.$$

Hence the rank of given matrix A is 3. **Ans.**

Example 21:

Fine the rank of the matrix $\begin{bmatrix} 1 & 3 & 4 & 3 \\ 3 & 9 & 12 & 9 \\ -1 & -3 & -4 & -3 \end{bmatrix}$

Solution:

In this matrix, a minor of order 3

$$= \begin{vmatrix} 1 & 3 & 4 \\ 3 & 9 & 12 \\ -1 & -3 & -4 \end{vmatrix} = 3 \begin{vmatrix} 1 & 3 & 4 \\ 1 & 3 & 4 \\ -1 & -3 & -4 \end{vmatrix}, \text{ taking common from } R_3$$

$= 0$, as R_1 and R_2 are identical.

In a similar way we can prove that all minors of order 3 are zero.

Now a minor of order 2.

$$= \begin{vmatrix} 1 & 3 \\ 3 & 9 \end{vmatrix} = 3 \begin{vmatrix} 1 & 3 \\ 1 & 3 \end{vmatrix}, \text{ taking out 3 common from } R_3$$

$= 0$, as rows are identical.

Similarly, all the minors of order 2 are zero.

Hence, we are left with minors of order unity, viz. the elements of the given matrix, which are not equal to zero.

Hence rank of the given matrix = 1. **Ans.**

Example 22:

Find the rank of the matrix $A = \begin{bmatrix} 1 & -1 & 3 & 6 \\ 1 & 3 & -3 & -4 \\ 5 & 3 & 3 & 11 \end{bmatrix}$

Solution:

The given matrix A possesses a minor of order 3 viz.

$$\begin{vmatrix} 1 & 3 & 6 \\ 1 & -3 & -4 \\ 5 & 3 & 11 \end{vmatrix} = \begin{vmatrix} 2 & 0 & 2 \\ 1 & -3 & -4 \\ 6 & 0 & 7 \end{vmatrix},$$ replacing R_1, R_2 by $R_2 + R_1$, $R_3 + R_2$

$$= -3 \begin{vmatrix} 2 & 2 \\ 6 & 7 \end{vmatrix} = -3\,(14 - 12) = -6 \neq 0$$

$\therefore$ $\quad p\,(A) \geq 3$ ...(i)

Also A does not posses any minor of order 4 *i.e.*,, 3 + 1, so

$p\,(A) \leq 3$ ...(ii)

$\therefore$ From (i) and (ii) we get p (A) = 3. **Ans.**

Example 23:

Find the rank of the matrix $A = \begin{bmatrix} 1 & 3 & 4 & 5 \\ 1 & 2 & 6 & 7 \\ 1 & 5 & 0 & 1 \end{bmatrix}$

Solution:

One minor of order three of A

$$= \begin{vmatrix} 1 & 4 & 5 \\ 1 & 6 & 7 \\ 1 & 0 & 1 \end{vmatrix} = \begin{vmatrix} 1 & 0 & 0 \\ 1 & 2 & 2 \\ 1 & -4 & -4 \end{vmatrix},$$ replacing C_2, C_3 by $C_2 - 4C_1$ and $C_3 - 5\,C_1$ respectively.

$$= \begin{vmatrix} 2 & 2 \\ -4 & -4 \end{vmatrix},$$ expanding with respect to R_1

$= 2\,(-4) - 2\,(-4) = 0.$

In a similar way we can prove that all the four minors of order three are zero.

Now a minor or order 2 is $\begin{vmatrix} 2 & 6 \\ 5 & 0 \end{vmatrix} = 2.0 - 6.5 \neq 0.$

Hence the rank of A is 2.

Example 24:

Find the rank of the matrix $A = \begin{bmatrix} 1 & 3 & 5 & 1 \\ 2 & 4 & 8 & 0 \\ 3 & 1 & 7 & 5 \end{bmatrix}$

Solution:

The given matrix A possesses a minor of order 3 viz.

$$\begin{vmatrix} 1 & 3 & 1 \\ 2 & 4 & 0 \\ 3 & 1 & 5 \end{vmatrix} = \begin{vmatrix} 0 & 0 & 1 \\ 2 & 4 & 0 \\ -2 & -14 & 5 \end{vmatrix}, \text{ repacing } C_1 \text{ and } C_2 \text{ by } C_1 - C_3 \text{ and } C_2 - 3C_2$$

$$= \begin{vmatrix} 2 & 4 \\ -2 & -14 \end{vmatrix}, \text{expanding with respect to } R_1$$

$$= 2\,(-14) - (4)\,(-2) = -\,28 + 8 \neq 0$$

$\therefore$ $\quad p\,(A) \geq 3$ ...(i)

Also A does not posses any minor of order 4 *i.e.*,, 3 + 1, so

$p\,(A) \leq 3$ ...(ii)

From (i) and (ii) we get p (A) = 3 *i.e.*, the rank of A is 3. **Ans.**

Example 25:

Determinant the rank of $A = \begin{bmatrix} 6 & 1 & 8 & 3 \\ 2 & 1 & 0 & 2 \\ 4 & -1 & -8 & -3 \end{bmatrix}$

Solution:

The given matrix A possesses a minor of order 3 viz.

$$\begin{vmatrix} 6 & 1 & 8 \\ 2 & 1 & 0 \\ 4 & -1 & -8 \end{vmatrix} = \begin{vmatrix} 10 & 0 & 0 \\ 6 & 0 & -8 \\ 4 & -1 & -8 \end{vmatrix}, \text{ replacing } R_1, R_2 \text{ by } R_1 + R_3, R_2 + R_3$$

$$= 10 \begin{vmatrix} 0 & -8 \\ -1 & -8 \end{vmatrix} = 10\,(0 - 8) = -\,80 \neq 0$$

$\therefore$ $\quad p\,(A) \geq 3$ ...(i)

Also A does not posses any minor of order 4 *i.e.*,, 3 + 1, so

$p\,(A) \leq 3$ ...(ii)

$\therefore$ From (i) and (ii) we get p (A) = 3. **Ans.**

3.2 NORMAL FORM OF A MATRIX

Every non-zero matrix A of order m × n can be reduced by application of elementary row and column operations into equivalent matrix of one of the following forms:

(i) $\begin{bmatrix} I_r & 0 \\ 0 & 0 \end{bmatrix}$, (ii) $\begin{bmatrix} I_r \\ 0 \end{bmatrix}$, (iii) $[I_r \ 0]$, (iv) $[I_r]$,

where I_r is r × r identity matrix and 0 is null matrix of any order.

These four forms are called **Normal** or **canonical form** of **A.**

Important Theorems (Without Proof)

Theorem 1:

If m × n matrix A is reduced to the canonical form or normal form

$\begin{bmatrix} I_r & 0 \\ 0 & 0 \end{bmatrix}$ *by application of elementary row or column operations, then r, the order to the identity sub-matrix I_r is the rank of the matrix A.*

Theorem 2:

If a non-singular matrix of order n × n is reduce to the identity matrix I_n (which is its canonical or normal form), then the rank of the matrix is n.

Example 1:

Find the rank of the matrix $A = \begin{bmatrix} 1 & 2 & 3 \\ 4 & 5 & 6 \\ 2 & 1 & 2 \end{bmatrix}$

Solution:

$$A \sim \begin{bmatrix} 1 & 0 & 0 \\ 4 & -3 & -6 \\ 2 & -3 & -4 \end{bmatrix}$$, replacing C_2, C_3 by $C_2 - 2C_1$, $C_3 - 3C_1$ respectively.

$$\sim \begin{bmatrix} 1 & 0 & 0 \\ 4 & -3 & -6 \\ 2 & -3 & -4 \end{bmatrix}$$, replacing R_2, R_3 by $R_2 - 4R_1$ $R_3 - 2R_1$ respectively

$$\sim \begin{bmatrix} 1 & 0 & 0 \\ 0 & 0 & -2 \\ 0 & -3 & 4 \end{bmatrix}$$, replacing R_3 by $R_2 - R_3$

$$\sim \begin{bmatrix} 1 & 0 & 0 \\ 0 & 0 & 1 \\ 0 & 1 & 2 \end{bmatrix}$$, replacing C_2, C_3 by $-\frac{1}{3}C_2$, $-\frac{1}{2}C_3$ respectively

$$\sim \begin{bmatrix} 1 & 0 & 0 \\ 0 & 0 & 1 \\ 0 & 1 & 0 \end{bmatrix}, \text{ replacing } R_3 \text{ by } R_3 - 2R_2$$

$$\sim \begin{bmatrix} 1 & 0 & 0 \\ 0 & 1 & 0 \\ 0 & 0 & 1 \end{bmatrix}, \text{ interchanging } C_2 \text{ and } C_3$$

$$\sim [\, I_3 \,]$$

Hence the rank of A is 3. **Ans.**

Example 2:

Find the rank of the matrix $A = \begin{bmatrix} -2 & -1 & -3 & -1 \\ 1 & 2 & 3 & -1 \\ 1 & 0 & 1 & 1 \\ 0 & 1 & 1 & -1 \end{bmatrix}$

Solution:

$$A \sim \begin{bmatrix} 0 & -1 & -1 & 1 \\ 0 & 2 & 2 & -2 \\ 1 & 0 & 1 & 1 \\ 0 & 1 & 1 & -1 \end{bmatrix}, \text{ replacing } R_1, R_2 \text{ by } R_1 + 2R_3, R_2 - R_3 \text{ respectively}$$

$$\sim \begin{bmatrix} 0 & 0 & 0 & 1 \\ 0 & 0 & 0 & -2 \\ 1 & 1 & 2 & 1 \\ 0 & 0 & 0 & -1 \end{bmatrix}, \text{ replacing } C_2, C_3 \text{ by } C_2 + C_4, C_3 + C_4 \text{ respectively}$$

$$\Rightarrow \quad A \sim \begin{bmatrix} 0 & 0 & 0 & 1 \\ 0 & 0 & 0 & 0 \\ 1 & 1 & 2 & 0 \\ 0 & 0 & 0 & 0 \end{bmatrix}, \text{ replacing } R_2, R_3, R_4 \text{ by } R_2 + 2R_1, R_3 - R_1, R_4 + R_1 \text{ respectively}$$

$$\sim \begin{bmatrix} 0 & 0 & 0 & 1 \\ 0 & 0 & 0 & 0 \\ 1 & 0 & 0 & 0 \\ 0 & 0 & 0 & 0 \end{bmatrix}, \text{ replacing } C_2, C_3 \text{ by } C_2 - C_1, C_3 - 2C_1 \text{ respectively.}$$

$$\sim \begin{bmatrix} 1 & 0 & 0 & 0 \\ 0 & 0 & 0 & 0 \\ 0 & 0 & 0 & 1 \\ 0 & 0 & 0 & 0 \end{bmatrix}, \text{ interchanging } R_1 \text{ and } R_4$$

$$\sim \begin{bmatrix} 1 & 0 & 0 & 0 \\ 0 & 0 & 0 & 0 \\ 0 & 1 & 0 & 0 \\ 0 & 0 & 0 & 0 \end{bmatrix}, \text{ interchanging } C_2 \text{ and } C_4$$

$$\sim \begin{bmatrix} 1 & 0 & 0 & 0 \\ 0 & 1 & 0 & 0 \\ 0 & 0 & 0 & 0 \\ 0 & 0 & 0 & 0 \end{bmatrix}, \text{ interchanging } R_2 \text{ and } R_3$$

$$\sim \begin{bmatrix} I_2 & 0 \\ 0 & 0 \end{bmatrix}$$

Hence the rank of A is 2. **Ans.**

Example 3:

Find the rank of the matrix $A = \begin{bmatrix} 1 & -1 & 2 & -3 \\ 4 & 1 & 0 & 2 \\ 0 & 3 & 0 & 4 \\ 0 & 1 & 0 & 2 \end{bmatrix}$

Solution:

$$A \sim \begin{bmatrix} 1 & 0 & 2 & 0 \\ 4 & 5 & 0 & 14 \\ 0 & 3 & 0 & 4 \\ 0 & 1 & 0 & 2 \end{bmatrix}, \text{ replacing } C_2, C_4 \text{ by } C_2 + C_1, C_4 + 3C_1 \text{ respectively}$$

$$\sim \begin{bmatrix} 1 & 0 & 1 & 0 \\ 4 & 5 & 0 & 7 \\ 0 & 3 & 0 & 2 \\ 0 & 1 & 0 & 1 \end{bmatrix}, \text{ replacing } C_3, C_4 \text{ by } \frac{1}{2}C_3, \frac{1}{2}C_4 \text{ respectively}$$

$$\Rightarrow \quad A \sim \begin{bmatrix} 0 & 0 & 1 & 0 \\ 4 & 5 & 0 & 2 \\ 0 & 3 & 0 & -1 \\ 0 & 1 & 0 & 0 \end{bmatrix}, \text{ replacing } C_1, C_4 \text{ by } C_1 - C_3, C_4 - C_2 \text{ respectively}$$

$$\sim \begin{bmatrix} 0 & 0 & 1 & 0 \\ 1 & 5 & 0 & 2 \\ 0 & 3 & 0 & -1 \\ 0 & 1 & 0 & 0 \end{bmatrix}, \text{ replacing } C_1 \text{ by } \frac{1}{2} C_1$$

$$\sim \begin{bmatrix} 0 & 0 & 1 & 0 \\ 1 & 0 & 0 & 0 \\ 0 & 3 & 0 & -1 \\ 0 & 1 & 0 & 0 \end{bmatrix}, \text{ replacing } C_2, C_4 \text{ by } C_2 - 5C_1, C_4 - 2C_1 \text{ respectively}$$

$$\sim \begin{bmatrix} 0 & 0 & 1 & 0 \\ 1 & 0 & 0 & 0 \\ 0 & 0 & 0 & 1 \\ 0 & 1 & 0 & 0 \end{bmatrix}, \text{ replacing } C_2, C_4 \text{ by } C_2 + 3C_4, -C_4 \text{ respectively}$$

$$\sim \begin{bmatrix} 1 & 0 & 0 & 0 \\ 0 & 1 & 0 & 0 \\ 0 & 0 & 1 & 0 \\ 0 & 0 & 0 & 1 \end{bmatrix}, \text{ rearranging columns}$$

$$\sim [\, I_4 \,].$$

$\therefore$ Rank of A is 4. **Ans.**

Example 4:

Find by reducing to normal form the rank of the matrix

$$A = \begin{bmatrix} 2 & 3 & -1 & -1 \\ 1 & -1 & -2 & -4 \\ 3 & 1 & 3 & -2 \\ 6 & 3 & 0 & -7 \end{bmatrix}$$

Solution:

$$A \sim \begin{bmatrix} 6 & 3 & 0 & -7 \\ 1 & -1 & -2 & -4 \\ 3 & 1 & 3 & -2 \\ 6 & 3 & 0 & -7 \end{bmatrix}, \text{ replacing } R_1 \text{ by } R_1 + R_2 + R_3$$

$$\sim \begin{bmatrix} 6 & 3 & 0 & -7 \\ 1 & -1 & -2 & -4 \\ 3 & 1 & 3 & -2 \\ 0 & 0 & 0 & 0 \end{bmatrix}, \text{ replacing } R_4 \text{ by } R_4 - R_1$$

$$\sim \begin{bmatrix} 0 & 9 & 12 & 17 \\ 1 & -1 & -2 & -4 \\ 0 & 4 & 9 & 10 \\ 0 & 0 & 0 & 0 \end{bmatrix}, \text{ replacing } R_1 \text{ and } R_3 \text{ by } R_1 - 6R_2,\ R_3 - 3R_2 \text{ respectively.}$$

$$\sim \begin{bmatrix} 0 & 9 & 12 & 17 \\ 1 & 0 & 0 & 0 \\ 0 & 4 & 9 & 10 \\ 0 & 0 & 0 & 0 \end{bmatrix}, \text{ replacing } C_2, C_3 \text{ and } C_4 \text{ by } C_2 + C_1,\ C_3 + 2C_1,\ C_4 + 4C_1 \text{ respectively.}$$

$$\Rightarrow \quad A \sim \begin{bmatrix} 1 & 0 & 0 & 0 \\ 0 & 9 & 12 & 17 \\ 0 & 4 & 9 & 10 \\ 0 & 0 & 0 & 0 \end{bmatrix}, \text{ interchanging } R_1 \text{ and } R_2$$

$$\sim \begin{bmatrix} 1 & 0 & 0 & 0 \\ 0 & 1 & -6 & -3 \\ 0 & 4 & 9 & 10 \\ 0 & 0 & 0 & 0 \end{bmatrix}, \text{ replacing } R_2 \text{ by } R_2 - 2R_3$$

$$\sim \begin{bmatrix} 1 & 0 & 0 & 0 \\ 0 & 1 & 0 & 0 \\ 0 & 4 & 33 & 22 \\ 0 & 0 & 0 & 0 \end{bmatrix}, \text{ replacing } C_3 \text{ and } C_4 \text{ by } C_3 + 6C_2 \text{ and } C_4 + 3C_2 \text{ respectively.}$$

$$\sim \begin{bmatrix} 1 & 0 & 0 & 0 \\ 0 & 1 & 0 & 0 \\ 0 & 4 & 1 & 1 \\ 0 & 0 & 0 & 0 \end{bmatrix},$$ replacing C_2 and C_4 by $\frac{1}{33}C_3$ and $\frac{1}{22}C_4$ respectively.

$$\sim \begin{bmatrix} 1 & 0 & 0 & 0 \\ 0 & 1 & 0 & 0 \\ 0 & 0 & 1 & 0 \\ 0 & 0 & 0 & 0 \end{bmatrix},$$ replacing C_2 and C_4 by $C_2 - 4C_3$ and $C_4 - C_3$ respectively.

$$\sim \begin{bmatrix} I_3 & 0 \\ 0 & 0 \end{bmatrix}$$

Hence the rank of the given matrix = 3. **Ans.**

Example 5:

Reduce the matrix A by the normal form and hence find the rank of the matrix A, where

$$A = \begin{bmatrix} 1 & -1 & 1 & -1 \\ 4 & 2 & -1 & 2 \\ 2 & 2 & -2 & 2 \end{bmatrix}$$

Solution:

$$A \sim \begin{bmatrix} 1 & 0 & 1 & 0 \\ 4 & 6 & -1 & 1 \\ 2 & 4 & -2 & 0 \end{bmatrix},$$ replacing C_2, C_4 by $C_2 + C_1, C_4 + C_3$ respectively.

$$\sim \begin{bmatrix} 1 & 0 & 1 & 0 \\ 5 & 6 & 0 & 1 \\ 4 & 4 & 0 & 0 \end{bmatrix},$$ replacing R_2, R_3 by $R_2 + R_1, R_3 + 2R_1$ respectively.

$$\Rightarrow \quad A \sim \begin{bmatrix} 0 & 0 & 1 & 0 \\ 5 & 3 & 0 & 1 \\ 4 & 2 & 0 & 0 \end{bmatrix},$$ replacing C_1, C_2 by $C_1 - C_3, \frac{1}{2}C_2$ respectively

$$\sim \begin{bmatrix} 0 & 0 & 1 & 0 \\ -1 & 3 & 0 & 1 \\ 0 & 2 & 0 & 0 \end{bmatrix},$$ replacing C_1 by $C_1 - 2C_2$

$$\sim \begin{bmatrix} 0 & 0 & 1 & 0 \\ 0 & 0 & 0 & 1 \\ 0 & 2 & 0 & 0 \end{bmatrix},$$ replacing C_1, C_2 by $C_1 + C_4, C_2 - 3C_1$ respectively.

$$\sim \begin{bmatrix} 1 & 0 & 0 & 0 \\ 0 & 0 & 0 & 1 \\ 0 & 1 & 0 & 0 \end{bmatrix},$$ replacing C_2 by $\frac{1}{2}C_2$ and interchanging C_1 and C_2

$$\sim \begin{bmatrix} 1 & 0 & 0 & 0 \\ 0 & 1 & 0 & 0 \\ 0 & 0 & 1 & 0 \end{bmatrix},$$ rearranging columns

$$\sim [\, I_3 \ 0 \,]$$

Hence the rank of A is 3. **Ans.**

Example 6:

Find the rank of the matrix $A = \begin{bmatrix} 1 & -1 & 2 & -3 \\ 4 & 1 & 0 & 2 \\ 0 & 3 & 0 & 4 \\ 0 & 1 & 0 & 2 \end{bmatrix}$

Solution:

$$A \sim \begin{bmatrix} 1 & 0 & 1 & 0 \\ 4 & 5 & 0 & 14 \\ 0 & 3 & 0 & 4 \\ 0 & 1 & 0 & 2 \end{bmatrix},$$ replacing C_2, C_3, C_4 by $C_2 + C_1, \frac{1}{2}C_3, C_4 + 3C_1$ respectively.

$$\sim \begin{bmatrix} 0 & 0 & 1 & 0 \\ 4 & 5 & 0 & 7 \\ 0 & 3 & 0 & 2 \\ 0 & 1 & 0 & 1 \end{bmatrix},$$ replacing C_1, C_4 by $C_1 - C_3, \frac{1}{2}C_4$ respectively.

$$\sim \begin{bmatrix} 0 & 0 & 1 & 0 \\ 1 & 5 & 0 & 2 \\ 0 & 3 & 0 & -1 \\ 0 & 1 & 0 & 0 \end{bmatrix},$$ replacing C_1, C_4 by $\frac{1}{4}C_1$, $C_4 - C_2$ respectively.

$$\sim \begin{bmatrix} 0 & 0 & 1 & 0 \\ 1 & 0 & 0 & 0 \\ 0 & 3 & 0 & -1 \\ 0 & 1 & 0 & 0 \end{bmatrix}, \text{ replacing } C_2, C_4 \text{ by } C_2 - 5C_1, C_4 - 2C_1 \text{ respectively.}$$

$$\Rightarrow \quad A \sim \begin{bmatrix} 1 & 0 & 0 & 0 \\ 0 & 0 & 1 & 0 \\ 0 & 0 & 0 & 1 \\ 0 & 1 & 0 & 0 \end{bmatrix}, \text{ replacing } C_2, C_4 \text{ by } C_2 + 3C_4, -C_4 \text{ respectively and interchanging } C_1, C_2$$

$$\sim \begin{bmatrix} 1 & 0 & 0 & 0 \\ 0 & 1 & 0 & 0 \\ 0 & 0 & 1 & 0 \\ 0 & 0 & 0 & 1 \end{bmatrix}, \text{ rearranging columns}$$

$$\sim [\, I_4 \,]$$

Hence the rank of A is 4. **Ans.**

Example 7:

Find the rank of the matrix $A = \begin{bmatrix} 1 & 3 & 4 & 7 \\ 2 & 4 & 5 & 8 \\ 3 & 1 & 2 & 4 \end{bmatrix}$

Solution:

$$A \sim \begin{bmatrix} 1 & 3 & 4 & 7 \\ 1 & 1 & 1 & 1 \\ 3 & 1 & 2 & 4 \end{bmatrix}, \text{ replacing } R_2 \text{ by } R_2 - R_1$$

$$\sim \begin{bmatrix} 1 & 2 & 3 & 6 \\ 1 & 0 & 0 & 0 \\ 3 & -2 & -1 & 1 \end{bmatrix}, \text{ replacing } C_2, C_3, C_4 \text{ by } C_2 - C_1, C_3 - C_1, C_4 - C_1 \text{ respectively.}$$

$$\sim \begin{bmatrix} 0 & 2 & 3 & 6 \\ 1 & 0 & 0 & 0 \\ 3 & -2 & -1 & 1 \end{bmatrix}, \text{ replacing } R_1, R_3 \text{ by } R_1 - R_2, R_3 - 3R_2$$

$$\sim \begin{bmatrix} 0 & 1 & 3 & 0 \\ 1 & 0 & 0 & 0 \\ 3 & -1 & -1 & 3 \end{bmatrix}, \text{ replacing } C_2, C_4 \text{ by } \frac{1}{2}C_2, C_4 - 2C_3 \text{ respectively.}$$

$$\Rightarrow \quad A \sim \begin{bmatrix} 0 & 1 & 0 & 0 \\ 1 & 0 & 0 & 0 \\ 0 & -1 & 2 & 3 \end{bmatrix}, \text{ replacing } C_3 \text{ by } C_3 - 3C_2$$

$$\sim \begin{bmatrix} 0 & 1 & 0 & 0 \\ 1 & 0 & 0 & 0 \\ 0 & 0 & 2 & 3 \end{bmatrix}, \text{ replacing } R_3 \text{ by } R_3 + R_1$$

$$\sim \begin{bmatrix} 0 & 1 & 0 & 0 \\ 1 & 0 & 0 & 0 \\ 0 & 0 & 1 & 1 \end{bmatrix}, \text{ replacing } C_3, C_4 \text{ by } \frac{1}{2}C_3 \text{ and } \frac{1}{3}C_4 \text{ respectively.}$$

$$\sim \begin{bmatrix} 0 & 1 & 0 & 0 \\ 1 & 0 & 0 & 0 \\ 0 & 0 & 1 & 0 \end{bmatrix}, \text{ replacing } C_4 \text{ by } C_4 - C_3$$

$$\sim \begin{bmatrix} 1 & 0 & 0 & 0 \\ 0 & 1 & 0 & 0 \\ 0 & 0 & 1 & 0 \end{bmatrix}, \text{ interchanging } R_1 \text{ and } R_2$$

$$\sim [\, I_3 \; 0 \,]$$

Hence the rank of A is 3. **Ans.**

Example 8:

Find the rank of $A = \begin{bmatrix} 0 & 1 & -3 & -1 \\ 1 & 0 & 1 & 1 \\ 3 & 1 & 0 & 2 \\ 1 & 1 & -2 & 0 \end{bmatrix}$

Solution:

$$A \sim \begin{bmatrix} 0 & 1 & -3 & -1 \\ 1 & 0 & 0 & 0 \\ 3 & 1 & -3 & -1 \\ 1 & 1 & -3 & -1 \end{bmatrix}, \text{ replacing } C_4, C_4 \text{ by } C_3 - C_1 \text{ and } C_4 - C_1 \text{ respectively}$$

$$\sim \begin{bmatrix} 0 & 1 & -3 & -1 \\ 1 & 0 & 0 & 0 \\ 3 & 0 & 0 & 0 \\ 1 & 0 & 0 & 0 \end{bmatrix}, \text{ replacing } R_3, R_4 \text{ by } R_3 - R_1 \text{ and } R_4 - R_1 \text{ respectively.}$$

$$\sim \begin{bmatrix} 0 & 1 & 0 & 0 \\ 1 & 0 & 0 & 0 \\ 3 & 0 & 0 & 0 \\ 1 & 0 & 0 & 0 \end{bmatrix}, \text{ replacing } C_3, C_4 \text{ by } C_3 + 3C_2, C_4 + C_2 \text{ respectively}$$

$$\sim \begin{bmatrix} 0 & 1 & 0 & 0 \\ 1 & 0 & 0 & 0 \\ 0 & 0 & 0 & 0 \\ 0 & 0 & 0 & 0 \end{bmatrix}, \text{ replacing } R_3, R_4 \text{ by } R_3 - 3R_2, R_4 - R_2 \text{ respectively.}$$

$$\sim \begin{bmatrix} 1 & 0 & 0 & 0 \\ 0 & 1 & 0 & 0 \\ 0 & 0 & 0 & 0 \\ 0 & 0 & 0 & 0 \end{bmatrix}, \text{ interchanging } C_1, C_2$$

$$\sim \begin{bmatrix} I_2 & 0 \\ 0 & 0 \end{bmatrix}$$

Hence the rank of A is 2. **Ans.**

Example 9:

Find the rank of the matrix $A = \begin{bmatrix} 1 & 0 & 2 & 1 \\ 0 & 1 & -2 & 1 \\ 1 & -1 & 4 & 0 \\ -2 & 2 & 8 & 0 \end{bmatrix}$

Solution:

$$A \sim \begin{bmatrix} 1 & 0 & 2 & 1 \\ 0 & 1 & -2 & 1 \\ 0 & -1 & 2 & -1 \\ 0 & 2 & 12 & 2 \end{bmatrix}, \text{ replacing } R_3, R_4 \text{ by } R_3 - R_2 \text{ and } R_4 + 2R_1 \text{ respectively}$$

$$\sim \begin{bmatrix} 1 & 0 & 2 & 1 \\ 0 & 1 & -2 & 1 \\ 0 & 0 & 0 & 0 \\ 0 & 0 & 16 & 0 \end{bmatrix}, \text{ replacing } R_3 \text{ and } R_4 \text{ by } R_3 + R_2 \text{ and } R_4 - 2R_2 \text{ respectively.}$$

$$\sim \begin{bmatrix} 1 & 0 & 2 & 1 \\ 0 & 1 & 0 & 1 \\ 0 & 0 & 0 & 0 \\ 0 & 0 & 16 & 0 \end{bmatrix}, \text{ replacing } C_2 \text{ by } C_3 + 2C_2$$

$$\sim \begin{bmatrix} 1 & 0 & 0 & 0 \\ 0 & 1 & 0 & 0 \\ 0 & 0 & 0 & 0 \\ 0 & 0 & 16 & 0 \end{bmatrix}, \text{ replacing } C_3, C_4 \text{ by } C_3 - 2C_1, C_4 - C_1 - C_2 \text{ respectively}$$

$$\sim \begin{bmatrix} 1 & 0 & 0 & 0 \\ 0 & 1 & 0 & 0 \\ 0 & 0 & 0 & 0 \\ 0 & 0 & 1 & 0 \end{bmatrix}, \text{ replacing } C_3 \text{ by } \frac{1}{16} C_2$$

$$\sim \begin{bmatrix} 1 & 0 & 0 & 0 \\ 0 & 1 & 0 & 0 \\ 0 & 0 & 1 & 0 \\ 0 & 0 & 0 & 0 \end{bmatrix}, \text{ interchanging } R_3 \text{ and } R_4$$

$$\sim \begin{bmatrix} I_3 & 0 \\ 0 & 0 \end{bmatrix}$$

Hence the rank of the given matrix = 3. **Ans.**

Example 10:

Use elementary transformation to reduce the following matrix A to triangular form and hence find the rank of A.

$$A = \begin{bmatrix} 5 & 3 & 14 & 4 \\ 0 & 1 & 2 & 1 \\ 1 & -1 & 2 & 0 \end{bmatrix}$$

Solution:

$$A \sim \begin{bmatrix} 5 & 3 & 8 & 1 \\ 0 & 1 & 0 & 0 \\ 1 & -1 & 4 & 1 \end{bmatrix}, \text{ replacing } C_3, C_4 \text{ by } C_3 - 2C_2, C_4 - C_2 \text{ respectively}$$

$$\Rightarrow \quad A \sim \begin{bmatrix} 5 & 8 & -12 & -4 \\ 0 & 1 & 0 & 0 \\ 1 & 0 & 0 & 0 \end{bmatrix}, \text{ replacing } C_2, C_3, C_4 \text{ by } C_2 + C_1, C_3 - 4C_1, C_4 - C_1$$

$$\sim \begin{bmatrix} 5 & 0 & 0 & -4 \\ 0 & 1 & 0 & 0 \\ 1 & 0 & 0 & 0 \end{bmatrix}, \text{ replacing } C_2, C_3 \text{ by } C_2 + 2C_4, C_3 - 3C_4 \text{ respectively}$$

$$\sim \begin{bmatrix} 5 & 0 & 0 & 1 \\ 0 & 1 & 0 & 0 \\ 1 & 0 & 0 & 0 \end{bmatrix}, \text{ replacing } C_4 \text{ by } -\frac{1}{4}C_4$$

$$\sim \begin{bmatrix} 0 & 0 & 0 & 1 \\ 0 & 1 & 0 & 0 \\ 1 & 0 & 0 & 0 \end{bmatrix}, \text{ replacing } C_1 \text{ by } C_1 - 5C_4$$

$$\sim \begin{bmatrix} 0 & 1 & 0 & 0 \\ 0 & 0 & 0 & 1 \\ 1 & 0 & 0 & 0 \end{bmatrix}, \text{ int erchanging } R_1, R_2$$

$$\sim \begin{bmatrix} 0 & 1 & 0 & 0 \\ 1 & 0 & 0 & 0 \\ 0 & 0 & 0 & 1 \end{bmatrix}, \text{ int erchanging } R_2, R_3$$

$$\sim \begin{bmatrix} 1 & 0 & 0 & 0 \\ 0 & 1 & 0 & 0 \\ 0 & 0 & 0 & 1 \end{bmatrix}, \text{ int erchanging } R_1, R_2$$

$$\sim \begin{bmatrix} 1 & 0 & 0 & 0 \\ 0 & 1 & 0 & 0 \\ 0 & 0 & 1 & 0 \end{bmatrix}, \text{ int erchanging } C_3, C_4$$

$$\sim [\, I_3 \; 0 \,]$$

Hence the rank of A is 3. **Ans.**

Example 11:

Find the rank of the matrix $A = \begin{bmatrix} 1 & 2 & -1 & 3 \\ 2 & 4 & -4 & 7 \\ -1 & -2 & -1 & 1 \end{bmatrix}$

Solution:

$$A \sim \begin{bmatrix} 1 & 0 & 0 & 0 \\ 2 & 0 & -2 & 1 \\ -1 & 0 & -2 & 5 \end{bmatrix}, \text{ replacing } C_2, C_3, C_4 \text{ by } C_2 - 2C_1, C_3 + C_1 \text{ and } C_4 - 3C_1 \text{ respectively}$$

$$\sim \begin{bmatrix} 1 & 0 & 0 & 0 \\ 2 & 0 & -2 & 1 \\ -3 & 0 & 0 & 4 \end{bmatrix}, \text{ replacing } R_3 \text{ by } R_3 - R_2$$

$$\sim \begin{bmatrix} 1 & 0 & 0 & 0 \\ 0 & 0 & -2 & 1 \\ 0 & 0 & 0 & 4 \end{bmatrix}, \text{ replacing } R_2, R_3 \text{ by } R_2 - 2R_1 \text{ and } R_3 + 3R_1 \text{ respectively}$$

$$\sim \begin{bmatrix} 1 & 0 & 0 & 0 \\ 0 & 0 & 1 & 1 \\ 0 & 0 & 0 & 1 \end{bmatrix}, \text{ replacing } C_3 \text{ by } -\frac{1}{2}C_3 \text{ and } R_3 \text{ by } \frac{1}{4}R_3$$

$$\sim \begin{bmatrix} 1 & 0 & 0 & 0 \\ 0 & 0 & 1 & 0 \\ 0 & 0 & 0 & 1 \end{bmatrix}, \text{ replacing } C_4 \text{ by } C_4 - C_3$$

$$\sim \begin{bmatrix} 1 & 0 & 0 & 0 \\ 0 & 1 & 0 & 0 \\ 0 & 0 & 0 & 1 \end{bmatrix}, \text{ interchanging } C_2 \text{ and } C_3$$

$$\sim \begin{bmatrix} 1 & 0 & 0 & 0 \\ 0 & 1 & 0 & 0 \\ 0 & 0 & 1 & 0 \end{bmatrix}, \text{ interchanging } C_3 \text{ and } C_4$$

$\sim [\, I_3 \; 0 \,]$ **(Note)**

Hence the rank of A is . **Ans.**

Example 12:

Determine by reducing to normal form the rank of the matrix

$$A = \begin{bmatrix} 8 & 1 & 3 & 6 \\ 0 & 3 & 2 & 2 \\ -8 & -1 & -3 & 4 \end{bmatrix}$$

Solution:

$$A \sim \begin{bmatrix} 1 & 1 & 3 & 3 \\ 0 & 3 & 2 & 1 \\ -1 & -1 & -3 & 2 \end{bmatrix}, \text{ replacing } C_1 \text{ by } \frac{1}{8}C_1 \text{ and } C_4 \text{ by } \frac{1}{2}C_4$$

$$\sim \begin{bmatrix} 1 & 0 & 0 & 0 \\ 0 & 3 & 2 & 1 \\ -1 & 0 & 0 & 5 \end{bmatrix}, \text{ replacing } C_2, C_3 \text{ and } C_4 \text{ by } C_2 - C_1, C_3 - 3C_1, C_4 - 3C_1 \text{ respectively.}$$

$$\sim \begin{bmatrix} 1 & 0 & 0 & 0 \\ 0 & 3 & 2 & 1 \\ 0 & 0 & 0 & 5 \end{bmatrix}, \text{ replacing } R_3 \text{ by } R_3 + R_1$$

$$\sim \begin{bmatrix} 1 & 0 & 0 & 0 \\ 0 & 1 & 1 & 1 \\ 0 & 0 & 0 & 5 \end{bmatrix}, \text{ repacing } C_2 \text{ by } \frac{1}{3}C_2 \text{ and } C_2 \text{ by } \frac{1}{2}C_3$$

$$\sim \begin{bmatrix} 1 & 0 & 0 & 0 \\ 0 & 1 & 0 & 0 \\ 0 & 0 & 0 & 5 \end{bmatrix}, \text{ repacing } C_3 \text{ and } C_4 \text{ by } C_3 - C_2 \text{ and } C_4 - C_2 \text{ respectively.}$$

$$\sim \begin{bmatrix} 1 & 0 & 0 & 0 \\ 0 & 1 & 0 & 0 \\ 0 & 0 & 5 & 0 \end{bmatrix}, \text{ int erchanging } C_3 \text{ and } C_4$$

$$\Rightarrow \quad A \sim \begin{bmatrix} 1 & 0 & 0 & 0 \\ 0 & 1 & 0 & 0 \\ 0 & 0 & 1 & 0 \end{bmatrix}, \text{ replacing } C_3 \text{ by } \frac{1}{5}C_3$$

$$\sim [\ I_3\ 0\ ]$$ **(Note)**

Hence the rank of A is . **Ans.**

Example 13:

Find the rank of the matrix $A = \begin{bmatrix} 1 & 1 & 1 \\ 2 & 2 & 2 \\ 3 & 3 & 3 \end{bmatrix}$

Solution:

$$A \sim \begin{bmatrix} 1 & 0 & 0 \\ 2 & 0 & 0 \\ 3 & 0 & 0 \end{bmatrix}, \text{ replacing } C_2, C_3 \text{ by } C_3 - C_1, C_3 - C_1 \text{ respectively}$$

$$\sim \begin{bmatrix} 1 & 0 & 0 \\ 0 & 0 & 0 \\ 0 & 0 & 0 \end{bmatrix}, \text{ replacing } R_2, R_3 \text{ by } R_2 - 2R_1 \; R_3 - 3R_1 \text{ respectively}$$

$$\sim \begin{bmatrix} I_1 & 0 \\ 0 & 0 \end{bmatrix}$$

Hence the rank of the matrix A is 1. **Ans.**

Example 14:

Find the rank of the matrix $A = \begin{bmatrix} 0 & 2 & 3 \\ 0 & 4 & 6 \\ 0 & 6 & 9 \end{bmatrix}$

Solution:

$$A \sim \begin{bmatrix} 0 & 2 & 1 \\ 0 & 4 & 2 \\ 0 & 6 & 3 \end{bmatrix}, \text{ replacing } C_3 \text{ by } C_3 - C_2$$

$$\sim \begin{bmatrix} 0 & 0 & 1 \\ 0 & 0 & 2 \\ 0 & 0 & 3 \end{bmatrix}, \text{ replacing } C_2 \text{ by } C_2 - 2C_3$$

$$\sim \begin{bmatrix} 0 & 0 & 1 \\ 0 & 0 & 0 \\ 0 & 0 & 0 \end{bmatrix}, \text{ replacing } R_2, R_3 \text{ by } R_2 - 2R_1, R_3 - 3R_1$$

$$\sim \begin{bmatrix} 1 & 0 & 0 \\ 0 & 0 & 0 \\ 0 & 0 & 0 \end{bmatrix}, \text{ interchanging } C_1 \text{ and } C_2$$

$$\sim \begin{bmatrix} I_1 & 0 \\ 0 & 0 \end{bmatrix}$$

Hence the rank of A is 1. **Ans.**

Example 15:

Reduce the matrix A to its normal form where

$A = \begin{bmatrix} 0 & 1 & 2 & -2 \\ 4 & 0 & 2 & 6 \\ 2 & 1 & 3 & 1 \end{bmatrix}$ *and hence determine its rank.*

Solution:

$$A \sim \begin{bmatrix} 0 & 1 & 0 & 0 \\ 4 & 0 & 2 & 8 \\ 2 & 1 & 1 & 4 \end{bmatrix},$$ replacing C_3, C_4 by $C_3 - 2C_2$ and $C_4 + C_3$ respectively

$$\sim \begin{bmatrix} 0 & 1 & 0 & 0 \\ 4 & 0 & 2 & 8 \\ 2 & 1 & 1 & 4 \end{bmatrix},$$ replacing R_2 by $R_2 - 2R_3$

$$\sim \begin{bmatrix} 0 & 1 & 0 & 0 \\ 0 & 0 & 0 & 0 \\ 2 & 0 & 1 & 4 \end{bmatrix},$$ replacing R_2, R_3 by $R_2 + 2R_1$ and $R_3 - R_1$ respectively.

$$\sim \begin{bmatrix} 0 & 1 & 0 & 0 \\ 0 & 0 & 0 & 0 \\ 0 & 0 & 1 & 0 \end{bmatrix},$$ replacing C_1, C_4 by $C_1 - 3C_3, C_4 - 4C_3$ respectively.

$$\sim \begin{bmatrix} 1 & 0 & 0 & 0 \\ 0 & 0 & 0 & 0 \\ 0 & 0 & 1 & 0 \end{bmatrix},$$ int erchanging C_1 and C_2

$$\sim \begin{bmatrix} 1 & 0 & 0 & 0 \\ 0 & 0 & 0 & 0 \\ 0 & 1 & 0 & 0 \end{bmatrix},$$ int erchanging C_2, C_3

$$\sim \begin{bmatrix} 1 & 0 & 0 & 0 \\ 0 & 1 & 0 & 0 \\ 0 & 0 & 0 & 0 \end{bmatrix}, \text{ interchanging } R_2, R_3$$

$$\sim \begin{bmatrix} I_2 & 0 \\ 0 & 0 \end{bmatrix}$$

Hence the rank of A is 2. **Ans.**

Example 16:

Find the rank of the matrix $A = \begin{bmatrix} 1 & 2 & 3 \\ 2 & 3 & 4 \\ 3 & 5 & 7 \end{bmatrix}$ *after reducing it to the normal form.*

Solution:

$$A \sim \begin{bmatrix} 1 & 1 & 1 \\ 2 & 1 & 1 \\ 3 & 2 & 2 \end{bmatrix}, \text{ replacing } C_2 \text{ and } C_3 \text{ by } C_2 - C_1 \text{ and } C_3 - C_2$$

$$\sim \begin{bmatrix} 1 & 1 & 1 \\ 1 & 0 & 0 \\ 1 & 1 & 1 \end{bmatrix}, \text{ replacing } R_2 \text{ and } R_3 \text{ by } R_2 - R_1 \text{ and } R_3 - R_2$$

$$\sim \begin{bmatrix} 1 & 1 & 1 \\ 1 & 0 & 0 \\ 0 & 0 & 0 \end{bmatrix}, \text{ replacing } R_3 \text{ by } R_3 - R_1$$

$$\sim \begin{bmatrix} 1 & 1 & 0 \\ 1 & 0 & 0 \\ 0 & 0 & 0 \end{bmatrix}, \text{ replacing } C_3 \text{ by } C_3 - C_2$$

$$\sim \begin{bmatrix} 0 & 1 & 0 \\ 1 & 0 & 0 \\ 0 & 0 & 0 \end{bmatrix}, \text{ replacing } C_1 \text{ by } C_1 - C_2$$

$$\sim \begin{bmatrix} 1 & 0 & 0 \\ 0 & 1 & 0 \\ 0 & 0 & 0 \end{bmatrix}, \text{ interchanging } C_1 \text{ and } C_2$$

$$\sim \begin{bmatrix} I_2 & 0 \\ 0 & 0 \end{bmatrix}$$

Hence the rank of A is 2. **Ans.**

Example 17:

Find the rank of the matrix $A = \begin{bmatrix} -1 & -2 & -1 \\ 6 & 12 & 6 \\ 5 & 10 & 5 \end{bmatrix}$

Solution:

$$A \sim \begin{bmatrix} -1 & 0 & 0 \\ 6 & 0 & 0 \\ 5 & 0 & 0 \end{bmatrix}, \text{ replcaing } C_2, C_3 \text{ by } C_2 - 2C_1, C_3 - C_1 \text{ respectively.}$$

$$\sim \begin{bmatrix} -1 & 0 & 0 \\ 0 & 0 & 0 \\ 0 & 0 & 0 \end{bmatrix}, \text{ replacing } R_2, R_3 \text{ by } R_2 + 6R_1, R_3 + 5R_1 \text{ respectively}$$

$$\sim \begin{bmatrix} 1 & 0 & 0 \\ 0 & 0 & 0 \\ 0 & 0 & 0 \end{bmatrix}, \text{ replacing } R_1 \text{ by } -R_1$$

$$\sim \begin{bmatrix} I_1 & 0 \\ 0 & 0 \end{bmatrix}$$

Hence the rank of the matrix A is 2. **Ans.**

3.3 ECHELON FORM OF A MATRIX

Definition: *If in a matrix,*

(i) all the non-zero rows, if any, precede the zero rows,

(ii) the number of zero proceeding the first non-zero element in a row is less than the number of such zero in the succeeding row,

(iii) *the first non-zero element in a row is unity, then it is in the Echelon form.*

Note: The number of non-zero rows of a matrix given in the Echelon form is its rank. **(Remember)**

Examples of a matrix in the Echelon Form :-

$$\begin{bmatrix} 1 & 3 & 4 & 5 & 6 \\ 0 & 1 & 2 & 3 & 4 \\ 0 & 0 & 1 & 2 & 3 \\ 0 & 0 & 0 & 0 & 0 \end{bmatrix}$$

In this matrix we observe that

(i) the first three non-zero rows precede the fourth row which is a zero row,

(ii) the number of zero is R_1, R_2 and R_3 are 5, 2 and 1 respectively which are in descending order,

(iii) the first non-zero term in each row is unity.

Hence all the three conditions of the Echelon form are satisfied.

Also there being three non-zero rows in this matrix, its rank is 3. This act can be proved by actually finding the rank of this matrix.

In this matrix, a minor of order 4.

$$= \begin{vmatrix} 1 & 3 & 4 & 5 \\ 0 & 1 & 2 & 3 \\ 0 & 0 & 1 & 2 \\ 0 & 0 & 0 & 0 \end{vmatrix} = 0, \text{ one row being of zero.}$$

In a similar way we can show that all minors of order 4 are zero.

Now a minor of order 3 $= \begin{vmatrix} 1 & 3 & 4 \\ 0 & 1 & 2 \\ 0 & 1 & 0 \end{vmatrix}$

$$= \begin{vmatrix} 1 & 2 \\ 0 & 1 \end{vmatrix}, \text{ expanding w.r. to } C_1$$

$$= 1 \neq 0$$

Hence the rank of this matrix = 3.

Example :

Find the rank of the matrix $A = \begin{bmatrix} 0 & 1 & 2 & 3 \\ 0 & 0 & 1 & -1 \\ 0 & 0 & 0 & 0 \end{bmatrix}$

Solution:

In the given matrix we observe that

(i) the first two non-zero rows precede the third row which is a zero row,

(ii) the number of zero in R_2, R_3 and R_1 are 4, 2 and 1 respectively which are in descending order, and

(iii) the first non-zero term in each row is unity.

Hence all three conditions of the Echelon form are satisfied.

Also there being two non-zero rows in this matrix, its rank is 2. **Ans.**

3.4 INVARIANCE OF RANK UNDER ELEMENTARY OPERATIONS

Theorem:

All equivalent matrices have the same ranks i.e. the rank of a matrix remains unaltered by the application of elementary row and column operations.

Proof:

Let r be the rank of an m × n matrix $A = [a_{ij}]$.

Case I: *If ith and jth rows are interchanged (which may be written symbolically as R_{ij} or $R_i \leftrightarrow R_j$) then it does not effect the rank.*

Let B denote the matrix obtained from the matrix A by the elementary operation $R_i \leftrightarrow R_j$ and let p be the rank of B.

Also if D be any (r + 1) rowed square sub-matrix of B, then $|D| = \pm |C|$, where C is a particular (r + 1) rowed submatrix of A.

As r is the rank of the matrix A so every (r + 1) rowed minor of A vanishes and therefore p, the rank osw *are multiplied by a non-zero number λ (which may be written symbolically, as $R_i \to \lambda R_j$, $\lambda \neq 0$) it does not effect the rank.*

Let B denote the matrix obtained from the matrix A by the elementary operation $R_i \to \lambda R_j$ and let p be the rank of B.

Let D be any (r + 1) rowed square submatrix of B and let C be the sub-matrix of A having the same position as D. Then either $|D| = |C|$ or $|D| = \lambda |C|$.

[Here | D | = | C | happens if the ith row of A is one of those rows which are removed to obtain D from B and | D | = λ | C | happens when the ith row is not removed while obtaining C from A].

Also as r is the rank of the matrix A so every (r + 1) rowed minor of A vanishes and therefore in particular | C | = 0 and consequently in both the above cases | D | = 0.

∴ p, the rank of B, cannot exceed r, the rank of A.

i.e., $p \leq r$

Also we can obtain A from B by the elementary operation $R_i \to \lambda^{-1} R_j$, therefore in that case interchanging the roles of A and B we shall gets $r \leq p$.

Hence r = p.

Case III: *If to the elements of the ith row are added the products by any non-zero numbers λ of the corresponding elements of jth row (which may be written symbolically as* $R_i \to R_j + \lambda R_j$, $\lambda \neq 0$*)*

Let B denote the matrix obtained from the matrix A by the elementary operation $R_i \to R_j + \lambda R_i$ and let p be the rank of B.

Let D be any (r + 1) rowed square submatrix of B and let C be the submatrix of A having the same position as D.

Now three sub-cases arise:-

(i) If A and B differ only in the ith row *i.e.,* if ith row of B is one of those rows which have been removed while obtaining C.

In this case D = C and therefore | D | = | C |.

∴ The rank of A is r, so | C | = 0 and consequently | D | = 0

(ii) If ith row of B has not been removed but jth row has been removed while obtaining D.

In this case $| D | = | C | + \lambda | C_0 |$, where C_0 is an (r + 1) rowed matrix which is obtained from C by replacing A_{ik} by a_{jk} *i.e.,* C_0 is obtained from C by performing the elementary operation R_{ij} or $R_i \leftrightarrow R_j$ and then removing those rows and columns of the new matrix which are removed to obtain D from B.

∴ $| C_0 |$ is negative of some (r + 1) rowed minor of A and as the rank of A is r so every (r + 1) rowed minor of A is zero *i.e.,* | C | = 0, $| C_0 | = 0$ and consequently | D | = 0.

(iii) If neither the ith row nor the jth row of B has been removed while obtained D.

Here | D | = | C | and so as before | D | = 0.

∴ Every (r + 1) rowed minor of B vanishes so p, the rank of B, cannot exceed r, the rank of A *i.e.*, p ≤ r.

Also we can obtain A from B by the elementary operation $R_i \to R_i - \lambda R_j$, therefore in that case interchanging the roles of A and B we shall get r ≤ p.

Hence r = p.

Thus we have observed that the rank of a matrix remains invariant under elementary row operations. Similarly it can be shown that the rank of a matrix remains invariant under elementary column operations too.

Note: By the application of the above theorem we can easily obtain the rank of a matrix for if we can obtain a matrix B by elementary operations on a matrix A and of the rank of B can be easily determined by inspection or simple calculations as given in previous articles in this chapter, then we can determine the rank of A.

3.5 SOME IMPORTANT THEOREMS

Theorem 1:

The rank of the product matrix AB of two matrices A and B is less than the rank of either of the matrices A and B.

Proof:

Let r_1 and r_2 be the rank of the matrices A and B.

∵ r_1 is the rank of A, therefore $A \sim \begin{bmatrix} M \\ O \end{bmatrix}$, where M is a submatrix of ran r_1 and contains r_1 rows.

Post-Multiplying it by B, we get

$$AB \sim \begin{bmatrix} M \\ O \end{bmatrix} B.$$

But $\begin{bmatrix} M \\ O \end{bmatrix}$ B can have r_1 non-zero rows at the most which are obtained on multiplying r_1 non-zero rows of M with columns of B.

∴ Rank of AB = Rank of $\begin{bmatrix} M \\ O \end{bmatrix} B \le r_1$

i.e., Rank of AB £ rank of A ...(i)

In a similar way we get B ~ [N O], where N is a submatrix of ran r_2 and contains r_2 columns.

Pre-multiplying it by A, we get

$$AB \sim A\,[N\ O].$$

But [N O] can have r_2 non-zero columns at the most which are obtained on multiplying the rows of A with r_2 non-zero columns of [N O].

∴ Rank of AB = Rank of A[N O] $\leq r_2$

i.e., Rank of AB ≤ rank of B ...(ii)

Hence the theorem from (i) and (ii).

Theorem 2:

The rank of a matrix is equal to the rank of the transposed matrix.

Proof:

Let A = $[a_{ij}]$ be any m × n matrix.

The transposed matrix A' = $[A_{ij}]$ is an n × m matrix.

Let the rank of A be r and let B be the r × r sub-matrix of A such that | B | ≠ 0.

Also we know that the value of a determinant remains unaltered if its rows and columns are interchanged.

i.e., | B' | = | B | ≠ 0, where B' is evidently a r × r sub-matrix of A'

∴ The rank of A' ≥ r.

Again if C be a (r + 1) × (r + 1) submatrix of A, then by definition of rank we must e all | C | = 0.

Also C' is a (r + 1) × (r + 1) submatrix of A' so we have

| C' | = | C | = 0, as explained above.

∴ We conclude that there cannot be any (r + 1) × (r + 1) submatrix of A' with non-zero determinant.

∴ The rank of A' ≥ r and it cannot be greater than r as above.

∴ The rank of A' is r which is also the rank of A. **Hence proved.**

Example 1:

Show that AA', has the same rank as A, where A' is the transpose of A.

Solution:

Let B = AA', then rank of B $\leq$ rank of A ...(i)

Also A^{-1} B = A', and so we have

rank A = rank A' $\leq$ rank of B ...(ii)

$\therefore$ From (i) and (ii), rank of A = rank of B.

Example 2:

Show that the rank of a matrix (A does not alter by pre or post) multiplying it with any non-singular matrix R.

Solution:

Let B = RA

Then rank of B = rank of RA $\leq$ rank A ...(i)

Also A = R^{-1} B, where R^{-1} is the inverse matrix of R.

$\therefore$ rank of A = rank of (R^{-1} B) $\leq$ rank of B. ...(ii)

$\therefore$ From (i) and (ii) we conclude that

rank of A = rank of B. **Hence proved.**

Example 3:

Prove that if A is a matrix of order m × n and if B is a non singular matrix of order n, then the product P = AB has the same rank as A.

Solution:

Here A ~ (m × n), B ~ (n × n)

P = AB ~ (m × n)

If m < n, rank of A $\leq$ m but rank of B = n

$\therefore$ rank of A < rank of B.

Now rank (P) = rank (AB $\leq$ rank A ...(i)

But we can write A = PB^{-1}

$\therefore$ rank of A = rank of (PB^{-1}) $\leq$ rank of P ...(ii)

$\therefore$ From (i) and (ii) we get rank of P = rank of A.

3.6 SWEEP OUT METHOD OF FINDING THE RANK OF A MATRIX

In the process of evaluation of the rank of a matrix by means of elementary row and column transformations, if certain rows of columns are zero-rows or zero-columns (*i.e.*, each element of these rows or columns are zero) then we can remove these rows or columns without any effect on the rank of the matrix. This method is generally called the *Sweep out* method.

Example 1:

Find the rank of the matrix $\begin{bmatrix} 0 & 1 & -3 & -1 \\ 1 & 0 & 1 & 1 \\ 3 & 1 & 0 & 2 \\ 1 & 1 & -2 & 0 \end{bmatrix}$

Solution:

Let $A = \begin{bmatrix} 0 & 1 & -3 & -1 \\ 1 & 0 & 1 & 1 \\ 3 & 1 & 0 & 2 \\ 1 & 1 & -2 & 0 \end{bmatrix}$

Now $A \sim \begin{bmatrix} 0 & 1 & -3 & -1 \\ 1 & 1 & -2 & 0 \\ 3 & 3 & -6 & 0 \\ 1 & 1 & -2 & 0 \end{bmatrix}$, replacing R_2 and R_3 by $R_2 + R_1$ and $R_3 + 2R_1$ respectively

$\sim \begin{bmatrix} 0 & 1 & -3 & -1 \\ 1 & 1 & -2 & 0 \\ 1 & 1 & -2 & 0 \\ 0 & 0 & 0 & 0 \end{bmatrix}$, replacing R_3 and R_4 by $\frac{1}{3}R_3$ and $R_4 - R_2$ respectively.

$\sim \begin{bmatrix} 0 & 1 & -3 & -1 \\ 1 & 0 & 0 & 0 \\ 0 & 0 & 0 & 0 \\ 0 & 0 & 0 & 0 \end{bmatrix}$, replacing R_3 by $R_3 - R_2$ and then C_2 and C_3 by $C_2 - C_1$ and $C_3 + 2C_2$ respectively

$\sim \begin{bmatrix} 0 & 1 & -3 & -1 \\ 1 & 0 & 0 & 0 \end{bmatrix}$

Now a minor of order 2 is $\begin{vmatrix} 0 & 1 \\ 1 & 0 \end{vmatrix} = -1 \neq 0.$

Hence its rank is 2. **Ans.**

Example 2:

Find the rank of the matrix $A = \begin{bmatrix} 6 & 1 & 3 & 8 \\ 4 & 2 & 6 & -1 \\ 10 & 3 & 9 & 7 \\ 16 & 4 & 12 & 15 \end{bmatrix}$

Solution:

$$A \sim \begin{bmatrix} 6 & 1 & 3 & 8 \\ 4 & 2 & 6 & -1 \\ 4 & 2 & 6 & -1 \\ 4 & 2 & 6 & -1 \end{bmatrix},$$ replacing R_3 and R_4 by $R_3 - R_1$ and $R_4 - 2R_1$ respectively

$$\therefore\ A \sim \begin{bmatrix} 6 & 1 & 3 & 8 \\ 4 & 2 & 6 & -1 \end{bmatrix}$$

Now a minor of order 2 is

$$\begin{vmatrix} 1 & 8 \\ 2 & -1 \end{vmatrix} = -1 - 16 = -17 \neq 0.$$

Hence its rank is 2. **Ans.**

4

Adjoint of a Matrix and Inverse of the Matrix

4.1 ADJOINT OF A MATRIX

Definition: *If* C_{ij} *be the cofactor of the element* a_{ij} *in* $|a_{ij}|$ *of the* $n \times n$ *matrix* $A = [a_{ij}]$, *then*

$$\text{adjoint of } A = \begin{bmatrix} C_{11} & C_{21} \cdots C_{n1} \\ C_{12} & C_{22} \cdots C_{n2} \\ \cdots\cdots & \cdots\cdots \\ \cdots\cdots & \cdots\cdots \\ C_{1n} & C_{2n} \cdots C_{nn} \end{bmatrix}$$

This is also written as Adj A.

or adjoint of A = transposed of C, where $C = \begin{bmatrix} C_{11} & C_{21} \cdots C_{1n} \\ C_{12} & C_{22} \cdots C_{2n} \\ \cdots\cdots & \cdots\cdots \\ C_{n1} & C_{n2} \cdots C_{nn} \end{bmatrix}$

While solving problems we generally use this definition.

Here students should note carefully that the cofactors of the elements of the first row of $[a_{ij}]$ are the elements of the first column of Adj. A.

Similarly, the cofactors of the elements of the first column of $|a_{ij}|$ are the element of first row of Adj. A.

Example 1:

Find the adjoint of matrix $A = \begin{bmatrix} 1 & 2 & 3 \\ 0 & 5 & 0 \\ 2 & 4 & 3 \end{bmatrix}$

Solution:

For the given matrix A, we have

$$C_{11} = \begin{vmatrix} 5 & 0 \\ 4 & 3 \end{vmatrix} = 16; \; C_{12} = -\begin{vmatrix} 0 & 0 \\ 2 & 3 \end{vmatrix} = 0; \; C_{13} = \begin{vmatrix} 0 & 5 \\ 2 & 4 \end{vmatrix} = -10;$$

$$C_{21} = -\begin{vmatrix} 2 & 3 \\ 4 & 3 \end{vmatrix} = -15; \; C_{22} = \begin{vmatrix} 1 & 3 \\ 2 & 3 \end{vmatrix} = -3; \; C_{23} = -\begin{vmatrix} 1 & 2 \\ 2 & 4 \end{vmatrix} = 0;$$

$$C_{31} = \begin{vmatrix} 2 & 3 \\ 5 & 0 \end{vmatrix} = -15; \; C_{32} = -\begin{vmatrix} 1 & 3 \\ 0 & 0 \end{vmatrix} = 0; \; C_{33} = \begin{vmatrix} 1 & 2 \\ 2 & 5 \end{vmatrix} = 5;$$

$$\therefore \quad C = \begin{bmatrix} 15 & 0 & -10 \\ 6 & -3 & 0 \\ -15 & 0 & 5 \end{bmatrix}$$

$$\therefore \quad \text{Adj. } A = C' \text{(i.e. transpose of C)} = \begin{bmatrix} 15 & 6 & -15 \\ 0 & -3 & 0 \\ -10 & 0 & 5 \end{bmatrix}$$

Ans.

Example 2:

Find the adjoint of $A = \begin{bmatrix} -1 & -2 & 3 \\ -2 & 1 & 1 \\ -4 & -5 & 2 \end{bmatrix}$

Solution:

For the given matrix A, we have

$$C_{11} = \begin{vmatrix} 1 & 1 \\ -5 & 2 \end{vmatrix} = 7; \; C_{12} = -\begin{vmatrix} -2 & 1 \\ -4 & 2 \end{vmatrix} = 0; \; C_{13} = \begin{vmatrix} -2 & 1 \\ -4 & -5 \end{vmatrix} = 14;$$

$$C_{21} = -\begin{vmatrix} -2 & 3 \\ -5 & 2 \end{vmatrix} = -11; \; C_{22} = \begin{vmatrix} -1 & 3 \\ -4 & 2 \end{vmatrix} = 10; \; C_{23} = -\begin{vmatrix} -1 & -2 \\ -4 & -5 \end{vmatrix} = 3;$$

$C_{31} = \begin{vmatrix} -2 & 3 \\ 1 & 1 \end{vmatrix} = -5;\ C_{32} = \begin{vmatrix} -1 & 3 \\ -2 & 1 \end{vmatrix} = -5;\ C_{33} = \begin{vmatrix} -1 & -2 \\ -2 & 1 \end{vmatrix} = -5;$

$\therefore\quad C = \begin{bmatrix} 7 & 0 & 14 \\ -11 & 10 & 3 \\ -5 & -5 & -5 \end{bmatrix}$

$\therefore\quad \text{Adj. } A = C' = \begin{bmatrix} 7 & -11 & -5 \\ 0 & 10 & -5 \\ 14 & 3 & -5 \end{bmatrix}$ **Ans.**

Example 3:

Find the adjoint of the matrix $A = \begin{bmatrix} 1 & 1 & 1 \\ 1 & 2 & -3 \\ 2 & -1 & 3 \end{bmatrix}$

Solution:

For the given matrix A, we have

$C_{11} = \begin{vmatrix} 2 & -3 \\ -1 & 3 \end{vmatrix} = 3;\ C_{12} = -\begin{vmatrix} 1 & -3 \\ 2 & 3 \end{vmatrix} = 9;\ C_{13} = \begin{vmatrix} 1 & 2 \\ 2 & -1 \end{vmatrix} = -5;$

$C_{21} = -\begin{vmatrix} 1 & 1 \\ -1 & 3 \end{vmatrix} = -4;\ C_{22} = \begin{vmatrix} 1 & 1 \\ 2 & 3 \end{vmatrix} = 1;\ C_{23} = -\begin{vmatrix} 1 & 1 \\ 2 & -1 \end{vmatrix} = 3;$

$C_{31} = \begin{vmatrix} 1 & 1 \\ 2 & -3 \end{vmatrix} = -5;\ C_{32} = -\begin{vmatrix} 1 & 1 \\ 1 & -3 \end{vmatrix} = 4;\ C_{33} = \begin{vmatrix} 1 & 1 \\ 1 & 2 \end{vmatrix} = 5;$

$\therefore\quad C = \begin{bmatrix} 3 & -9 & -5 \\ -4 & 1 & 3 \\ -5 & 4 & 1 \end{bmatrix}$

$\therefore\quad \text{Adj. } A = C' = \begin{bmatrix} 3 & -4 & -5 \\ -9 & 1 & 4 \\ -5 & 3 & 1 \end{bmatrix}$ **Ans.**

Example 4:

Find the adjoint of the matrix $A = \begin{bmatrix} 1 & 0 & -1 \\ 3 & 4 & 5 \\ 0 & -6 & -7 \end{bmatrix}$

Solution:

For the given matrix A, we have

$$C_{11} = \begin{vmatrix} 4 & 5 \\ -6 & -7 \end{vmatrix} = 2; \; C_{12} = -\begin{vmatrix} 3 & 5 \\ 0 & -7 \end{vmatrix} = 21; \; C_{13} = \begin{vmatrix} 3 & 4 \\ 0 & -6 \end{vmatrix} = 18;$$

$$C_{21} = -\begin{vmatrix} 0 & -1 \\ -6 & -7 \end{vmatrix} = -6; \; C_{22} = \begin{vmatrix} 1 & -1 \\ 0 & -7 \end{vmatrix} = -7; \; C_{23} = -\begin{vmatrix} 1 & -0 \\ 0 & -0 \end{vmatrix} = 6;$$

$$C_{31} = \begin{vmatrix} 0 & -1 \\ 4 & 5 \end{vmatrix} = 5; \; C_{32} = -\begin{vmatrix} 1 & -1 \\ 3 & 5 \end{vmatrix} = -8; \; C_{33} = \begin{vmatrix} 1 & 0 \\ 3 & 4 \end{vmatrix} = 4;$$

$$\therefore \quad C = \begin{bmatrix} 2 & 21 & -18 \\ -6 & -7 & 6 \\ 4 & -8 & 4 \end{bmatrix}$$

$$\therefore \quad \text{Adj. } A = C' = \begin{bmatrix} 2 & -6 & 4 \\ 21 & -7 & 6 \\ 4 & -8 & 4 \end{bmatrix}$$

Ans.

4.2 THEOREMS ON ADJOINT OF A MATRIX

Theorem 1:

If $A = [a_{ij}]$ *be an* $n \times n$ *matrix, then*

$$|\text{Adj } A| = |A|^{n-1}, \text{ if } |A| \neq 0$$

Proof:

We know that $|A| . |B| = |AB|$

$$\therefore \quad |A| . |\text{Adj } A| = |A . \text{Adj } A|$$

$$= \begin{bmatrix} |A| & 0 & 0 & .. & 0 \\ 0 & |A| & 0 & .. & 0 \\ 0 & 0 & |A| & .. & 0 \\ .. & .. & .. & .. & .. \\ 0 & 0 & 0 & .. & |A| \end{bmatrix},$$

$\Rightarrow$ $|A| \,.\, |\text{Adj } A| = \{|A|\}^n$ **(Note)**

Dividing both sides by $|A|$, since $|A| \neq 0$, we get

$|\text{Adj } A| = |A|^{n-1}$ **Hence proved.**

Theorem 2:

If A and B are two n × n matrices, then

Adj (AB) = (Adj B) . (Adj A).

Proof:

We know A . (Adj A) = | A | . 1

So we have (AB) . (Adj AB) = | AB | . 1 ...(i)

Now (AB) . (Adj B). (Adj A)

= A . B , Adj B . Adj A

= A . (B . Adj B) . (Adj A) **(Note)**

= A . | B | . I . Adj A, $\because$ B . Adj B = | B | . I

= A . | B | . Adj A, $\because$ I . Adj A = Adj A as I . A = A always

= | B | . A . Adj A **(Note)**

= | B | . | A | . I, $\because$ A . Adj A = | A | . I

= | A | . | B | . I

= | AB | . I $\because$ | A | . | B | = | AB | ...(ii)

$\therefore$ From (i) and (ii) we get

(AB) . (Adj AB) = (AB) . (Adj B) . (Adj A)

$\Rightarrow$ Adj (AB) = (Adj B) . (Adj A). **Hence proved.**

Theorem 3:

If $A = [a_{ij}]$ *be an n × n matrix, then*

A . (Adj A) = (Adj A) . A = | A | . I, where I is on n × n identity matrix.

Proof:

We know Adj A = $[C'_{ik}]$,

where C_{kj} is the cofactor of a_{kj} in | A | and $C'_{jk} = C_{kj}$,

Therefore $A \,.\, (\text{Adj } A) = [a_{ij}]\,[C'_{ik}]$

$$= [B_{ik}], \text{ say,} \qquad \ldots\text{(ii)}$$

where $$B_{ik} = \sum_{j=1}^{n} a_{ij}\, C'_{jk}$$

$$= \sum_{j=1}^{n} a_{ij}\, C'_{kj} \qquad \because C'_{jk} = C_{kj}$$

$$\left.\begin{array}{l} = |A|, \text{ if } i = k \\ = 0, \quad \text{if } i \neq k \end{array}\right\}$$

$\therefore$ From (i), (i, k)th element of A . (Adj. A) = | A |, or 0 according as i = k or i ≠ k.

i.e.,, all diagonal terms of A . (Adj . A) are | A | and non-diagonal terms are zero.

i.e., $$A \,.\, (\text{Adj} \,.\, A) = \begin{bmatrix} |A| & 0 & 0 & .. & 0 \\ 0 & |A| & 0 & .. & 0 \\ 0 & 0 & |A| & .. & 0 \\ .. & .. & .. & .. & .. \\ 0 & 0 & 0 & .. & |A| \end{bmatrix}$$

$$= |A| \begin{bmatrix} 1 & 0 & 0 & .. & 0 \\ 0 & 1 & 0 & .. & 0 \\ 0 & 0 & 1 & .. & 0 \\ .. & .. & .. & .. & .. \\ 0 & 0 & 0 & .. & 1 \end{bmatrix}.$$

$$= |A| \,.\, I. \qquad \ldots\text{(i)}$$

Similarly we can prove that (Adj. A) . A = | A | . I ...(ii)

Hence from (i) and (ii), we get

$$A \,.\, (\text{Adj } A) = (\text{Adj } A) \,.\, A = |A| \,.\, I$$

$$\Rightarrow \quad A \,.\, \frac{(\text{Adj } A)}{|A|} = \frac{(\text{Adj } A)}{|A|} \,.\, A - I$$

$$\Rightarrow \quad A^{-1} = \frac{(\text{Adj } A)}{|A|}, \quad \because AA^{-1} = I = A^{-1} A$$

i.e.,, the inverse of $A = \frac{\text{Adj } A}{|A|}$. ...(iii)

Note: The result (iii) gives us another method of finding the inverse of a given matrix.

Example 1:

Find the adjoin and inverse of the matrix $A = \begin{bmatrix} \cos\alpha & -\sin\alpha \\ \sin\alpha & \cos\alpha \end{bmatrix}$

Solution:

Here $|A| = \begin{bmatrix} \cos\alpha & -\sin\alpha \\ \sin\alpha & \cos\alpha \end{bmatrix}$

$= \cos^2\alpha + \sin^2\alpha = 1 \neq 0$ ...(i)

Also we have

$C_{11} = \cos\alpha$, $C_{12} = -\sin\alpha$, $C_{21} = -(-\sin\alpha) = \sin\alpha$ and $C_{22} = \cos\alpha$

(Note)

$\therefore \quad C = \begin{bmatrix} \cos\alpha & -\sin\alpha \\ \sin\alpha & \cos\alpha \end{bmatrix}$

$\therefore \quad \text{Adj } A = C' = \begin{bmatrix} \cos\alpha & \sin\alpha \\ -\sin\alpha & \cos\alpha \end{bmatrix}$

$\therefore \quad A^{-1} = \frac{\text{Adj } A}{|A|} = \begin{bmatrix} \cos\alpha & \sin\alpha \\ -\sin\alpha & \cos\alpha \end{bmatrix}$

substituting values from (i) and (ii). **Ans.**

Example 2:

Find the adjoint and inverse of the matrix $A = \begin{bmatrix} \cos\theta & -\sin\theta & 0 \\ \sin\theta & \cos\theta & 0 \\ 0 & 0 & 1 \end{bmatrix}$

Solution:

For the given matrix A, we have

$C_{11} = \begin{vmatrix} \cos\theta & 0 \\ 0 & 1 \end{vmatrix} = \cos\theta$; $C_{12} = -\begin{vmatrix} \sin\theta & 0 \\ 0 & 1 \end{vmatrix} = -\sin\theta$;

$$C_{13} = \begin{vmatrix} \sin\theta & \cos\theta \\ 0 & 0 \end{vmatrix} = \theta;$$

$$C_{21} = -\begin{vmatrix} -\sin\theta & 0 \\ 0 & 1 \end{vmatrix} = \sin\theta; \; C_{22} = \begin{vmatrix} \cos\theta & 0 \\ 0 & 1 \end{vmatrix} = \cos\theta;$$

$$C_{23} = -\begin{vmatrix} \cos\theta & -\sin\theta \\ 0 & 0 \end{vmatrix} = \theta;$$

$$C_{31} = \begin{vmatrix} -\sin\theta & 0 \\ \cos\theta & 0 \end{vmatrix} = 0; \; C_{32} = -\begin{vmatrix} \cos\theta & 0 \\ \sin\theta & 0 \end{vmatrix} = 0;$$

$$C_{33} = \begin{vmatrix} \cos\theta & -\sin\theta \\ \sin\theta & \cos\theta \end{vmatrix} = 1;$$

$$\therefore \quad C = \begin{bmatrix} \cos\theta & -\sin\theta & 0 \\ \sin\theta & \cos\theta & 0 \\ 0 & 0 & 1 \end{bmatrix}$$

$$\therefore \quad \text{Adj. } A = C' = \begin{bmatrix} \cos\theta & \sin\theta & 0 \\ -\sin\theta & \cos\theta & 0 \\ 0 & 0 & 1 \end{bmatrix}$$

$$\text{Also } |A| = \begin{vmatrix} \cos\theta & -\sin\theta & 0 \\ \sin\theta & \cos\theta & 0 \\ 0 & 0 & 1 \end{vmatrix} = \begin{vmatrix} \cos\theta & -\sin\theta \\ \sin\theta & \cos\theta \end{vmatrix}$$

$$= \cos^2\theta + \sin^2\theta = 1 \neq 0$$

$$\therefore \quad \text{The inverse of } A = \frac{\text{Adj. } A}{|A|} = \begin{vmatrix} \cos\theta & \sin\theta & 0 \\ -\sin\theta & \cos\theta & 0 \\ 0 & 0 & 1 \end{vmatrix}$$ **Ans.**

Example 3:

Find the inverse of A $= \begin{bmatrix} 0 & 1 & 2 \\ 1 & 2 & 3 \\ 3 & 1 & 1 \end{bmatrix}$

Solution:

For the given matrix A, we have

$$C_{11} = \begin{vmatrix} 2 & 3 \\ 1 & 1 \end{vmatrix} = -1;\ C_{12} = -\begin{vmatrix} 1 & 3 \\ 3 & 1 \end{vmatrix} = 8;\ C_{13} = \begin{vmatrix} 1 & 2 \\ 3 & 1 \end{vmatrix} = -5;$$

$$C_{21} = -\begin{vmatrix} 1 & 2 \\ 1 & 1 \end{vmatrix} = 1;\ C_{22} = \begin{vmatrix} 0 & 2 \\ 3 & 1 \end{vmatrix} = -6;\ C_{23} = -\begin{vmatrix} 0 & 1 \\ 3 & 1 \end{vmatrix} = 3;$$

$$C_{31} = \begin{vmatrix} 1 & 2 \\ 2 & 3 \end{vmatrix} = -1;\ C_{32} = -\begin{vmatrix} 0 & 2 \\ 1 & 3 \end{vmatrix} = 2;\ C_{33} = \begin{vmatrix} 0 & 1 \\ 1 & 2 \end{vmatrix} = -1;$$

$$\therefore \quad C = \begin{bmatrix} -1 & 8 & -5 \\ 1 & -6 & 3 \\ -1 & 2 & -1 \end{bmatrix}$$

$$\therefore \quad \text{Adj. } A = C' = \begin{bmatrix} -1 & 1 & -1 \\ 8 & -6 & 2 \\ -5 & 3 & -1 \end{bmatrix}$$

and $$|A| = \begin{vmatrix} 0 & 1 & 2 \\ 1 & 2 & 3 \\ 3 & 1 & 1 \end{vmatrix} = \begin{vmatrix} 0 & 1 & 0 \\ 1 & 2 & -1 \\ 3 & 1 & 1 \end{vmatrix}$$, replacing C_3 by $C_3 - 2C_2$

$$\Rightarrow \quad |A| = -\begin{vmatrix} 1 & -1 \\ 3 & -1 \end{vmatrix} = -2 \neq 0$$

$$\therefore \quad \text{Inverse of } A = \frac{\text{Adj. } A}{|A|}$$

$$= -\frac{1}{2}\begin{bmatrix} -1 & 1 & -1 \\ 8 & -6 & 2 \\ -5 & 3 & -1 \end{bmatrix} = \frac{1}{2}\begin{bmatrix} 1 & -1 & 1 \\ -8 & 6 & -2 \\ 5 & -3 & 1 \end{bmatrix}$$ **Ans.**

Example 4:

Given $A = \begin{bmatrix} 1 & -2 & -1 \\ 2 & 3 & 1 \\ 0 & 5 & -2 \end{bmatrix}$, *compute det. A, Adj. A and* A^{-1}.

Solution:

Here $|A| = \begin{vmatrix} 1 & -2 & -1 \\ 2 & 3 & 1 \\ 0 & 5 & -2 \end{vmatrix}$

$= \begin{vmatrix} 1 & -2 & -1 \\ 0 & 7 & 3 \\ 0 & 5 & -2 \end{vmatrix}$, replacing R_2 by $R_2 - 2R_1$

$= \begin{vmatrix} 7 & 3 \\ 5 & -2 \end{vmatrix}$, expanding with respect to C_1

$= 7(-2) - 5(3) = -(29) \neq 0.$ **Ans.**

For we have

$C_{11} = \begin{vmatrix} 3 & 1 \\ 5 & -2 \end{vmatrix} = -11;\ C_{12} = -\begin{vmatrix} 2 & 1 \\ 0 & -2 \end{vmatrix} = 4;\ C_{13} = \begin{vmatrix} 2 & 3 \\ 0 & 5 \end{vmatrix} = 10;$

$C_{21} = -\begin{vmatrix} -2 & -1 \\ 3 & -2 \end{vmatrix} = -9;\ C_{22} = \begin{vmatrix} 1 & -1 \\ 0 & -2 \end{vmatrix} = -2;\ C_{23} = -\begin{vmatrix} 1 & -2 \\ 0 & 5 \end{vmatrix} = -5;$

$C_{31} = \begin{vmatrix} -2 & -1 \\ 3 & 1 \end{vmatrix} = 1;\ C_{32} = -\begin{vmatrix} 1 & -1 \\ 2 & 1 \end{vmatrix} = -3;\ C_{33} = \begin{vmatrix} 1 & -2 \\ 2 & 3 \end{vmatrix} = 7;$

$\therefore \quad C = \begin{bmatrix} -11 & 4 & 10 \\ -9 & -2 & -5 \\ 1 & -3 & 7 \end{bmatrix}$

$\therefore$ Adj. $A = C' = \begin{bmatrix} -11 & -9 & 1 \\ 4 & -2 & -3 \\ 10 & -5 & 7 \end{bmatrix}$ **Ans.**

$\therefore \quad A^{-1} = \frac{\text{Adj. } A}{|A|}$

$= -\frac{1}{29}\begin{bmatrix} -11 & -9 & 1 \\ 4 & -2 & -3 \\ 10 & -5 & 7 \end{bmatrix} = \frac{1}{29}\begin{bmatrix} 11 & -9 & -1 \\ -4 & 2 & 3 \\ -10 & 5 & -7 \end{bmatrix}$ **Ans.**

Example 5:

Verify the theorem $A \cdot (Adj \cdot A) = (Adj \cdot A) \cdot A = |A| I_2$, *when*

$A = \begin{bmatrix} 1 & 2 \\ 3 & -5 \end{bmatrix}$.

Solution:

Here $|A| = \begin{vmatrix} 1 & 2 \\ 3 & -5 \end{vmatrix} = -5 - 6 = -11$; $I_2 = \begin{bmatrix} 1 & 0 \\ 0 & 1 \end{bmatrix}$ ...(i)

Also for the given matrix A, we have

$C_{11} = -5,\ C_{12} = -3;\ C_{21} = -2,\ C_{22} = 1$

$\therefore \quad C = \begin{bmatrix} -5 & -3 \\ -2 & 1 \end{bmatrix}$ and as Adj. $A = C' = \begin{bmatrix} -5 & -2 \\ -3 & 1 \end{bmatrix}$

$$\therefore \quad A \cdot (\text{Adj. } A) = \begin{bmatrix} 1 & 2 \\ 3 & -5 \end{bmatrix} \cdot \begin{bmatrix} -5 & -2 \\ -3 & 1 \end{bmatrix}$$

$$= \begin{bmatrix} 1.(-5) + 2(-3) & 1(-2) + 2.1 \\ 3(-5) - 5(-3) & 3(-2) - 5.1 \end{bmatrix}$$

$$= \begin{bmatrix} -11 & 0 \\ 0 & -11 \end{bmatrix}$$

$$\text{And } (\text{Adj. } A) \cdot A = \begin{bmatrix} -5 & -2 \\ -3 & 1 \end{bmatrix} \cdot \begin{bmatrix} 1 & 2 \\ 3 & -5 \end{bmatrix}$$

$$= \begin{bmatrix} -5.1 - 2.3 & -5.2 - 2(-5) \\ -3.1 + 1.3 & -3.2 + 1(-5) \end{bmatrix} = \begin{bmatrix} -11 & 0 \\ 0 & -11 \end{bmatrix}$$

$$\therefore \quad A \cdot (\text{Adj. } A) = (\text{Adj. } A) \cdot A$$

$$= \begin{bmatrix} -11 & 0 \\ 0 & -11 \end{bmatrix} = -11 \begin{bmatrix} 1 & 0 \\ 0 & 1 \end{bmatrix}$$

$= |A| I_2$, from (i) **Hence verified.**

Example 6:

Find the inverse of $A = \begin{bmatrix} 1 & 2 & 3 \\ 1 & 3 & 4 \\ 1 & 4 & 3 \end{bmatrix}$

Solution:

Here $| A | = \begin{bmatrix} 1 & 2 & 3 \\ 1 & 3 & 4 \\ 1 & 4 & 3 \end{bmatrix} = \begin{bmatrix} 1 & 2 & 3 \\ 0 & 1 & 1 \\ 0 & 2 & 0 \end{bmatrix}$

replacing R_2, R_3 by $R_3 - R_1$, $R_3 - R_1$

$$\Rightarrow \qquad | A | = \begin{vmatrix} 1 & 1 \\ 2 & 0 \end{vmatrix} = -2 \qquad \ldots(i)$$

Also for the matrix A, we have

$$C_{11} = \begin{vmatrix} 3 & 4 \\ 4 & 3 \end{vmatrix} = -7;\ C_{12} = -\begin{vmatrix} 1 & 4 \\ 1 & 3 \end{vmatrix} = 1;\ C_{13} = \begin{vmatrix} 1 & 3 \\ 1 & 4 \end{vmatrix} = 1;$$

$$C_{21} = -\begin{vmatrix} 2 & 3 \\ 4 & 3 \end{vmatrix} = 6;\ C_{22} = \begin{vmatrix} 1 & 3 \\ 1 & 3 \end{vmatrix} = 0;\ C_{23} = -\begin{vmatrix} 1 & 2 \\ 1 & 4 \end{vmatrix} = -2;$$

$$C_{31} = \begin{vmatrix} 2 & 3 \\ 3 & 4 \end{vmatrix} = -1;\ C_{32} = -\begin{vmatrix} 1 & 3 \\ 1 & 4 \end{vmatrix} = -1;\ C_{33} = \begin{vmatrix} 1 & 2 \\ 1 & 3 \end{vmatrix} = 5;$$

$$\therefore \quad C = \begin{bmatrix} -7 & 1 & 1 \\ 6 & 0 & -2 \\ -1 & -1 & 1 \end{bmatrix}$$

$$\therefore \quad \text{Adj. } A = C' = \begin{bmatrix} -7 & 6 & -1 \\ 1 & 0 & -1 \\ 1 & -2 & 1 \end{bmatrix}$$

$$\therefore \quad A^{-1} = \frac{\text{Adj. } A}{| A |} = -\frac{1}{2}\begin{bmatrix} -7 & 6 & -1 \\ 1 & 0 & -1 \\ 1 & -2 & 1 \end{bmatrix}$$

$$= \begin{bmatrix} \frac{7}{2} & -3 & \frac{1}{2} \\ -\frac{1}{2} & 0 & \frac{1}{2} \\ -\frac{1}{2} & 1 & -\frac{1}{2} \end{bmatrix} \qquad \textbf{Ans.}$$

Example 7:

Find the inverse of $A = \begin{bmatrix} 1 & 2 & 3 \\ 2 & 4 & 5 \\ 3 & 5 & 6 \end{bmatrix}$

Solution:

For the given matrix A, we have

$C_{11} = \begin{vmatrix} 4 & 5 \\ 5 & 6 \end{vmatrix} = -1;\ C_{12} = -\begin{vmatrix} 2 & 5 \\ 3 & 6 \end{vmatrix} = 3;\ C_{13} = \begin{vmatrix} 2 & 4 \\ 3 & 5 \end{vmatrix} = -2;$

$C_{21} = -\begin{vmatrix} 2 & 3 \\ 5 & 6 \end{vmatrix} = 3;\ C_{22} = \begin{vmatrix} 1 & 3 \\ 3 & 6 \end{vmatrix} = -3;\ C_{23} = -\begin{vmatrix} 1 & 2 \\ 3 & 5 \end{vmatrix} = 1;$

$C_{31} = \begin{vmatrix} 2 & 3 \\ 4 & 5 \end{vmatrix} = -2;\ C_{32} = -\begin{vmatrix} 1 & 3 \\ 2 & 5 \end{vmatrix} = 1;\ C_{33} = \begin{vmatrix} 1 & 2 \\ 2 & 4 \end{vmatrix} = 0;$

$\therefore\ C = \begin{bmatrix} -1 & 3 & -2 \\ 3 & -3 & 1 \\ -2 & 1 & 0 \end{bmatrix}$

$\therefore\ \text{Adj. } A = C' = \begin{bmatrix} -1 & 3 & -2 \\ 3 & -3 & 1 \\ -2 & 1 & 0 \end{bmatrix}$

Also $|A| = \begin{bmatrix} 1 & 2 & 3 \\ 2 & 4 & 5 \\ 3 & 5 & 6 \end{bmatrix} = \begin{bmatrix} 1 & 0 & 0 \\ 2 & 0 & -1 \\ 3 & -1 & -3 \end{bmatrix}$, replacing C_2, C_3 by $C_2 - 2C_1, C_3 - 3C_1$

$= \begin{vmatrix} 0 & -1 \\ -1 & -3 \end{vmatrix} = -1$

$\therefore\ A^{-1} = \dfrac{\text{Adj. } A}{|A|} = -\begin{bmatrix} -1 & 3 & -2 \\ 3 & -3 & 1 \\ -2 & 1 & 0 \end{bmatrix}$ **(Note)**

$= \begin{bmatrix} 1 & -3 & 2 \\ -3 & 3 & -1 \\ 2 & -1 & 0 \end{bmatrix}$ **Ans.**

Example 8:

Find the adjoint of the matrix A and evaluate A^{-1}, where

$$A = \begin{bmatrix} 2 & 2 & 2 \\ 2 & 5 & 5 \\ 2 & 5 & 11 \end{bmatrix}$$

Solution:

Hence for the matrix A, we have

$$C_{11} = \begin{vmatrix} 5 & 5 \\ 5 & 11 \end{vmatrix} = 30; \; C_{12} = -\begin{vmatrix} 2 & 5 \\ 2 & 11 \end{vmatrix} = -12; \; C_{13} = \begin{vmatrix} 2 & 5 \\ 2 & 5 \end{vmatrix} = 0;$$

$$C_{21} = -\begin{vmatrix} 2 & 2 \\ 5 & 11 \end{vmatrix} = -12; \; C_{22} = \begin{vmatrix} 2 & 2 \\ 2 & 11 \end{vmatrix} = 18; \; C_{23} = -\begin{vmatrix} 2 & 2 \\ 2 & 5 \end{vmatrix} = -6;$$

$$C_{31} = \begin{vmatrix} 2 & 2 \\ 5 & 5 \end{vmatrix} = 0; \; C_{32} = -\begin{vmatrix} 2 & 2 \\ 2 & 5 \end{vmatrix} = -6; \; C_{33} = \begin{vmatrix} 2 & 2 \\ 2 & 5 \end{vmatrix} = 6;$$

$$\therefore \quad C = \begin{bmatrix} 30 & -12 & 0 \\ -12 & 18 & -6 \\ 0 & -6 & 6 \end{bmatrix}$$

$$\therefore \quad \text{Adj. A} = C' = \begin{bmatrix} 30 & -12 & 0 \\ -12 & 18 & -6 \\ 0 & -6 & 6 \end{bmatrix}$$

Ans.

Also $|A| = \begin{bmatrix} 2 & 2 & 2 \\ 2 & 5 & 5 \\ 2 & 5 & 11 \end{bmatrix} = \begin{bmatrix} 2 & 0 & 0 \\ 2 & 3 & 3 \\ 2 & 3 & 9 \end{bmatrix}$, applying $C_2 - C_1$, $C_3 - C_1$

$$= 2\begin{vmatrix} 3 & 3 \\ 3 & 9 \end{vmatrix} 2\,[27 - 9] = 36$$

$$\therefore \quad A^{-1} = \frac{\text{Adj. A}}{|A|} = \frac{1}{36}\begin{bmatrix} 30 & -12 & 0 \\ -12 & 18 & -6 \\ 0 & -6 & 6 \end{bmatrix}$$

$$= \frac{1}{36} \times 6 \begin{bmatrix} 5 & -2 & 0 \\ -2 & 3 & -1 \\ 0 & -1 & 1 \end{bmatrix} = \begin{bmatrix} 5/6 & -1/3 & 0 \\ -1/3 & 1/2 & -1/6 \\ 0 & -1/6 & 1/6 \end{bmatrix}$$ **Ans.**

Example 9:

Find the inverse of $A = \begin{bmatrix} 1 & 2 & 3 \\ 0 & 5 & 0 \\ 2 & 0 & 3 \end{bmatrix}$

Solution:

Here $|A| = \begin{vmatrix} 1 & 2 & 3 \\ 0 & 5 & 0 \\ 2 & 0 & 3 \end{vmatrix} = 5 \begin{vmatrix} 1 & 3 \\ 2 & 3 \end{vmatrix} = -15$ ...(i)

Also for the matrix A, we have

$C_{11} = \begin{vmatrix} 5 & 0 \\ 0 & 3 \end{vmatrix} = 15$; $C_{12} = -\begin{vmatrix} 0 & 0 \\ 2 & 3 \end{vmatrix} = 0$; $C_{13} = \begin{vmatrix} 0 & 5 \\ 2 & 0 \end{vmatrix} = -10$;

$C_{21} = -\begin{vmatrix} 2 & 3 \\ 0 & 3 \end{vmatrix} = -6$; $C_{22} = \begin{vmatrix} 1 & 3 \\ 2 & 3 \end{vmatrix} = -3$; $C_{23} = -\begin{vmatrix} 1 & 2 \\ 2 & 0 \end{vmatrix} = 4$;

$C_{31} = \begin{vmatrix} 2 & 3 \\ 5 & 0 \end{vmatrix} = 15$; $C_{32} = -\begin{vmatrix} 1 & 3 \\ 0 & 0 \end{vmatrix} = 0$; $C_{33} = \begin{vmatrix} 1 & 2 \\ 0 & 5 \end{vmatrix} = 5$;

$\therefore\ C = \begin{bmatrix} 15 & 0 & -10 \\ -6 & -3 & 4 \\ -15 & 0 & 5 \end{bmatrix}$

$\therefore\ A^{-1} = \frac{\text{Adj. A}}{|A|} = -\frac{1}{15} \begin{bmatrix} 15 & -6 & -15 \\ 0 & -3 & 0 \\ -10 & 4 & 5 \end{bmatrix}$ **Ans.**

Example 10:

Find the inverse of $A = \begin{bmatrix} 1 & 2 & 4 \\ 5 & 7 & 8 \\ 9 & 10 & 12 \end{bmatrix}$

Solution:

Here $|A| = \begin{vmatrix} 1 & 2 & 4 \\ 5 & 7 & 8 \\ 9 & 10 & 12 \end{vmatrix} = \begin{vmatrix} 1 & 0 & 0 \\ 5 & -3 & -12 \\ 9 & -8 & -24 \end{vmatrix}$

replacing C_2, C_3 by $C_2 - 2C_1, C_3 - 4C_1$

$$\Rightarrow \quad |A| = \begin{vmatrix} -3 & -12 \\ -8 & -24 \end{vmatrix} = 72 - 96 = -24 \qquad \text{...(i)}$$

Also for the matrix A, we have

$C_{11} = \begin{vmatrix} 7 & 8 \\ 10 & 12 \end{vmatrix} = 4$; $C_{12} = -\begin{vmatrix} 5 & 8 \\ 9 & 12 \end{vmatrix} = 12$; $C_{13} = \begin{vmatrix} 5 & 7 \\ 9 & 10 \end{vmatrix} = -13$;

$C_{21} = -\begin{vmatrix} 2 & 4 \\ 10 & 12 \end{vmatrix} = 16$; $C_{22} = \begin{vmatrix} 1 & 4 \\ 9 & 12 \end{vmatrix} = -24$; $C_{23} = -\begin{vmatrix} 1 & 2 \\ 9 & 10 \end{vmatrix} = 8$;

$C_{31} = \begin{vmatrix} 2 & 4 \\ 7 & 8 \end{vmatrix} = -12$; $C_{32} = -\begin{vmatrix} 1 & 4 \\ 5 & 8 \end{vmatrix} = 12$; $C_{33} = \begin{vmatrix} 1 & 2 \\ 5 & 7 \end{vmatrix} = -3$;

$$\therefore \quad C = \begin{bmatrix} 4 & 12 & -13 \\ 16 & -24 & 8 \\ -12 & 12 & -3 \end{bmatrix}$$

$$\therefore \quad \text{Adj. A} = C' = \begin{bmatrix} 4 & 16 & -12 \\ 12 & -24 & 12 \\ -13 & 8 & -3 \end{bmatrix}$$

$$\therefore \quad A^{-1} = \frac{\text{Adj. A}}{|A|} = \frac{1}{24}\begin{bmatrix} 4 & 16 & -12 \\ 12 & -24 & 12 \\ -13 & 8 & -3 \end{bmatrix}, \text{ from (i), (ii)}$$

$$= \begin{bmatrix} -\frac{4}{24} & -\frac{16}{24} & \frac{12}{24} \\ -\frac{12}{24} & 1 & -\frac{12}{24} \\ \frac{13}{24} & -\frac{8}{24} & \frac{8}{24} \end{bmatrix} = \begin{bmatrix} -\frac{1}{6} & -\frac{2}{3} & \frac{1}{2} \\ -\frac{1}{2} & 1 & -\frac{1}{2} \\ \frac{13}{24} & -\frac{1}{3} & \frac{1}{3} \end{bmatrix} \qquad \textbf{Ans.}$$

Example 11:

If $A = \begin{bmatrix} 1 & 4 & 0 \\ -1 & 2 & 2 \\ 0 & 0 & 2 \end{bmatrix}$, *find* A^{-1}.

Solution:

Here $|A| = \begin{vmatrix} 1 & 4 & 0 \\ -1 & 2 & 2 \\ 0 & 0 & 2 \end{vmatrix} = \begin{vmatrix} 1 & 4 & 0 \\ 0 & 6 & 2 \\ 0 & 0 & 2 \end{vmatrix}$, replacing R_2 by $R_2 + R_2$

$$= \begin{vmatrix} 6 & 2 \\ 0 & 2 \end{vmatrix} = 12$$

Also for the matrix A, we have

$C_{11} = \begin{vmatrix} 2 & 2 \\ 0 & 2 \end{vmatrix} = 4$; $C_{12} = -\begin{vmatrix} -1 & 2 \\ 0 & 2 \end{vmatrix} = 2$; $C_{13} = \begin{vmatrix} -1 & 2 \\ 0 & 0 \end{vmatrix} = 0$;

$C_{21} = -\begin{vmatrix} 4 & 0 \\ 0 & 2 \end{vmatrix} = -8$; $C_{22} = \begin{vmatrix} 1 & 0 \\ 0 & 2 \end{vmatrix} = 2$; $C_{23} = -\begin{vmatrix} 1 & 4 \\ 0 & 0 \end{vmatrix} = 0$;

$C_{31} = \begin{vmatrix} 4 & 0 \\ 2 & 2 \end{vmatrix} = 8$; $C_{32} = -\begin{vmatrix} 1 & 0 \\ -1 & 2 \end{vmatrix} = -2$; $C_{33} = \begin{vmatrix} 1 & 4 \\ -1 & 2 \end{vmatrix} = 6$;

$$\therefore \quad C = \begin{bmatrix} 4 & 2 & 0 \\ -8 & 2 & 0 \\ 8 & -2 & 6 \end{bmatrix}$$

$$\therefore \quad \text{Adj. A} = C' = \begin{bmatrix} 4 & -8 & 8 \\ 2 & 2 & -2 \\ 0 & 0 & 6 \end{bmatrix} = 2\begin{bmatrix} 2 & -4 & 4 \\ 1 & 1 & -1 \\ 0 & 0 & 3 \end{bmatrix}$$

$$\therefore \quad A^{-1} = \frac{\text{Adj. A}}{|A|} = -\frac{1}{2} \times 2\begin{bmatrix} 2 & -4 & 4 \\ 1 & 1 & -1 \\ 0 & 0 & 3 \end{bmatrix} = \frac{1}{6}\begin{bmatrix} 2 & -4 & 4 \\ 1 & 1 & -1 \\ 0 & 0 & 3 \end{bmatrix}$$

Example 12:

If $A = \begin{bmatrix} 1 & 0 & -1 \\ 3 & 4 & 5 \\ 0 & -6 & -7 \end{bmatrix}$, *find adj. A and* A^{-1}

Solution:

Here $|A| = \begin{vmatrix} 1 & 0 & -1 \\ 3 & 4 & 5 \\ 0 & -6 & -7 \end{vmatrix} = \begin{vmatrix} 1 & 0 & -1 \\ 0 & 4 & 8 \\ 0 & -6 & -7 \end{vmatrix}$, replacing R_2 by $R_2 - 3R_1$

$\Rightarrow \quad |A| = \begin{vmatrix} 4 & 8 \\ -6 & -7 \end{vmatrix} = -28 + 48 = 20$

Also for the matrix A, we have

$C_{11} = \begin{vmatrix} 4 & 5 \\ -6 & -7 \end{vmatrix} = 2;\ C_{12} = -\begin{vmatrix} 3 & 5 \\ 0 & -7 \end{vmatrix} = 21;\ C_{13} = \begin{vmatrix} 3 & 4 \\ 0 & -6 \end{vmatrix} = -18;$

$C_{21} = -\begin{vmatrix} 0 & -1 \\ -6 & -7 \end{vmatrix} = 6;\ C_{22} = \begin{vmatrix} 1 & -1 \\ 0 & -7 \end{vmatrix} = -7;\ C_{23} = -\begin{vmatrix} 1 & 0 \\ 0 & -6 \end{vmatrix} = 6;$

$C_{31} = \begin{vmatrix} 0 & -1 \\ 4 & 5 \end{vmatrix} = 4;\ C_{32} = -\begin{vmatrix} 1 & -1 \\ 3 & 5 \end{vmatrix} = -8;\ C_{33} = \begin{vmatrix} 1 & 0 \\ 3 & 4 \end{vmatrix} = 4;$

$\therefore \quad C = \begin{bmatrix} 2 & 21 & -18 \\ 6 & -7 & 6 \\ 4 & -8 & 4 \end{bmatrix}$

$\therefore \quad \text{Adj. } A = C' = \begin{bmatrix} 2 & 6 & 4 \\ 21 & -7 & -8 \\ -18 & 6 & 4 \end{bmatrix}$

$\therefore \quad A^{-1} = \frac{\text{Adj. } A}{|A|} = \frac{1}{20}\begin{bmatrix} 2 & 6 & 4 \\ 21 & -7 & -8 \\ -18 & 6 & 4 \end{bmatrix}$ **Ans.**

Example 13:

Find the reciprocal of the matrix $A = \begin{bmatrix} 2 & 1 & 2 \\ 2 & 2 & 1 \\ 1 & 2 & 2 \end{bmatrix}$

Solution:

Here $|A| = \begin{vmatrix} 2 & 1 & 2 \\ 2 & 2 & 1 \\ 1 & 2 & 2 \end{vmatrix} = \begin{vmatrix} 0 & -3 & -2 \\ 0 & -2 & -3 \\ 1 & 2 & 2 \end{vmatrix}$, applying $R_1 - 2R_3$, $R_2 - 2R_3$

$$= \begin{vmatrix} -3 & -2 \\ -2 & -3 \end{vmatrix} = 9 - 4 = 5$$

Also we have

$C_{11} = \begin{vmatrix} 2 & 1 \\ 2 & 2 \end{vmatrix} = 2$; $C_{12} = -\begin{vmatrix} 2 & 1 \\ 1 & 2 \end{vmatrix} = -3$; $C_{13} = \begin{vmatrix} 2 & 2 \\ 1 & 2 \end{vmatrix} = 2$;

$C_{21} = -\begin{vmatrix} 1 & 2 \\ 2 & 2 \end{vmatrix} = 2$; $C_{22} = \begin{vmatrix} 2 & 2 \\ 1 & 2 \end{vmatrix} = 2$; $C_{23} = -\begin{vmatrix} 2 & 1 \\ 1 & 2 \end{vmatrix} = -3$;

$C_{31} = \begin{vmatrix} 1 & 2 \\ 2 & 1 \end{vmatrix} = -3$; $C_{32} = -\begin{vmatrix} 2 & 2 \\ 2 & 1 \end{vmatrix} = 2$; $C_{33} = \begin{vmatrix} 2 & 1 \\ 2 & 2 \end{vmatrix} = 2$;

$\therefore \quad C = \begin{bmatrix} 2 & -3 & 2 \\ 2 & 2 & -3 \\ -3 & 2 & 2 \end{bmatrix}$

$\therefore \quad \text{Adj. } A = C' = \begin{bmatrix} 2 & 2 & -3 \\ -3 & 2 & 2 \\ 2 & -3 & 2 \end{bmatrix}$

$\therefore \quad$ Reciprocal of $A = A^{-1}$

$$= \frac{\text{Adj. } A}{|A|} = \frac{1}{5}\begin{bmatrix} 2 & 2 & -3 \\ -3 & 2 & 2 \\ 2 & -3 & 2 \end{bmatrix}$$

Ans.

Example 14:

Find the inverse of $A = \begin{bmatrix} 1 & 2 & 1 \\ 3 & 2 & 3 \\ 1 & 1 & 2 \end{bmatrix}$

Solution:

For the given matrix A, we have

$$C_{11} = \begin{vmatrix} 2 & 3 \\ 1 & 2 \end{vmatrix} = 1;\ C_{12} = -\begin{vmatrix} 3 & 3 \\ 1 & 2 \end{vmatrix} = -3;\ C_{13} = \begin{vmatrix} 3 & 2 \\ 1 & 1 \end{vmatrix} = 1;$$

$$C_{21} = -\begin{vmatrix} 2 & 1 \\ 1 & 2 \end{vmatrix} = -3;\ C_{22} = \begin{vmatrix} 1 & 1 \\ 1 & 2 \end{vmatrix} = 1;\ C_{23} = -\begin{vmatrix} 1 & 2 \\ 1 & 1 \end{vmatrix} = 1;$$

$$C_{31} = \begin{vmatrix} 2 & 1 \\ 1 & 3 \end{vmatrix} = 4;\ C_{32} = -\begin{vmatrix} 1 & 1 \\ 3 & 3 \end{vmatrix} = 0;\ C_{33} = \begin{vmatrix} 1 & 2 \\ 3 & 2 \end{vmatrix} = -4;$$

$$\therefore \quad C = \begin{bmatrix} 1 & -3 & 1 \\ -3 & 1 & 1 \\ 4 & 0 & -4 \end{bmatrix}$$

$$\therefore \quad \text{Adj. } A = C' = \begin{bmatrix} 1 & -3 & 4 \\ -3 & 1 & 0 \\ 1 & 1 & -4 \end{bmatrix}$$

and $|A| = \begin{vmatrix} 1 & 2 & 1 \\ 3 & 2 & 3 \\ 1 & 1 & 2 \end{vmatrix} = \begin{vmatrix} 1 & 2 & 0 \\ 3 & 2 & 0 \\ 1 & 1 & 1 \end{vmatrix}$, replcaing C_2 by $C_3 - C_1$

$$= \begin{vmatrix} 1 & 2 \\ 3 & 2 \end{vmatrix} = -4$$

Ans.

Example 15:

Find the adjoint and rank of $A = \begin{bmatrix} 1 & 2 & 3 \\ 2 & 3 & 2 \\ 3 & 3 & 4 \end{bmatrix}$

Solution:

Here $|A| = \begin{vmatrix} 1 & 2 & 3 \\ 2 & 3 & 2 \\ 3 & 3 & 4 \end{vmatrix} = \begin{vmatrix} 1 & 0 & 0 \\ 2 & -1 & -4 \\ 3 & -3 & -5 \end{vmatrix}$

replacing C_2, C_3 by $C_2 - 2C_1, C_3 - 3C_1$

$\Rightarrow \quad |A| = \begin{vmatrix} -1 & -4 \\ -3 & -5 \end{vmatrix} = (-1)(-5) - (-4)(-3) = -7$ ...(i)

Also for the matrix A, we have

$C_{11} = \begin{vmatrix} 3 & 2 \\ 3 & 4 \end{vmatrix} = 6; \; C_{12} = -\begin{vmatrix} 2 & 2 \\ 3 & 4 \end{vmatrix} = -2; \; C_{13} = \begin{vmatrix} 2 & 3 \\ 3 & 3 \end{vmatrix} = -3;$

$C_{21} = -\begin{vmatrix} 2 & 3 \\ 3 & 4 \end{vmatrix} = 1; \; C_{22} = \begin{vmatrix} 1 & 3 \\ 3 & 4 \end{vmatrix} = -5; \; C_{23} = -\begin{vmatrix} 1 & 2 \\ 3 & 3 \end{vmatrix} = 3;$

$C_{31} = \begin{vmatrix} 1 & 3 \\ 3 & 2 \end{vmatrix} = -5; \; C_{32} = -\begin{vmatrix} 1 & 3 \\ 2 & 2 \end{vmatrix} = 4; \; C_{33} = \begin{vmatrix} 1 & 2 \\ 2 & 3 \end{vmatrix} = -1;$

$\therefore \quad C = \begin{bmatrix} 6 & -2 & -3 \\ 1 & -5 & 3 \\ -5 & 4 & -1 \end{bmatrix}$

$\therefore \quad \text{Adj. } A = C' = \begin{bmatrix} 6 & 1 & -5 \\ -2 & -5 & 4 \\ -3 & 3 & -1 \end{bmatrix}$ ...(ii) **Ans.**

$\therefore \quad A^{-1} = \dfrac{\text{Adj. } A}{|A|} = \dfrac{\begin{bmatrix} 6 & 1 & -5 \\ -2 & -5 & 4 \\ -3 & 3 & -1 \end{bmatrix}}{-7}$, from (i), (ii)

$= \dfrac{1}{7}\begin{bmatrix} -6 & -1 & 5 \\ 2 & 5 & -4 \\ 3 & -3 & 1 \end{bmatrix} = \begin{bmatrix} -\frac{6}{7} & -\frac{1}{7} & \frac{5}{7} \\ \frac{2}{7} & \frac{5}{7} & -\frac{4}{7} \\ \frac{3}{7} & -\frac{3}{7} & \frac{1}{7} \end{bmatrix}$ **Ans.**

Example 16:

Find A^{-1}, if $A = \begin{bmatrix} 2 & 3 & 1 \\ 1 & 2 & 3 \\ 3 & 1 & 2 \end{bmatrix}$

Solution:

Here $|A| = \begin{vmatrix} 2 & 3 & 1 \\ 1 & 2 & 3 \\ 3 & 1 & 2 \end{vmatrix} = \begin{vmatrix} 0 & 0 & 1 \\ -5 & -7 & 3 \\ -1 & -5 & 2 \end{vmatrix}$,

replacing C_1 and C_2 by $C_1 - 2C_3$ and $C_2 - 3C_3$ respectively.

$= \begin{vmatrix} -5 & -7 \\ -1 & -5 \end{vmatrix}$, expanding with respect to R_1

$= (-5)(-5) - (-7)(-1) = 25 - 7 = 18.$

Also we have

$C_{11} = \begin{vmatrix} 2 & 3 \\ 1 & 2 \end{vmatrix} = 1; \; C_{12} = -\begin{vmatrix} 1 & 3 \\ 3 & 2 \end{vmatrix} = 7; \; C_{13} = \begin{vmatrix} 1 & 2 \\ 3 & 1 \end{vmatrix} = -5;$

$C_{21} = -\begin{vmatrix} 3 & 1 \\ 1 & 2 \end{vmatrix} = 5; \; C_{22} = \begin{vmatrix} 2 & 1 \\ 3 & 2 \end{vmatrix} = 1; \; C_{23} = -\begin{vmatrix} 2 & 3 \\ 3 & 2 \end{vmatrix} = 7;$

$C_{31} = \begin{vmatrix} 3 & 1 \\ 2 & 3 \end{vmatrix} = 7; \; C_{32} = -\begin{vmatrix} 2 & 1 \\ 1 & 3 \end{vmatrix} = -5; \; C_{33} = \begin{vmatrix} 2 & 3 \\ 1 & 1 \end{vmatrix} = 1;$

$\therefore \; C = \begin{bmatrix} 1 & 7 & -5 \\ -5 & 1 & 7 \\ 7 & -5 & 1 \end{bmatrix}$

$\therefore \; \text{Adj. A} = C' = \begin{bmatrix} 1 & -5 & 7 \\ 7 & 1 & -5 \\ -5 & 7 & 1 \end{bmatrix}$

$\therefore \; A^{-1} = \dfrac{\text{Adj. A}}{|A|} = \dfrac{1}{18}\begin{bmatrix} 1 & -5 & 7 \\ 7 & 1 & -5 \\ -5 & 7 & 1 \end{bmatrix}$ **Ans.**

Example 17:

If A' denotes the transpose of a matrix A and

$$A = \begin{bmatrix} 1 & -2 & 3 \\ 0 & -1 & 4 \\ -2 & 2 & 1 \end{bmatrix} \textit{find } (A')^{-1}$$

Solution:

$$A' = \begin{bmatrix} 1 & 0 & -2 \\ -2 & -1 & 2 \\ 3 & 4 & 1 \end{bmatrix}, \text{ by definition of transpose of a matrix}$$

$$= B \text{ (say)}$$

$$\text{Now } |B| = \begin{vmatrix} 1 & 0 & -2 \\ -2 & -1 & 2 \\ 3 & 4 & 1 \end{vmatrix} = \begin{vmatrix} 1 & 0 & 0 \\ -2 & -1 & -2 \\ 3 & 4 & 7 \end{vmatrix}, \text{ replacing } C_3 \text{ by } C_3 + 2C_1$$

$$= \begin{vmatrix} -1 & -2 \\ 4 & 7 \end{vmatrix}, \text{expanding with respect to } R_1$$

$$= (-1)(7) - (-2)(4) = -7 + 8 = 1 \neq 0.$$

Also we have

$$C_{11} = \begin{vmatrix} -1 & 2 \\ 4 & 1 \end{vmatrix} = -9;\ C_{12} = -\begin{vmatrix} -2 & 2 \\ 3 & 1 \end{vmatrix} = 8;\ C_{13} = \begin{vmatrix} -2 & 1 \\ 3 & 4 \end{vmatrix} = -5;$$

$$C_{21} = -\begin{vmatrix} 0 & -2 \\ 4 & 1 \end{vmatrix} = -8;\ C_{22} = \begin{vmatrix} 1 & -2 \\ 3 & 1 \end{vmatrix} = 7;\ C_{23} = -\begin{vmatrix} 1 & 0 \\ 3 & 4 \end{vmatrix} = -4;$$

$$C_{31} = \begin{vmatrix} 0 & -2 \\ -1 & 2 \end{vmatrix} = -2;\ C_{31} = -\begin{vmatrix} 1 & -2 \\ -2 & 2 \end{vmatrix} = 2;\ C_{33} = \begin{vmatrix} 1 & 0 \\ -2 & -1 \end{vmatrix} = -1;$$

$$\therefore\ C = \begin{bmatrix} -9 & 8 & -5 \\ -8 & 7 & -4 \\ -2 & 2 & -1 \end{bmatrix}$$

$$\therefore \quad \text{Adj. B} = C' = \begin{bmatrix} -9 & -8 & -2 \\ 8 & 7 & 2 \\ -5 & -4 & -1 \end{bmatrix}$$

$$\therefore \quad B^{-1} = \frac{\text{Adj. B}}{|B|} = \begin{bmatrix} -9 & -8 & -2 \\ 8 & 7 & 2 \\ -5 & -4 & -1 \end{bmatrix}$$

$$\Rightarrow \quad (A')^{-1} = B^{-1} = \begin{bmatrix} -9 & -8 & -2 \\ 8 & 7 & 2 \\ -5 & -4 & -1 \end{bmatrix}$$

Ans.

Example 18:

How will you use the notion of determinant to compute the inverse of a non-singular square matrix? Compute the inverse of the matrix

$$A = \begin{bmatrix} 1 & 2 & 3 \\ 4 & 5 & 6 \\ 7 & 8 & 10 \end{bmatrix}$$

Solution:

We have for the matrix A,

$$C_{11} = \begin{vmatrix} 5 & 6 \\ 8 & 10 \end{vmatrix} = 2; \; C_{12} = -\begin{vmatrix} 4 & 6 \\ 7 & 10 \end{vmatrix} = 2; \; C_{13} = \begin{vmatrix} 4 & 5 \\ 7 & 8 \end{vmatrix} = -3;$$

$$C_{21} = -\begin{vmatrix} 2 & 3 \\ 8 & 10 \end{vmatrix} = 4; \; C_{22} = \begin{vmatrix} 1 & 3 \\ 7 & 10 \end{vmatrix} = -11; \; C_{23} = -\begin{vmatrix} 1 & 2 \\ 7 & 8 \end{vmatrix} = 6;$$

$$C_{31} = \begin{vmatrix} 2 & 3 \\ 5 & 6 \end{vmatrix} = -3; \; C_{32} = -\begin{vmatrix} 1 & 3 \\ 4 & 6 \end{vmatrix} = 6; \; C_{33} = \begin{vmatrix} 1 & 2 \\ 4 & 5 \end{vmatrix} = -3;$$

$$\therefore \quad C = \begin{bmatrix} 2 & 2 & -3 \\ 4 & -11 & 6 \\ -3 & 6 & -3 \end{bmatrix}$$

$$\therefore \text{ Adj. A} = C' = \begin{bmatrix} 2 & 4 & -3 \\ 2 & -11 & 6 \\ -3 & 6 & -3 \end{bmatrix}$$

Also $|A| = \begin{vmatrix} 1 & 2 & 3 \\ 4 & 5 & 6 \\ 7 & 8 & 10 \end{vmatrix} = \begin{vmatrix} 1 & 0 & 0 \\ 4 & -3 & -6 \\ 7 & -6 & -11 \end{vmatrix}$, replacing C_2, C_3 by $C_2 - 2C_1, C_3 - 3C_1$ respectively

$$\therefore A^{-1} = \frac{\text{Adj. A}}{|A|} = \frac{1}{3}\begin{bmatrix} 2 & 4 & -3 \\ 2 & -11 & 6 \\ -3 & 6 & -3 \end{bmatrix}$$

$$= \begin{bmatrix} -\frac{2}{3} & -\frac{4}{3} & 1 \\ -\frac{2}{3} & \frac{11}{3} & -2 \\ 1 & -2 & 1 \end{bmatrix}$$

Ans.

Example 19:

Find the inverse of the matrix A, where $A = \begin{bmatrix} 1 & 0 & -4 \\ -2 & 2 & 5 \\ 3 & -1 & 2 \end{bmatrix}$

Solution:

Here $|A| = \begin{vmatrix} 1 & 0 & -4 \\ -2 & 2 & 5 \\ 3 & -1 & 2 \end{vmatrix} = \begin{vmatrix} 1 & 0 & 0 \\ -2 & 2 & -3 \\ 3 & -1 & 14 \end{vmatrix}$ applying $C_3 + 4C_1$

$$= \begin{vmatrix} 2 & -3 \\ -1 & 14 \end{vmatrix} = 28 - 3 = 25 \neq 0$$

Also we have

$C_{11} = \begin{vmatrix} 2 & 5 \\ -1 & 2 \end{vmatrix} = 9$; $C_{12} = -\begin{vmatrix} -2 & 5 \\ 3 & 2 \end{vmatrix} = 19$; $C_{13} = \begin{vmatrix} -2 & 2 \\ 3 & -1 \end{vmatrix} = -4$;

$C_{21} = -\begin{vmatrix} 0 & -4 \\ -1 & 2 \end{vmatrix} = 4$; $C_{22} = \begin{vmatrix} 1 & -4 \\ 3 & 2 \end{vmatrix} = 14$; $C_{23} = -\begin{vmatrix} 1 & 0 \\ 3 & -1 \end{vmatrix} = 1$;

$C_{31} = \begin{vmatrix} 0 & -4 \\ 2 & 5 \end{vmatrix} = 8;\ C_{32} = -\begin{vmatrix} 1 & -4 \\ -2 & 5 \end{vmatrix} = 3;\ C_{33} = \begin{vmatrix} 1 & 0 \\ -2 & 2 \end{vmatrix} = 2;$

$$\therefore\ C = \begin{bmatrix} 9 & 19 & -4 \\ 4 & 14 & 1 \\ 8 & 3 & 2 \end{bmatrix}$$

$$\therefore\ \text{Adj. } A = C' = \begin{bmatrix} 9 & 4 & 8 \\ 19 & 14 & 3 \\ -4 & 1 & 2 \end{bmatrix}$$

$$\therefore\ A^{-1} = \frac{\text{Adj. } A}{|A|} = \frac{1}{25}\begin{bmatrix} 9 & 4 & 8 \\ 19 & 14 & 3 \\ -4 & 1 & 2 \end{bmatrix}$$

Ans.

Example 20:

If $A = \begin{bmatrix} 1 & 1 & 1 \\ 2 & 2 & 3 \\ 1 & 4 & 9 \end{bmatrix}$, *find* A^{-1}.

Solution:

Here $|A| = \begin{vmatrix} 1 & 1 & 1 \\ 2 & 2 & 3 \\ 1 & 4 & 9 \end{vmatrix} = \begin{vmatrix} 1 & 0 & 0 \\ 2 & 0 & 1 \\ 1 & 3 & 8 \end{vmatrix}$

replacing C_2, C_3 by $C_2 - C_1$ and $C_3 - C_1$

$$= \begin{vmatrix} 0 & 1 \\ 3 & 8 \end{vmatrix} = 0 \qquad \text{...(i)}$$

Also for the matrix A, we have

$C_{11} = \begin{vmatrix} 2 & 3 \\ 4 & 9 \end{vmatrix} = 6;\ C_{12} = -\begin{vmatrix} 2 & 3 \\ 1 & 9 \end{vmatrix} = -15;\ C_{13} = \begin{vmatrix} 2 & 2 \\ 1 & 4 \end{vmatrix} = 6;$

$$C_{21} = -\begin{vmatrix} 1 & 1 \\ 4 & 9 \end{vmatrix} = -5;\ C_{22} = \begin{vmatrix} 1 & 1 \\ 1 & 9 \end{vmatrix} = 8;\ C_{23} = -\begin{vmatrix} 1 & 1 \\ 1 & 4 \end{vmatrix} = -3;$$

$$C_{31} = \begin{vmatrix} 1 & 1 \\ 2 & 3 \end{vmatrix} = 1;\ C_{32} = -\begin{vmatrix} 1 & 1 \\ 2 & 3 \end{vmatrix} = -1;\ C_{33} = \begin{vmatrix} 1 & 1 \\ 2 & 2 \end{vmatrix} = 0;$$

$$\therefore \quad C = \begin{bmatrix} 6 & -15 & 6 \\ -5 & 8 & -3 \\ 1 & -1 & 0 \end{bmatrix}$$

$$\therefore \text{ Adj. } A = C' = \begin{bmatrix} 6 & -5 & 1 \\ -15 & 8 & -1 \\ 6 & -3 & 0 \end{bmatrix}$$

$$\therefore \ A^{-1} = \frac{\text{Adj. } A}{|A|} = -\frac{1}{3}\begin{bmatrix} 6 & -5 & 1 \\ -15 & 8 & -1 \\ 6 & -3 & 0 \end{bmatrix}$$

$$= \begin{bmatrix} -2 & 5/3 & -1/3 \\ 5 & -8/3 & 1/2 \\ -2 & 1 & 0 \end{bmatrix}$$

Ans.

4.3 EXISTENCE OF INVERSE

Theorem:

The necessary and sufficient condition that a square matrix may possess an inverse is that it be non-singular.

Proof:

The condition is necessary: If A is an n × n matrix and B is its inverse then by definition of the inverse we have

$$AB = I_n$$

Taking the determinants of both sides we get

$$|AB| = |I_n| \qquad \ldots(i)$$

But $\quad |AB| = |A| . |B|$

and $|I_n| = 1$, where I_n is the n × n identity matrix

∴ From (i) we get $|A| . |B| = 1$.

which implies that $|A| \neq 0$.

∴ The matrix A is non-singular.

The condition is sufficient: If A is an n × n non-singular matrix and there be another matrix B defined by

$$B = \frac{1}{|A|}(\text{Adj } A)$$

Then
$$AB = A\frac{1}{|A|}(\text{Adj } A)$$

$$= \frac{1}{|A|}(A.\text{Adj } A)$$

$$= \frac{1}{|A|}.|A|\, I_n$$

$$= I_n$$

∴ $AB = BA = I_n$

∴ B is the inverse of A and it exists.

4.4 SOME IMPORTANT THEOREMS

Theorem 1:

The inverse of the transposed conjugate of a non-singular matrix A is the transposed conjugate of the inverse of A singular matrix A is the transposed conjugate of the inverse of A i.e. $(A^\theta)^{-1} = (A^{-1})^\theta$

Proof:

If A is a non-singular matrix, then A is invertible and we have

$$AA^{-1} = I = A^{-1} A$$

⇒ $$(AA^{-1})^\theta = I^\theta = (A^{-1} A)^\theta$$

⇒ $(A^{-1})^\theta A^\theta = I = A^\theta (A^{-1})^\theta$, since $(AB)^\theta = B^\theta A^\theta$, $I^\theta = I$.

∴ A^θ is invertible and we have $(A^\theta)^{-1} = (A^{-1})^\theta$. **Hence proved.**

Theorem 2:

If a non-singular matrix A is symmetric, then A^{-1} *is also symmetric.*

Proof:

If A is symmetric, then $A = A'$ ...(i)

Also by definition if A is non-singular, then

$$A^{-1} A = I$$

$\Rightarrow$ $A^{-1} A = I'$, since $I' = I$

$= (AA^{-1})$, since $I = A^{-1} A = AA^{-1}$

$= (A^{-1})' A'$, since $(AB)' = B'A'$

i.e., $A^{-1} A = (A^{-1})' A$, since $A = A'$, from (i)

$\Rightarrow$ $A^{-1} = (A^{-1})'$, by right cancellation law.

Hence A^{-1} is symmetric by definition. **Hence proved.**

Theorem 3:

The inverse of the inverse of a matrix is the matrix itself. i.e. $(A^{-1})^{-1} = A$, where A is the inverse.

Proof:

Let A be the given matrix. Then its inverse is A^{-1}.

Also by definition $AA^{-1} = I = A^{-1} A$.

$\therefore A^{-1}$ is invertible and we have $(A^{-1})^{-1} = A$

i.e., the inverse of the inverse of A is A itself. **Hence proved.**

Theorem 4:

If A, B are any two $n \times n$ matrices such that $AB = O$, where O is the null matrix, then at least one of them is singular.

Proof:

Since A, B are two $n \times n$ matrices.

so $AB = O$, where O is the null matrix

$\Rightarrow$ $|A| . |B| = 0$ **(Note)**

$$\Rightarrow \begin{cases} \text{either } |A| = 0, \text{ which means A is singular} \\ \text{or } \quad |B| = 0, \text{ which means B is singular} \\ \text{or both } |A| \text{ and } |B| \text{ are zero which means both A and B are singular.} \end{cases}$$

Hence at least one of A and B is singular.

Theorem 5:

If A is a non-singular matrix of order n such that AX = AY, then X = Y.

Proof:

If A is a non-singular matrix, then A^{-1} exists

Given $AX = AY$

$\Rightarrow$ $A^{-1}(AX) = A^{-1}(AY)$

$\Rightarrow$ $(A^{-1}A)X = (A^{-1}A)Y$

$\Rightarrow$ $IX = IY \quad \because A^{-1}A = I$

$\Rightarrow$ X = Y, by left cancellation law. **Hence proved.**

Theorem 6:

If r be the rank of a matrix A of order m × n; A_r be the normal form of A, R be the product of elementary matrices of order m and S be the product of elementary matrices of order n, then A_r = RAS.

Proof:

Since R and S are non-singular (*i.e.,* their inverses exist), therefore

$R^{-1}A, S^{-1} = A$, where R^{-1} and S^{-1} are the inverses of R and S respectively.

$\Rightarrow$ $A = B A_r C$, where $B = R^{-1}, C = S^{-1}$

$\Rightarrow$ $A_r = B^{-1} A C^{-1}$ **(Note)**

Now if A is a non-singular matrix of order n, then r = n and

$A_r = I_n$

Hence $A = B I_n C$,

which is of the form A = B, since B and C are the product of elementary matrices.

Cor.:

If two matrices A and B are of the same order m × n and same rank, then there exists non-singular square matrices P, Q such that B = PAQ.

Proof:

From above theorem we find that

$A = CA_rD, B = C_1 A_r D_1$

where C, C_1 are product of elementary matrices of order m and D, D_1 of order n.

From $A = C\,A_r\,D$, we get $A_r = C^{-1}\,A\,D^{-1}$

Substituting this is $B = C_1\,A_r\,D_1$, we get

$$B = C_1\,(C^{-1}\,A\,D^{-1})\,D_1 = (C_1\,C^{-1})\,A\,(D^{1}\,D_1)$$

which is of the form B = PAQ.

Example 1:

Find two non-singular matrices P and Q such that PAQ is in the normal form, where $A = \begin{bmatrix} 1 & 1 & 1 \\ 1 & -1 & -1 \\ 3 & 1 & 1 \end{bmatrix}$

Solution:

Here we find that A is a 3 × 3 matrix

$$\therefore \quad [A]_{3\times 3} = I_3\,A\,I_3$$

$$\Rightarrow \begin{bmatrix} 1 & 1 & 1 \\ 1 & -1 & -1 \\ 3 & 1 & 1 \end{bmatrix} = \begin{bmatrix} 1 & 0 & 0 \\ 0 & 1 & 0 \\ 0 & 0 & 1 \end{bmatrix}.A.\begin{bmatrix} 1 & 0 & 0 \\ 0 & 1 & 0 \\ 0 & 0 & 1 \end{bmatrix}$$

$$\Rightarrow \begin{bmatrix} 1 & 1 & 1 \\ 2 & 0 & 0 \\ 2 & 0 & 0 \end{bmatrix} = \begin{bmatrix} 1 & 0 & 0 \\ 1 & 1 & 0 \\ -1 & 0 & 1 \end{bmatrix}.A.\begin{bmatrix} 1 & 0 & 0 \\ 0 & 1 & 0 \\ 0 & 0 & 1 \end{bmatrix},$$

applying $R_2 + R_1$, $R_3 - R_1$

$$\Rightarrow \begin{bmatrix} 1 & 1 & 1 \\ 2 & 0 & 0 \\ 0 & 0 & 0 \end{bmatrix} = \begin{bmatrix} 1 & 0 & 0 \\ 1 & 1 & 0 \\ -2 & -1 & 1 \end{bmatrix}.A.\begin{bmatrix} 1 & 0 & 0 \\ 0 & 1 & 0 \\ 0 & 0 & 1 \end{bmatrix}, \text{ applying } R_3 - R_2$$

$$\Rightarrow \begin{bmatrix} 1 & 1 & 1 \\ 1 & 0 & 0 \\ 0 & 0 & 0 \end{bmatrix} = \begin{bmatrix} 1 & 0 & 0 \\ 1/2 & 1/2 & 0 \\ -2 & -1 & 1 \end{bmatrix}.A.\begin{bmatrix} 1 & 0 & 0 \\ 0 & 1 & 0 \\ 0 & 0 & 1 \end{bmatrix}, \text{ applying } R_2\left(\frac{1}{2}\right)$$

$$\Rightarrow \begin{bmatrix} 0 & 1 & 1 \\ 1 & 0 & 0 \\ 0 & 0 & 0 \end{bmatrix} = \begin{bmatrix} 1/2 & -1/2 & 0 \\ 1/2 & 1/2 & 0 \\ -2 & -1 & 1 \end{bmatrix}.A.\begin{bmatrix} 1 & 0 & 0 \\ 0 & 1 & 0 \\ 0 & 0 & 1 \end{bmatrix}, \text{ applying } R_1 - R_2$$

$$\Rightarrow \quad \begin{bmatrix} 1 & 0 & 0 \\ 0 & 1 & 1 \\ 0 & 0 & 0 \end{bmatrix} = \begin{bmatrix} 1/2 & -1/2 & 0 \\ 1/2 & -1/2 & 0 \\ -2 & -1 & 1 \end{bmatrix} . A . \begin{bmatrix} 1 & 0 & 0 \\ 0 & 1 & 0 \\ 0 & 0 & 1 \end{bmatrix},$$

interchanging R_1 and R_2

$$\Rightarrow \quad \begin{bmatrix} 1 & 0 & 0 \\ 0 & 1 & 0 \\ 0 & 0 & 0 \end{bmatrix} = \begin{bmatrix} 1/2 & 1/2 & 0 \\ 1/2 & -1/2 & 0 \\ -2 & -1 & 1 \end{bmatrix} . A . \begin{bmatrix} 1 & 0 & 0 \\ 0 & 1 & -1 \\ 0 & 0 & 1 \end{bmatrix},$$

applying $C_3 - C_2$ **(Note)**

Since L.H.S. is in the normal form, so we have

$$P = \begin{bmatrix} 1/2 & 1/2 & 0 \\ 1/2 & -1/2 & 0 \\ -2 & -1 & 1 \end{bmatrix}, Q = \begin{bmatrix} 1 & 0 & 0 \\ 0 & 1 & -1 \\ 0 & 0 & 1 \end{bmatrix},$$

Ans.

Example 2:

Find the non-singular matrices R and S, such that RAS is the normal form, where $A = \begin{bmatrix} 2 & 2 & -6 \\ -1 & 2 & 2 \end{bmatrix}$.

Solution:

Here we find that A is a 2×3 matrix.

$$\therefore \quad [A]_{2 \times 3} = I_2 \, A \, I_3$$

$$\Rightarrow \quad \begin{bmatrix} 2 & 2 & -6 \\ -1 & 2 & 2 \end{bmatrix} = \begin{bmatrix} 1 & 0 \\ 0 & 1 \end{bmatrix} . A . \begin{bmatrix} 1 & 0 & 0 \\ 0 & 1 & 0 \\ 0 & 0 & 1 \end{bmatrix}$$

Now we are to bring L.H.S. to the normal form by applying elementary row and column operations.

$$\therefore \begin{bmatrix} 1 & 1 & -3 \\ -1 & 2 & 2 \end{bmatrix} = \begin{bmatrix} 1/2 & 0 \\ 0 & 1 \end{bmatrix} . A . \begin{bmatrix} 1 & 0 & 0 \\ 0 & 1 & 0 \\ 0 & 0 & 1 \end{bmatrix}, \text{ by } R_1\left(\frac{1}{2}\right)$$

$$\Rightarrow \quad \begin{bmatrix} 1 & 1 & -3 \\ 0 & 3 & -1 \end{bmatrix} = \begin{bmatrix} 1/2 & 0 \\ 1/2 & 1 \end{bmatrix} . A . \begin{bmatrix} 1 & 0 & 0 \\ 0 & 1 & 0 \\ 0 & 0 & 1 \end{bmatrix}, \text{ by } R_2 + R_1$$

$$\Rightarrow \begin{bmatrix} 1 & 0 & 0 \\ 0 & 3 & -1 \end{bmatrix} = \begin{bmatrix} 1/2 & 0 \\ 1/2 & 1 \end{bmatrix} .A. \begin{bmatrix} 1 & -1 & 3 \\ 0 & 1 & 0 \\ 0 & 0 & 1 \end{bmatrix}, \text{ by } C_2 - C_1 \text{ and } C_3 + 3C_1$$

$$\Rightarrow \begin{bmatrix} 1 & 0 & 0 \\ 0 & 3 & -1 \end{bmatrix} = \begin{bmatrix} 1/2 & 0 \\ 1/2 & 1 \end{bmatrix} .A. \begin{bmatrix} 1 & -1/3 & 3 \\ 0 & 1/3 & 0 \\ 0 & 0 & 1 \end{bmatrix}, \text{ replacing } C_2 \text{ by } \frac{1}{3}C_2$$

$$\Rightarrow \begin{bmatrix} 1 & 0 & 0 \\ 0 & 1 & 0 \end{bmatrix} = \begin{bmatrix} 1/2 & 0 \\ 1/2 & 1 \end{bmatrix} .A. \begin{bmatrix} 1 & -1/3 & 8/3 \\ 0 & 1/3 & 1/3 \\ 0 & 0 & 1 \end{bmatrix}, \text{ by } C_3 + C_2$$

Since L.H.S. is in the normal form, so

$$R = \begin{bmatrix} 1/2 & 0 \\ 1/2 & 1 \end{bmatrix} \text{ and } S = \begin{bmatrix} 1 & -1/3 & 8/3 \\ 0 & 1/3 & 1/3 \\ 0 & 0 & 0 \end{bmatrix}$$

Ans.

SOLVED EXAMPLES

Example 1:

Find the rank of an m × n matrix, every element of which is unity.

Solution:

Let an m × n matrix be $A = \begin{bmatrix} 1 & 1 & 1 & \dots & 1 \\ 1 & 1 & 1 & \dots & 1 \\ \dots & \dots & \dots & \dots & \dots \\ 1 & 1 & 1 & \dots & 1 \end{bmatrix}$

Then we find that every square submatrix of A higher than 1 × 1 will be a matrix each element of which is unity and therefore the value of the determinant will be always zero, since its rows and columns are identical. But the square sub-matrices of order 1 × 1 are [1] and the determinants of these are $| A | \neq 0$.

Hence the rank of A is 1. **Ans.**

Example 2:

Prove that the rank of a matrix remains unaltered by the application of elementary row and column operations.

Or

Prove that two equivalent matrices have the same rank.

Solution:

Let an m × n matrix A be given by

$$A = \begin{bmatrix} a_{11} & a_{12} & \cdots & a_{1n} \\ a_{21} & a_{22} & \cdots & a_{2n} \\ \cdots & \cdots & \cdots & \cdots \\ a_{m1} & a_{m2} & \cdots & a_{mn} \end{bmatrix}$$

Let M be any minor of order r belonging to the first r rows of | A |.

Now firstly if we interchange any two rows or columns of A, then the minor M either remains unaltered or changes sign.

Secondly if we multiply one row or column of A by a number λ, then either the minor M remains unaltered or changes into λ M.

Thirdly if we replace any row R_i (or column C_i) by $R_i + \lambda R_j$ (or $C_i + \lambda C_j$), then either the minor M remains unaltered or change into a sum or difference of two of the original minors.

Let B the matrix obtained from A by the application of any one of the above three elementary row or column operations.

Thus if all the minors of order r in | A | are zero, then all the minors of order r in | B | are also zero.

∴ rank of B ≤ rank of A ...(i)

Similarly if all the minors of order r in | B | are zero, then all the minors of order r in | A | are also zero.

∴ rank of A ≤ rank of B ...(ii)

∴ From (i) and (ii) we get

rank of A = rank of B. **Hence proved.**

Example 3:

If A is of order m × n, R is a non-singular matrix of order m, show that

Rank of RA = Rank of A.

Solution:

Let $A = E\, A_r\, F$ and $R = E_1$

Then $RA = E_1\,(E\, A_r\, F) = E_1\, E\, A_r\, F$

i.e., RA has been expressed as the result of elementary operations on A_r.

Thus Rank of (RA) = Rank A_r = Rank A.

Example 4:

Find the reciprocal (or inverse) of the matrix

$$S = \begin{bmatrix} 0 & 1 & 1 \\ 1 & 0 & 1 \\ 1 & 1 & 0 \end{bmatrix}$$ *and show that the transform of*

the matrix $A = \frac{1}{2}\begin{bmatrix} b+c & c-a & b-a \\ c-b & c+a & a-b \\ b-c & a-c & a+b \end{bmatrix}$ *by S i.e.*

Solution:

In the usual way we can show that

$$S^{-1} = \text{inverse of } S = \frac{1}{2}\begin{bmatrix} -1 & 1 & 1 \\ 1 & -1 & 1 \\ 1 & 1 & -1 \end{bmatrix}$$

$$\therefore \quad SA = \begin{bmatrix} 0 & 1 & 1 \\ 1 & 0 & 1 \\ 1 & 1 & 0 \end{bmatrix} \times \frac{1}{2}\begin{bmatrix} b+c & c-a & b-a \\ c-b & c+a & a-b \\ b-c & a-c & a+b \end{bmatrix}$$

$$= \frac{1}{2}\begin{bmatrix} 0 & 2a & 2a \\ 2b & 0 & 2b \\ 2c & 2c & 0 \end{bmatrix}$$, multiplying the matrices in the usual way.

$$\Rightarrow \quad SA = \begin{bmatrix} 0 & a & a \\ b & 0 & b \\ c & c & 0 \end{bmatrix}$$

$$\therefore \quad SAS^{-1} = \begin{bmatrix} 0 & a & a \\ b & 0 & b \\ c & c & 0 \end{bmatrix} \times \frac{1}{2}\begin{bmatrix} -1 & 1 & 1 \\ 1 & -1 & 1 \\ 1 & 1 & -1 \end{bmatrix}$$

$$= \frac{1}{2}\begin{bmatrix} 2a & 0 & 0 \\ 0 & 2b & 0 \\ 0 & 0 & 2c \end{bmatrix},$$

multiplying the two matrices in the usual way.

$$= \begin{bmatrix} a & 0 & 0 \\ 0 & b & 0 \\ 0 & 0 & c \end{bmatrix}, \text{ which is a diagonal matrix.}$$

Example 5:

Find the rank of the matrix $A = \begin{bmatrix} 4 & 5 & 8 \\ 5 & 6 & 7 \\ 7 & 8 & 9 \end{bmatrix}$

Solution:

The determinant of order 3 formed by A

$$= \begin{vmatrix} 4 & 5 & 8 \\ 5 & 6 & 7 \\ 7 & 8 & 9 \end{vmatrix} = \begin{vmatrix} 4 & 1 & 2 \\ 5 & 1 & 1 \\ 7 & 1 & 1 \end{vmatrix}, \text{ replacing } C_2, C_3 \text{ by } C_2 - C_1, C_3 - C_2 \text{ respectively}$$

$$= \begin{vmatrix} 0 & 1 & 0 \\ 1 & 1 & -2 \\ 3 & 1 & -2 \end{vmatrix}, \text{ replacing } C_1, C_3 \text{ by } C_1 - 4C_2 \text{ and } C_3 - 3C_2 \text{ respectively}$$

$$= -\begin{vmatrix} 1 & -2 \\ 3 & -2 \end{vmatrix}, \text{ expanding with respect to } R_1$$

$$= -[-2 + 6] = -4 \neq 0$$

$\therefore \quad p(A) \geq 3$ ...(i)

Also the matrix A does not possess any minor of order 4 *i.e.*, 3 + 1 so

$p(A) \leq 3$ ...(ii)

$\therefore$ From (i) and (ii) we get $p(A) = 3$. **Ans.**

Example 6:

Find the rank of the matrix

$$A = \begin{bmatrix} 1^2 & 2^2 & 3^2 & 4^2 \\ 2^2 & 3^2 & 4^2 & 5^2 \\ 3^2 & 4^2 & 5^2 & 6^2 \\ 4^2 & 5^2 & 6^2 & 7^2 \end{bmatrix}$$

Solution:

$$\text{Given } A = \begin{bmatrix} 1 & 4 & 9 & 16 \\ 4 & 9 & 16 & 25 \\ 9 & 16 & 25 & 36 \\ 16 & 25 & 36 & 49 \end{bmatrix}$$

$$\therefore\ A \sim \begin{bmatrix} 1 & 0 & 0 & 0 \\ 4 & -7 & -20 & -39 \\ 9 & -20 & -56 & -108 \\ 16 & -39 & -108 & -107 \end{bmatrix},$$ replacing C_2, C_3, C_4 by $C_2 - 4C_1, C_2 - 9C_1$ and $C_4 - 16C_1$ respectively.

$$\sim \begin{bmatrix} 1 & 0 & 0 & 0 \\ 0 & -7 & -20 & -39 \\ 1 & -6 & -16 & -30 \\ 0 & -11 & -28 & -51 \end{bmatrix},$$ replacing R_2, R_3, R_4 by $R_2 - 4R_1, R_3 - 2R_2$ and $R_4 - 4R_2$ respectively

$$\sim \begin{bmatrix} 1 & 0 & 0 & 0 \\ 0 & -7 & -20 & -39 \\ 0 & -6 & -16 & -30 \\ 0 & -4 & -8 & -12 \end{bmatrix},$$ replacing R_3, R_4 by $R_3 - R_1$ and $R_4 - R_2$ respectively

$$\sim \begin{bmatrix} 1 & 0 & 0 & 0 \\ 0 & 1 & 4 & 9 \\ 0 & 2 & 8 & 18 \\ 0 & 4 & 8 & 12 \end{bmatrix},$$ replacing R_2, R_3, R_4 by $-(R_2 - R_3) -(R_3 - R_4)$ and $- R_4$ respectively.

$$\sim \begin{bmatrix} 1 & 0 & 0 & 0 \\ 0 & 1 & 4 & 9 \\ 0 & 0 & 0 & 0 \\ 0 & 0 & -8 & -24 \end{bmatrix},$$ replacing R_3, R_4 by $R_3 - 2R_2$ $R_4 - 4R_2$ respectively.

$$\sim \begin{bmatrix} 1 & 0 & 0 & 0 \\ 0 & 1 & 0 & 0 \\ 0 & 0 & 0 & 0 \\ 0 & 0 & -8 & -24 \end{bmatrix}, \text{ replacing } C_3, C_4 \text{ by } C_3 - 4C_2, C_4 - 9C_2 \text{ respectively.}$$

$$\sim \begin{bmatrix} 1 & 0 & 0 & 0 \\ 0 & 1 & 0 & 0 \\ 0 & 0 & 0 & 0 \\ 0 & 0 & 1 & 3 \end{bmatrix}, \text{ replacing } R_4 \text{ by } -\frac{1}{8}R_4$$

$$\sim \begin{bmatrix} 1 & 0 & 0 & 0 \\ 0 & 1 & 0 & 0 \\ 0 & 0 & 1 & 3 \\ 0 & 0 & 0 & 0 \end{bmatrix}, \text{ interchanging } R_3 \text{ and } R_4$$

$$\sim \begin{bmatrix} 1 & 0 & 0 & 0 \\ 0 & 1 & 0 & 0 \\ 0 & 0 & 1 & 0 \\ 0 & 0 & 0 & 0 \end{bmatrix}, \text{ replacing } C_4 \text{ by } C_4 - 3C_2$$

$$\sim \begin{bmatrix} I_2 & 0 \\ 0 & 0 \end{bmatrix}$$

∴ The rank of the matrix A is 3. **Ans.**

Example 7:

If $A = \begin{bmatrix} a_1 & 0 & 0 \\ 0 & a_2 & 0 \\ 0 & 0 & a_3 \end{bmatrix}$, *where none of a's is zero, then show that A is invertible. Also evaluate* A^{-}.

Solution:

$$|A| = \begin{vmatrix} a_1 & 0 & 0 \\ 0 & a_2 & 0 \\ 0 & 0 & a_3 \end{vmatrix} = a_1\, a_2\, a_3 \text{ on evaluating}$$

i.e., $|A| \neq 0$. Hence A is invertible.

Also

$C_{11} = \begin{vmatrix} a_2 & 0 \\ 0 & a_3 \end{vmatrix} = a_2 a_3$; $C_{12} = -\begin{vmatrix} 0 & 0 \\ 0 & a_3 \end{vmatrix} = 0$; $C_{13} = \begin{vmatrix} 0 & a_2 \\ 0 & 0 \end{vmatrix} = 0$;

$C_{21} = -\begin{vmatrix} 0 & 0 \\ 0 & a_3 \end{vmatrix} = 0$; $C_{22} = \begin{vmatrix} a_1 & 0 \\ 0 & a_3 \end{vmatrix} = a_1 a_3$; $C_{23} = -\begin{vmatrix} a_1 & 0 \\ 0 & 0 \end{vmatrix} = 0$;

$C_{31} = \begin{vmatrix} 0 & 0 \\ a_2 & 0 \end{vmatrix} = 0$; $C_{32} = -\begin{vmatrix} a_1 & 0 \\ 0 & 0 \end{vmatrix} = 0$; $C_{33} = \begin{vmatrix} a_1 & 0 \\ 0 & a_2 \end{vmatrix} = a_1 a_3$;

$$\therefore \quad C = \begin{bmatrix} a_2a_3 & 0 & 0 \\ 0 & a_3a_1 & 0 \\ 0 & 0 & a_1a_2 \end{bmatrix}$$

$$\therefore \quad \text{Adj. } A = C' = \begin{bmatrix} a_2a_3 & 0 & 0 \\ 0 & a_3a_1 & 0 \\ 0 & 0 & a_1a_2 \end{bmatrix}$$

$$A^{-1} = \frac{\text{Adj. } A}{|A|} = \frac{1}{a_1 a_2 a_3}\begin{bmatrix} a_2a_3 & 0 & 0 \\ 0 & a_3a_1 & 0 \\ 0 & 0 & a_1a_2 \end{bmatrix}$$

$$= \begin{bmatrix} 1/a_1 & 0 & 0 \\ 0 & 1/a_2 & 0 \\ 0 & 0 & 1/a_3 \end{bmatrix}$$

Ans.

Example 8:

If A is invertible show that $\overline{A}$ is invertible

Solution:

If A is invertible, then we know that

$$AA^{-1} = I = A^{-1}A$$

$$\Rightarrow \quad \overline{(AA^{-1})} = \overline{I} = \overline{(A^{-1}A)}$$

$$\Rightarrow \quad \overline{A}\,\overline{(A^{-1})} = I = \overline{(A^{-1})}\,\overline{A}. \quad \because \overline{AB} = \overline{A}.\overline{B}.$$

Hence $\overline{A}$ is invertible and we have $(\overline{A})^{-1} = \overline{(A^{-1})}$.

Hence proved.

Example 9:

Find the rank of $A = \begin{bmatrix} 1 & 2 & 3 & 1 \\ 2 & 4 & 6 & 2 \\ 1 & 2 & 3 & 2 \end{bmatrix}$

Solution:

$$A \sim \begin{bmatrix} 1 & 2 & 3 & 1 \\ 0 & 0 & 0 & 0 \\ 0 & 0 & 0 & 1 \end{bmatrix},$$ replacing R_2, R_3 by $R_2 - 2R_1$ and $R_3 - R_1$ respecitvely

$$\sim \begin{bmatrix} 1 & 2 & 3 & 0 \\ 0 & 0 & 0 & 0 \\ 0 & 0 & 0 & 1 \end{bmatrix},$$ replacing R_1 by $R_1 - R_2$

$$\sim \begin{bmatrix} 1 & 0 & 0 & 0 \\ 0 & 0 & 0 & 0 \\ 0 & 0 & 0 & 1 \end{bmatrix},$$ replacing C_2, C_3 by $C_2 - 2C_1$ and $C_3 - 3C_1$ respectively

$$\sim \begin{bmatrix} 1 & 0 & 0 & 0 \\ 0 & 0 & 0 & 0 \\ 0 & 1 & 0 & 0 \end{bmatrix},$$ int erchaing C_2 and C_3

$$\sim \begin{bmatrix} 1 & 0 & 0 & 0 \\ 0 & 1 & 0 & 0 \\ 0 & 0 & 0 & 0 \end{bmatrix},$$ int erchaing R_2 and R_3

$$\sim \begin{bmatrix} I_2 & 0 \\ 0 & 0 \end{bmatrix}$$

$\therefore$ The rank of matrix A is 2. **Ans.**

Example 10:

If two non-singular symmetric matrices A and B be such that AB = BA (i.e. commute under multiplication), then prove that $A^{-1}B$ *and* $A^{-1}B^{-2}$ *are symmetric.*

Solution:

Now we are given that AB = BA

$\therefore$ We have $A^{-1}AB = A^{-1}BA$ premultiplying by A^{-1}

$\Rightarrow$ $IB = A^{-1}BA, \quad A^{-1}A = 1$

$\Rightarrow$ $B = A^{-1}BA, \quad \because \quad IB = B$

$\Rightarrow$ $BA^{-1} = A^{-1}BAA^{-1}$, post multiplying by A^{-1}

$= A^{-1}BI = A^{-1}B,$ ...(i)

since $AA^{-1} = I$ and $BI = B$.

Again $(A^{-1}B)' = B'(A^{-1})', \quad \because \quad (AB)' = B'A'$

$= B'(A')^{-1} \quad \because \quad (A^{-1})' = (A')^{-1}$

$= BA^{-1}, \quad \because \quad A' = A,$

$B' = B$ as A and B are symmetric

i.e., $(A^{-1}B)' = A^{-1}B$, from (i).

Hence $A^{-1}B$ is symmetric.

Similarly $(A^{-1}B^{-1})' = (B^{-1})'(A^{-1})$, as $(CD)' = D'C'$

$\Rightarrow$ $(A^{-1}B^{-1}) = ((B')^{-1}(A')^{-1}$

$= B^{-1}A^{-1} \quad \because \quad A' = A, B' = B$

$= (AB)^{-1}, \quad \because \quad (AB)^{-1} = B^{-1}A^{-1}$

$= (BA)^{-1}, \quad \because \quad AB = BA$ (given)

$\Rightarrow$ $(A^{-1}B^{-1})' = A^{-1}B^{-1}$.

Hence $A^{-1}B^{-1}$ is symmetric.

Example 11:

If $A = \begin{bmatrix} 3 & -3 & 4 \\ 2 & -3 & 4 \\ 0 & -1 & 1 \end{bmatrix}$, *show that* $A^3 = A^{-1}$.

Solution:

Here $|A| = \begin{vmatrix} 3 & -3 & 4 \\ 2 & -3 & 4 \\ 0 & -1 & 1 \end{vmatrix}$,

$$= \begin{vmatrix} 1 & 0 & 0 \\ 2 & -3 & 4 \\ 0 & -1 & 1 \end{vmatrix}, \text{ replacing } R_1 \text{ by } R_1 - R_2$$

$$\Rightarrow \quad |A| = \begin{vmatrix} -3 & 4 \\ -1 & 1 \end{vmatrix} = -3 + 4 = 1 \qquad \ldots(i)$$

Also for the matrix A, we have

$$C_{11} = \begin{vmatrix} -3 & 4 \\ -1 & 1 \end{vmatrix} = 1; \; C_{12} = -\begin{vmatrix} 2 & 4 \\ 0 & 1 \end{vmatrix} = -2; \; C_{13} = \begin{vmatrix} 2 & -3 \\ 0 & -1 \end{vmatrix} = -2;$$

$$C_{21} = -\begin{vmatrix} -3 & 4 \\ -1 & 1 \end{vmatrix} = -1; \; C_{22} = \begin{vmatrix} 3 & 4 \\ 0 & 1 \end{vmatrix} = 3; \; C_{23} = -\begin{vmatrix} 3 & -3 \\ 0 & -1 \end{vmatrix} = 3;$$

$$C_{31} = \begin{vmatrix} -3 & 4 \\ -3 & 4 \end{vmatrix} = 0; \; C_{32} = -\begin{vmatrix} 3 & 4 \\ 2 & 4 \end{vmatrix} = -4; \; C_{33} = \begin{vmatrix} 3 & -3 \\ 2 & -3 \end{vmatrix} = -3;$$

$$\therefore \quad C = \begin{bmatrix} 1 & -2 & -2 \\ -1 & 3 & 3 \\ 0 & -4 & -3 \end{bmatrix}.$$

$$\therefore \text{ Adj. } A = C' = \begin{bmatrix} 1 & -1 & 0 \\ -2 & 3 & -4 \\ -2 & 3 & -3 \end{bmatrix} \qquad \ldots(ii)$$

$$\therefore \quad A^{-1} = \frac{\text{Adj. } A}{|A|} = \begin{bmatrix} 1 & -1 & 0 \\ -2 & 3 & -4 \\ -2 & 3 & -3 \end{bmatrix}, \text{ from (i), (ii)} \qquad \ldots(iii)$$

$$\text{Alos } A^2 = \begin{bmatrix} 3 & -3 & 4 \\ 2 & -3 & 4 \\ 0 & -1 & 1 \end{bmatrix} \times \begin{bmatrix} 3 & -3 & 4 \\ 2 & -3 & 4 \\ 0 & -1 & 1 \end{bmatrix}$$

$$= \begin{bmatrix} 9-6+0 & -9+9-4 & 12-12+4 \\ 6-6+0 & -6+9-4 & 8-12+4 \\ 0-2+0 & 0+3-1 & 0-4+1 \end{bmatrix} = \begin{bmatrix} 3 & -4 & 4 \\ 0 & -1 & 0 \\ -2 & 2 & -3 \end{bmatrix}$$

$$\therefore \quad A^2 = A^2.A = \begin{bmatrix} 3 & -4 & 4 \\ 0 & -1 & 0 \\ -2 & 2 & -3 \end{bmatrix} \times \begin{bmatrix} 3 & -3 & 4 \\ 2 & -3 & 4 \\ 0 & -1 & 1 \end{bmatrix}$$

$$= \begin{bmatrix} 9-8+0 & -9+12-4 & 12-16+4 \\ 0-2+0 & 0+3+0 & 0-4+0 \\ -6+4+0 & 6-6+3 & -8+8-3 \end{bmatrix} = \begin{bmatrix} 1 & -1 & 0 \\ -2 & 3 & -4 \\ -2 & 3 & -3 \end{bmatrix}$$

i.e., $A^3 = A^{-1}$, from (iii) **Hence proved.**

Example 12:

Show that the matrix $A = \begin{bmatrix} 1 & a & \alpha & a\alpha \\ 1 & b & \beta & b\beta \\ 1 & c & \gamma & c\gamma \end{bmatrix}$ *is of rank 3 provided no two of a, b, c are equal and no two of* α, β γ *are equal.*

Solution:

$$A \sim \begin{bmatrix} 1 & a & \alpha & a\alpha \\ 0 & b-a & \beta-\alpha & b\beta - a\alpha \\ 0 & c-a & \gamma-\alpha & c\gamma - a\alpha \end{bmatrix},$$

replacing R_2, R_3 by $R_2 - R_1$, $R_3 - R_1$ respectively.

$$\sim \begin{bmatrix} 1 & 0 & 0 & 0 \\ 0 & b-a & \beta-\alpha & b\beta - a\alpha \\ 0 & c-a & \gamma-\alpha & c\gamma - a\alpha \end{bmatrix},$$

replacing C_2, C_3, C_4 by $C_2 - aC_1$, $C_3 - \alpha C_1$ and $C_4 - a\alpha C_1$ respectively

$$\Rightarrow \quad A \sim \begin{bmatrix} 1 & 0 & 0 & 0 \\ 0 & b-a & \beta-\alpha & b\beta - a\alpha \\ 0 & c-a & \gamma-\alpha & c\gamma - a\alpha \end{bmatrix}, \text{ replacing } C_4 \text{ by } C_4 - \alpha C_2$$

$$\sim \begin{bmatrix} 1 & 0 & 0 & 0 \\ 0 & b-a & \beta-\alpha & 0 \\ 0 & c-a & \gamma-\alpha & \begin{matrix} c\gamma - a\alpha \\ -b\gamma + b\alpha \end{matrix} \end{bmatrix}, \text{ replacing } C_4 \text{ by } C_4 - bC_3$$

$$\sim \begin{bmatrix} 1 & 0 & 0 & 0 \\ 0 & b-a & \beta-\alpha & 0 \\ 0 & c-a & \gamma-\alpha & (c-b)(\gamma-\alpha) \end{bmatrix} = B \text{ (say)},$$

Now a minor of order 3 of B

$$= \begin{vmatrix} 1 & 0 & 0 \\ 0 & b-a & 0 \\ 0 & c-a & (c-b)(\gamma-\alpha) \end{vmatrix} = \begin{vmatrix} b-a & 0 \\ c-a & (c-b)(\gamma-\alpha) \end{vmatrix},$$

expanding with respect to R_1

$= (b - a)(c - b)(\gamma - \alpha) \neq 0$, as no two of a, b, c and no two of α, β, γ are equal (given).

$\therefore \qquad \rho(B) \geq 3 \qquad$ ...(i)

Also the matrix B does not posses any minor of order 4 *i.e.,*, of order 3 + 1, so $\rho(B) \leq 3$ ...(ii)

$\therefore$ From (i) and (ii) we get $\rho(B) = 3$

and therefore $\rho(A) = 3$, as $A \sim B$. **Hence proved.**

Example 5:

Find the rank of the matrix $A = \begin{bmatrix} 4 & 5 & 8 \\ 5 & 6 & 7 \\ 7 & 8 & 9 \end{bmatrix}$

Solution:

The determinant of order 3 formed by A

$$= \begin{vmatrix} 4 & 5 & 8 \\ 5 & 6 & 7 \\ 7 & 8 & 9 \end{vmatrix} = \begin{vmatrix} 4 & 1 & 2 \\ 5 & 1 & 1 \\ 7 & 1 & 1 \end{vmatrix},$$ replacing C_2, C_3 by $C_2 - C_1, C_3 - C_2$ respectively

$$= \begin{vmatrix} 0 & 1 & 0 \\ 1 & 1 & -2 \\ 3 & 1 & -2 \end{vmatrix},$$ replacing C_1, C_3 by $C_1 - 4C_2$ and $C_3 - 3C_2$ respectively

$$= -\begin{vmatrix} 1 & -2 \\ 3 & -2 \end{vmatrix},$$ expanding with respect to R_1

$= -[-2+6] = -4 \neq 0$

$\therefore \quad p(A) \geq 3 \quad \ldots(i)$

Also the matrix A does not possess any minor of order 4 *i.e.*, 3 + 1 so

$p(A) \leq 3 \quad \ldots(ii)$

$\therefore$ From (i) and (ii) we get p(A) = 3. **Ans.**

EXERCISES

1. *Show that the inverse of*

$$\begin{bmatrix} 2 & -1 & 1 \\ -15 & 6 & -5 \\ 5 & -2 & 2 \end{bmatrix} \text{ is } \begin{bmatrix} -2 & 0 & 1 \\ -5 & 1 & 5 \\ 0 & 0 & 3 \end{bmatrix}$$

2. *Show that the adjoint of*

$$\begin{bmatrix} 1 & 1 & 3 \\ 0 & 1 & -1 \\ 2 & 0 & -4 \end{bmatrix} \text{ is } \begin{bmatrix} -4 & 4 & -4 \\ -2 & -10 & 1 \\ -2 & 2 & 1 \end{bmatrix}$$

3. *Show that the inverse of*

$$\begin{bmatrix} 3 & -2 & -1 \\ -4 & 1 & -1 \\ 2 & 0 & 1 \end{bmatrix} \text{ is } \begin{bmatrix} 1 & 2 & 3 \\ 2 & 5 & 7 \\ -2 & -4 & -5 \end{bmatrix}$$

4. *If* $A = \begin{bmatrix} a_1 & b_1 & c_1 \\ a_2 & b_2 & c_2 \\ a_3 & b_3 & c_3 \end{bmatrix}$, *prove that* $AA^{-1} = I_2$.

5. *Prove that* $Adj.\ (A') = (Adj.\ A)'$.

6. *Compute rank of the matrix* $\begin{bmatrix} 1 & 2 & 3 \\ 4 & 5 & 6 \\ 7 & 8 & 9 \end{bmatrix}$

7. $\begin{bmatrix} 1 & 2 & 3 \\ 2 & 4 & 9 \\ 3 & 6 & 10 \end{bmatrix}$ **Ans. 2**

8. $\begin{bmatrix} 1 & 1 & 1 \\ a & b & c \\ a^2 & b^2 & c^2 \end{bmatrix}$, *where a, b, c are all real.*

Ans. 2

9. $\begin{bmatrix} 2 & 1 & 3 & 5 \\ 4 & 2 & 1 & 3 \\ 8 & 4 & 7 & 13 \\ 8 & 4 & -3 & -1 \end{bmatrix}$

Ans. 2

10. $\begin{bmatrix} 4 & 5 & 6 \\ 5 & 6 & 7 \\ 7 & 8 & 9 \end{bmatrix}$

Ans. 2

11. $\begin{bmatrix} 2 & 1 & 3 \\ 4 & 7 & 13 \\ 4 & -3 & -1 \end{bmatrix}$

Ans. 2

12. $\begin{bmatrix} 1 & 2 & 1 & 0 \\ -2 & 4 & 3 & 0 \\ 1 & 0 & 2 & -8 \end{bmatrix}$

Ans. 3

13. $\begin{bmatrix} 3 & 1 & -5 & -1 \\ 1 & -2 & 1 & -5 \\ 1 & 5 & -7 & 2 \end{bmatrix}$

Ans. 3

14. $\begin{bmatrix} -4 & -3 & -3 \\ 1 & 0 & 1 \\ 4 & 4 & 3 \end{bmatrix}$ **Ans.** $\begin{bmatrix} -4 & -3 & -3 \\ 1 & 0 & 1 \\ 4 & 4 & 3 \end{bmatrix}$

15. $\begin{bmatrix} 2 & 3 & -1 \\ 0 & 1 & -1 \\ 2 & 1 & 2 \end{bmatrix}$ **Ans.** $\begin{bmatrix} 3 & -7 & -2 \\ 2 & 6 & 2 \\ -2 & 4 & 2 \end{bmatrix}$

16. $\begin{bmatrix} 2 & -1 & 3 \\ 0 & 2 & 0 \\ 2 & 1 & 1 \end{bmatrix}$ **Ans.** 2 $\begin{bmatrix} 1 & 2 & -3 \\ 0 & -2 & 0 \\ -2 & -2 & 2 \end{bmatrix}$

17. $\begin{bmatrix} 0 & 1 & 1 \\ 1 & 2 & 0 \\ 3 & -1 & 4 \end{bmatrix}$ **Ans.** $\begin{bmatrix} 8 & -5 & 2 \\ -4 & -3 & 1 \\ -7 & 3 & -1 \end{bmatrix}$

18. $\begin{bmatrix} 1 & 2 & 3 \\ 2 & 3 & 4 \\ 3 & 4 & 5 \end{bmatrix}$ **Ans.** $\begin{bmatrix} -1 & 2 & -1 \\ 2 & -4 & 2 \\ -1 & 2 & -1 \end{bmatrix}$

19. $\begin{bmatrix} 0 & 1 & 1 \\ 1 & 0 & 1 \\ 1 & 1 & 1 \end{bmatrix}$ **Ans.** $\begin{bmatrix} -1 & 0 & 1 \\ 0 & -1 & 1 \\ 1 & 1 & -1 \end{bmatrix}$

20. $\begin{bmatrix} 1 & 0 & 0 \\ 1 & 1 & 0 \\ 1 & 0 & 1 \end{bmatrix}$ **Ans.** $\begin{bmatrix} 1 & 0 & 0 \\ -1 & 1 & 0 \\ -1 & 0 & 1 \end{bmatrix}$

21. $\begin{bmatrix} 1 & 2 & 5 \\ 2 & 3 & 1 \\ -1 & 1 & 1 \end{bmatrix}$ **Ans.** $(1/21)\begin{bmatrix} 2 & 3 & -13 \\ -3 & 6 & 9 \\ 5 & -3 & -1 \end{bmatrix}$

22. $\begin{bmatrix} 1 & 2 & 3 \\ 3 & 4 & 5 \\ 6 & 7 & 8 \end{bmatrix}$ **Ans.** Not possible as $|A| = 0$.

23. $\begin{bmatrix} 1 & 3 & -2 \\ -3 & 0 & -6 \\ 2 & 5 & 0 \end{bmatrix}$ **Ans.** $(1/24)\begin{bmatrix} 30 & -10 & -18 \\ -12 & 4 & 12 \\ -15 & 6 & 4 \end{bmatrix}$

24. $\begin{bmatrix} 1 & 1 & 1 \\ 1 & 2 & 1 \\ 1 & 1 & 2 \end{bmatrix}$ **Ans.** $\begin{bmatrix} 3 & -1 & -1 \\ -1 & 1 & 0 \\ -1 & 0 & 1 \end{bmatrix}$

25. *Show that the inverse of*

$$\begin{bmatrix} 4 & 3 & 3 \\ -1 & 0 & -1 \\ -4 & -4 & -3 \end{bmatrix} \text{ is } \begin{bmatrix} 4 & 3 & 3 \\ 1 & 0 & 1 \\ 4 & 4 & 3 \end{bmatrix}$$

5

SOLUTION OF LINEAR EQUATIONS BY MATRIX METHOD

5.1 MATRIX OF COEFFICIENTS OF A SYSTEM OF EQUATIONS

Definition: Let the system of m simultaneous equations in n unknowns $x_1, x_2, x_3, \ldots, x_n$ be

$$a_{11}x_1 + x_{12}x_2 + a_{13}x_3 + \ldots + a_{1n}x_n = k_1$$
$$a_{21}x_1 + x_{22}x_2 + a_{23}x_3 + \ldots + a_{2n}x_n = k_2$$
$$a_{31}x_1 + x_{32}x_2 + a_{33}x_3 + \ldots + a_{3n}x_n = k_3$$

....

$$a_{m1}x_1 + x_{m2}x_2 + a_{m3}x_3 + \ldots + a_{mn}x_n = k_m$$

or written in a compact form

$$\sum_{j=1}^{n} a_{ij}\, x_j = k_j, \text{ where } i = 1, 2, \ldots m \qquad \ldots(i)$$

Then the matrix $A = [a_{ij}] = \begin{bmatrix} a_{11} & a_{12} \ldots a_{1n} \\ a_{22} & a_{23} \ldots a_{2n} \\ \ldots & \ldots \ldots \ldots \\ a_{m1} & a_{m2} \ldots a_{mn} \end{bmatrix}$

of order m × n is known as the *matrix of coefficient of the system of equations* given by (i).

The determinant of the matrix A, [if there be n equations in (i)] viz.

$$|A| = \begin{bmatrix} a_{11} & a_{12} \cdots a_{1n} \\ a_{21} & a_{22} \cdots a_{2n} \\ \cdots & \cdots \cdots \cdots \\ a_{n1} & a_{n2} \cdots a_{nn} \end{bmatrix}$$

is called the *determinant of coefficients of the system of equations* given by (i).

Note: If all the k's are zero, then the system of equations given by (i) is said to be *homogeneous* and if at least one of k's is not zero, then the above system of equations is said to be *non-homogeneous.*

5.2 SYSTEM OF EQUATIONS IN THE MATRIX FORM

The system of equations can be written in the matrix form as

$$\begin{pmatrix} a_{11} & a_{12} \cdots a_{1j} \cdots a_{1n} \\ a_{21} & a_{22} \cdots a_{2j} \cdots a_{2n} \\ \cdots\cdots & \cdots\cdots\cdots\cdots \\ a_{i1} & a_{i2} \cdots a_{ij} \cdots a_{in} \\ \cdots\cdots & \cdots\cdots\cdots\cdots \\ a_{m1} & a_{m2} \cdots a_{mj} \cdots a_{mn} \end{pmatrix} \begin{pmatrix} x_1 \\ x_2 \\ \cdots \\ \cdots \\ \cdots \\ x_n \end{pmatrix} = \begin{pmatrix} k_1 \\ k_2 \\ \cdots \\ \cdots \\ \cdots \\ k_m \end{pmatrix} \quad \textbf{(Note)} \quad \ldots(i)$$

or in a more compact form it may be written as

$$AX = K,$$

where $A = [a_{ij}]$ *i.e.,* the matrix of coefficients of the system of equations given by (i);

X = the transposed matrix of $[x_1\ x_2\ x_3 \ldots x_n]$ and

K = the transposed matrix of $[k_1\ k_2\ k_3 \ldots k_m]$.

Hence, students should note that the product AX is a matrix of order $m \times 1$, as A is a matrix of order $m \times n$ and X is a matrix of order $n \times 1$. And K is also a matrix of order $m \times 1$.

5.3 CONSISTENT AND INCONSISTENT EQUATIONS

Consider the system of equations.

If the above system has a solution (*i.e.,* a set of values of x_1, x_2, x_3 . . ., x_n, satisfy simultaneously these m equation), then the equations are said to be *consistent* otherwise the equations are said to be *inconsistent.*

A consistent system of equations has either one solution or infinitely many solutions.

5.4 SOLUTION OF NON-HOMOGENOUS SIMULTANEOUS EQUATIONS

Solution of , when $m = n$ and the matrix A is non-singular.

We know that the matrix form of the given equations is

$$AX = K \qquad \text{...(i)}$$

Also we know that if A is non-singular, its inverse matrix *i.e.,* A^{-1} exists such that

$$A^{-1}A = I, \qquad \text{...(ii)}$$

where I is the identity matrix.

Hence by multiplying both sides of (i) by A^{-1}, we have

$$A^{-1} AX = A^{-1} K$$

$\Rightarrow$ $IX = A^{-1} K$, from (ii)

$\Rightarrow$ $X = A^{-1} K$, which is the required solution of the given equations and is unique.

Example 1:

Solve the equations:

$$x + y + z = 9;\ 2x + 5y + 72 = 52;\ 2x + y - z = 0.$$

Solution:

Let $A = \begin{bmatrix} 1 & 1 & 1 \\ 2 & 5 & 7 \\ 2 & 1 & -1 \end{bmatrix}$; $K = \begin{bmatrix} 9 \\ 52 \\ 0 \end{bmatrix}$ and assume that there

exists a matrix $X = \begin{bmatrix} x \\ y \\ z \end{bmatrix}$ such that $AX = K$.

Then $\begin{bmatrix} 1 & 1 & 1 \\ 2 & 5 & 7 \\ 2 & 1 & -1 \end{bmatrix}\begin{bmatrix} x \\ y \\ z \end{bmatrix} = \begin{bmatrix} 9 \\ 32 \\ 0 \end{bmatrix}$

$\Rightarrow$ $\begin{bmatrix} 1 & 1 & 1 \\ 0 & 3 & 5 \\ 0 & -1 & -3 \end{bmatrix}\begin{bmatrix} x \\ y \\ z \end{bmatrix} = \begin{bmatrix} 9 \\ 34 \\ -18 \end{bmatrix}$, by the elementary row operation $R_2 \to R_2 - 2R_1$ and $R_3 \to R_3 - 2R_1$

$$\Rightarrow \begin{bmatrix} 1 & 1 & 1 \\ 0 & 0 & -4 \\ 0 & -1 & -3 \end{bmatrix} \begin{bmatrix} x \\ y \\ z \end{bmatrix} = \begin{bmatrix} -9 \\ -20 \\ -18 \end{bmatrix},$$ by the elementary row operation $R_2 \rightarrow R_2 + 3R_3$

$\Rightarrow$ $x + y + z = 9;\ 4x = -20,\ -y - 3z = -18$

$\Rightarrow$ $z = 5;\ y = 18 - 3z = 18 - 3\,(5) = 3$

and $x = 9 - y - z = 9 - 3 - 5 = 1$

$\Rightarrow$ $x = 1,\ y = 3,\ z = 5$ **Ans.**

Example 2:

Solve the equations:

$x_1 + 7x_2 + x_3 = 4,\ x_1 - x_2 + x_3 = 5,\ 2x_1 + 3x_2 + x_3 = 1$

Solution:

Let $A = \begin{bmatrix} 1 & 2 & 1 \\ 1 & -1 & 1 \\ 2 & 3 & -1 \end{bmatrix}$; $K = \begin{bmatrix} 4 \\ 5 \\ 1 \end{bmatrix}$ and assume that there exists a matrix $X = \begin{bmatrix} x_1 \\ x_2 \\ x_3 \end{bmatrix}$ such that AX = K.

Then $\begin{bmatrix} 1 & 2 & 1 \\ 1 & -1 & 1 \\ 2 & 3 & -1 \end{bmatrix} \begin{bmatrix} x_1 \\ x_2 \\ x_3 \end{bmatrix} = \begin{bmatrix} -4 \\ 5 \\ -1 \end{bmatrix}$

$$\Rightarrow \begin{bmatrix} 0 & 3 & 0 \\ 1 & -1 & 1 \\ 0 & 5 & -3 \end{bmatrix} \begin{bmatrix} x_1 \\ x_2 \\ x_3 \end{bmatrix} = \begin{bmatrix} -1 \\ 5 \\ -9 \end{bmatrix},$$ by the elementary row operation $R_1 \rightarrow R_1 - R_2$, $R_3 \rightarrow R_3 - 2R_1$

$\Rightarrow$ $3x_2 = -1,\ x_1 - x_2 + x_3 = 5,\ 5x_2 - 3x_3 = -9$

$\Rightarrow$ $x_2 = -1/3,\ x_1 + (1/3) + x_3 = 5,\ 5\,(-1/3) - 3x_3 = -9$

$\Rightarrow$ $x_2 = -1/3,\ x_1 + x_3 = 14/3,\ x_3 = 22/9$

$\Rightarrow$ $x_2 = -1/3,\ x_1 + (22/9) = 14/9,\ x_3 = 22/9$

$\Rightarrow$ $x_1 = 20/9;\ x_2 = -1/9,\ x_3 = 22/9$ **Ans**

Example 3:

Find the matrix X from the equations AX = B, where

$$A = \begin{bmatrix} 1 & -1 & 0 \\ 0 & 1 & -1 \\ 1 & 1 & 1 \end{bmatrix} \text{ and } B = \begin{bmatrix} 2 \\ 1 \\ 7 \end{bmatrix}$$

Solution:

Let $X = \begin{bmatrix} x \\ y \\ z \end{bmatrix}$, then from AX = B, we have,

$$\begin{bmatrix} 1 & -1 & 0 \\ 0 & 1 & -1 \\ 1 & 1 & 1 \end{bmatrix}\begin{bmatrix} x \\ y \\ z \end{bmatrix} = \begin{bmatrix} 2 \\ 1 \\ 7 \end{bmatrix}$$

$\Rightarrow \begin{bmatrix} 1 & -1 & 0 \\ 0 & 1 & -1 \\ 1 & 2 & 0 \end{bmatrix}\begin{bmatrix} x \\ y \\ z \end{bmatrix} = \begin{bmatrix} 2 \\ 1 \\ 8 \end{bmatrix}$, by the elementary row operation $R_3 \to R_3 + R_2$

$\Rightarrow \begin{bmatrix} 1 & -1 & 0 \\ 0 & 1 & -1 \\ 3 & 0 & 0 \end{bmatrix}\begin{bmatrix} x \\ y \\ z \end{bmatrix} = \begin{bmatrix} 2 \\ 1 \\ 12 \end{bmatrix}$, by the elementary row operation $R_3 \to R_3 + 2R_1$

$\Rightarrow \quad x - y = 2,\ y - z = 1,\ 3x = 12$ **(Note)**

$\Rightarrow \quad y = x - 2,\ z = y - 1,\ x = 4$

$\Rightarrow \quad y = 4 - 2 = 2,\ z = 2 - 1 = 1,\ x = 4$

$\Rightarrow \quad x = 4,\ y = 2,\ z = 1.$

$\therefore \quad X = \begin{bmatrix} x \\ y \\ z \end{bmatrix} = \begin{bmatrix} 4 \\ 2 \\ 1 \end{bmatrix}$ **Ans.**

Example 4:

Solve by matrix method

$$x - 2y + 3z = 2,\ 2x - 3z = 0,\ x + y + z = 0$$

Solution:

The given equations are

$$x - 2y + 3z = 2$$
$$2x + 0y - 3z = 0$$
$$x + y + z = 0.$$

Let $A = \begin{bmatrix} 1 & -2 & 3 \\ 2 & 0 & -3 \\ 1 & 1 & 1 \end{bmatrix}$ and $K = \begin{bmatrix} 2 \\ 0 \\ 0 \end{bmatrix}$ and assume that there exists a matrix $X = \begin{bmatrix} x \\ y \\ z \end{bmatrix}$ such that $AX = K$.

Then $\begin{bmatrix} 1 & -2 & 3 \\ 2 & 0 & -3 \\ 1 & 1 & 1 \end{bmatrix} \begin{bmatrix} x \\ y \\ z \end{bmatrix} = \begin{bmatrix} 2 \\ 0 \\ 0 \end{bmatrix}$

$\Rightarrow \begin{bmatrix} 3 & -2 & 0 \\ 2 & 0 & -3 \\ 2 & 2 & 2 \end{bmatrix} \begin{bmatrix} x \\ y \\ z \end{bmatrix} = \begin{bmatrix} 2 \\ 0 \\ 0 \end{bmatrix}$, by the elementary row operations $R_1 \to R_1 + R_2$ and $R_3 \to 2R_3$

$\Rightarrow \begin{bmatrix} 3 & -2 & 0 \\ 2 & 0 & -3 \\ 5 & 0 & 2 \end{bmatrix} \begin{bmatrix} x \\ y \\ z \end{bmatrix} = \begin{bmatrix} 2 \\ 0 \\ 2 \end{bmatrix}$, by the elementary row operation $R_3 \to R_3 + R_1$

$\Rightarrow \begin{bmatrix} 3 & -2 & 0 \\ 2/3 & 0 & -1 \\ 5/2 & 0 & 1 \end{bmatrix} \begin{bmatrix} x \\ y \\ z \end{bmatrix} = \begin{bmatrix} 2 \\ 0 \\ 1 \end{bmatrix}$, by the elementary row operations $R_2 \to \frac{1}{3} R_2$ and $R_3 \to \frac{1}{2} R_3$

$\Rightarrow \begin{bmatrix} 3 & -2 & 0 \\ 2/3 & 0 & -1 \\ 19/6 & 0 & 2 \end{bmatrix} \begin{bmatrix} x \\ y \\ z \end{bmatrix} = \begin{bmatrix} 2 \\ 0 \\ 1 \end{bmatrix}$, by the elementary row operation $R_3 \to R_3 + R_2$

$\Rightarrow \quad 3x - 2y = 2, \ \frac{2}{3}x - z = 0, \ (19/6)\, x = 1$ **(Note)**

$\Rightarrow \quad x = (6/19), \ 2y = 3x - 2, \ z = \frac{2}{3} x$

$\Rightarrow \quad x = (6/19), \ 2y = (18/19) - 2 = -(20/19), \ z = \frac{2}{3} \times (6/19) = (4/19)$

$\Rightarrow \quad x = (6/19), \ y = -(10/19), \ z = (4/19).$ **Ans.**

Example 5:

Using matrix method, solve the following equations:

$$2x - y + 3z = 9,\ x + y + z = 6 \text{ and } x - y + z = 2.$$

Solution:

The given equations are

$$2x - y + 3z = 9$$
$$x + y + z = 6$$
$$x - y + z = 2$$

Let $A = \begin{bmatrix} 2 & -1 & 3 \\ 1 & 1 & 1 \\ 1 & -1 & 1 \end{bmatrix}$; $K = \begin{bmatrix} 9 \\ 6 \\ 2 \end{bmatrix}$ and assume that there exists a matrix $X = \begin{bmatrix} x \\ y \\ z \end{bmatrix}$ such that $AX = K$.

Then $\begin{bmatrix} 2 & -1 & 3 \\ 1 & 1 & 1 \\ 1 & -1 & 1 \end{bmatrix}\begin{bmatrix} x \\ y \\ z \end{bmatrix} = \begin{bmatrix} 9 \\ 6 \\ 2 \end{bmatrix}$

$\Rightarrow \quad \begin{bmatrix} 3 & 0 & 4 \\ 1 & 1 & 1 \\ 2 & 0 & 2 \end{bmatrix}\begin{bmatrix} x \\ y \\ z \end{bmatrix} = \begin{bmatrix} 15 \\ 6 \\ 8 \end{bmatrix}$, by the elementary row operations $R_1 \to R_1 + R_2$, $R_3 \to R_3 + R_2$

$\Rightarrow \quad 3x + 4z = 15,\ x + y + z = 6,\ 2x + 2z = 8$

$\Rightarrow \quad 3x + 4z = 15,\ x + y + z = 6,\ x + z = 4$

$\Rightarrow \quad y = 6 - (x + z) = 6 - 4, \quad \because \quad x + z = 4$

$\Rightarrow \quad y = 2$

Also $\quad 3x + 4z = 15$ gives

$3x + 4(4 - x) = 15, \quad \because \quad x + z = 4$

$\Rightarrow \quad 3x + 16 - 4x = 15 \quad \Rightarrow \quad x = 1$

$\therefore \quad z = 4 - x = 4 - 1 = 3$

$\Rightarrow \quad x = 1,\ y = 2,\ z = 3.$ **Ans.**

5.5 TO COMPUTE THE INVERSE OF A SQUARE MATRIX WITH THE HELP OF THE LINEAR EQUATIONS

Let a system of n linear equations in n unknowns $x_1, x_2, x_3, \ldots, x_n$ be

$$a_{11}x_1 + x_{12}x_2 + \ldots + a_{1n}x_n = k_1$$
$$a_{21}x_1 + x_{22}x_2 + \ldots + a_{2n}x_n = k_2$$
$$a_{31}x_1 + x_{32}x_2 + \ldots + a_{3n}x_n = k_3$$

....

....

$$a_{n1}x_1 + x_{n2}x_2 + \ldots + a_{nn}x_n = k_n$$

Then this system can be written in the matrix form as

$$AX = K, \qquad \ldots(i)$$

where $A = \begin{bmatrix} a_{11} & a_{12} & \cdots & a_{1n} \\ a_{21} & a_{22} & \cdots & a_{2n} \\ a_{31} & a_{32} & \cdots & a_{3n} \\ \cdots & \cdots & \cdots & \cdots \\ a_{n1} & a_{n2} & \cdots & a_{nn} \end{bmatrix}$;

$X = \begin{bmatrix} x_1 \\ x_2 \\ \cdots \\ x_n \end{bmatrix}$ and $K = \begin{bmatrix} k_1 \\ k_2 \\ \cdots \\ k_n \end{bmatrix}$

If $|A| \neq 0$, then the matrix A is non-singular and the inverse of A exists. Hence pre multiplying (i) by A^{-1}, we have

$$A^{-1}AX = A^{-1}K$$

$\Rightarrow$ $(A^{-1}A)X = A^{-1}K$

$\Rightarrow$ $IX = A^{-1}K$ $\qquad \because A^{-1}A = I$

$\Rightarrow$ $X = A^{-1}K$, which gives the value of A^{-1}.

Example 1:

Find the inverse of the matrix $A = \begin{bmatrix} 1 & 2 & 3 \\ 0 & 5 & 0 \\ 2 & 4 & 3 \end{bmatrix}$

Solution:

The matrix equation $AX = K$ is here equivalent to the equations

$$x_1 + 2x_2 + 3x_2 = k_1 \qquad \ldots(i)$$

$$0.x_1 + 5x_2 + 0.x_3 = k_2 \qquad ...(ii)$$

$$2x_1 + 4x_2 + 3x_3 = k_3 \qquad ...(iii)$$

From (ii) we get $x_2 = \frac{1}{5}k_2 = 0.k_1 + \frac{1}{5}k_2 + 0.k_3$ ...(iv)

Subtracting (i) from (iii), we get

$$x_1 + 2x_2 = k_3 - k_1$$

$\Rightarrow \quad x_1 = k_3 - k_1 - 2x_2 = k_3 - k_1 - 2\ \frac{1}{5}k_2$, from (iv)

$\Rightarrow \quad x_1 = -k_1 - (2/5)k_2 + k_3$ ...(v)

Also from (i), $2x_3 = k_1 - x_1 - 2x_2$

$= k_1 + k_1 + \frac{2}{5}k_2 - k_3 - \frac{2}{5}k_2$, from (iv) and (v)

$\Rightarrow \quad x_3 = \frac{2}{3}k_1 + 0.k_2 - \frac{1}{3}k_3$ ...(vi)

From (iv), (v) and (vi) we have

$$\begin{bmatrix} x_1 \\ x_2 \\ x_3 \end{bmatrix} = \begin{bmatrix} -1 & -2/3 & 1 \\ 0 & 1/5 & 0 \\ 2/3 & 0 & -1/3 \end{bmatrix} \begin{bmatrix} k_1 \\ k_2 \\ k_3 \end{bmatrix}$$

i.e., $\quad X = A^{-1} K$

$$\therefore\ A^{-1} = \begin{bmatrix} -1 & -2/5 & 1 \\ 0 & 1/5 & 0 \\ 2/3 & 0 & -1/3 \end{bmatrix}$$

Ans.

Example 2:

Find the inverse of the matrix $A = \begin{bmatrix} 2 & 5 & 3 \\ 3 & 1 & 2 \\ 1 & 2 & -1 \end{bmatrix}$

And apply the results to solve the equations:

$2x + 5y + 3z = 9,\ 3x + y + 2z = 3,\ x + 2y - z = 6.$

Solution:

The matrix equation AX = K is here equivalent so the equations

$$2x + 5y + 3z = k_1 \quad ...(i)$$
$$3x + y + 2z = k_2 \quad ...(ii)$$
and $$x + 2y - z = k_3 \quad ...(iii)$$

Multiplying (iii) by 3 and adding to (i), we get

$$5x + 11y = k_1 + 3k_3 \quad ...(iv)$$

Multiplying (iii) by 2 and adding to (ii), we get

$$5x + 5y = k_2 + 2k_3 \quad ...(v)$$

Subtracting (v) from (iv) we get $6y = k_1 - k_2 + k_3$

$$\Rightarrow \quad y = \frac{1}{6}(k_1 - k_2 + k_3) \quad ...(vi)$$

$\therefore$ From (v) we get $5x = k_2 + 2k_3 - 5y$

$$\Rightarrow \quad x = \frac{1}{5}[k_2 + 2k_3 - (5/6)(k_1 - k_2 + k_3)], \text{ from (vi)}$$

$$\Rightarrow \quad x = (1/30)[-5k_1 + 11k_2 + 7k_3] \quad ...(vii)$$

Also from (iii) we get $z = x + 2y - k_3$

$$\Rightarrow \quad z = (1/30)[-5k_1 + 11k_2 + 7k_3] + \frac{1}{3}[k_1 - k_2 - k_3] - k_3$$

$$= (1/30)[5k_1 + k_2 - 13k_3] \quad ...(viii)$$

$\therefore$ From (vi), (vii) and (viii), we get

$$\begin{bmatrix} x \\ y \\ z \end{bmatrix} = \begin{bmatrix} -(5/30) & (11/30) & (7/30) \\ 1/6 & -1/6 & 1/6 \\ (5/30) & (1/30) & -(13/30) \end{bmatrix} \begin{bmatrix} k_1 \\ k_2 \\ k_3 \end{bmatrix}$$

$$\therefore A^{-1} = \begin{bmatrix} -(5/30) & (11/30) & (7/30) \\ 1/6 & -1/6 & 1/6 \\ (5/30) & (1/30) & -(13/30) \end{bmatrix}, = 1/6 \begin{bmatrix} -1 & (11/5) & (7/5) \\ 1 & -1 & 1 \\ 1 & 1/5 & -(13/5) \end{bmatrix}$$

Also the given equations can be written in the matrix form as where

$$\therefore \quad A^{-1} = \begin{bmatrix} -2 & 5 & (7/30) \\ 3 & 1 & 1/6 \\ 1 & 2 & -(13/30) \end{bmatrix}, \ X = \begin{bmatrix} x \\ y \\ z \end{bmatrix}, K = \begin{bmatrix} 9 \\ 3 \\ 6 \end{bmatrix}$$

Now from (x) we also have $X = A^{-1} K$

i.e. $$\begin{bmatrix} x \\ y \\ z \end{bmatrix} = \frac{1}{6}\begin{bmatrix} -1 & (11/5) & (7/5) \\ 1 & -1 & 1 \\ 1 & (1/5) & -(13/5) \end{bmatrix}\begin{bmatrix} 9 \\ 3 \\ 6 \end{bmatrix}, \text{ from (ix)}$$

$$\Rightarrow \quad x = \frac{1}{8}[-1(9) + (11/5)(3) + (7/5)(0)] = \frac{1}{6}(6) = 1;$$

$$y = \frac{1}{6}[1(9) - 1(3) + 1(6)] = \frac{1}{6}(12) = 2;$$

$$z = \frac{1}{6}[1(9) + \frac{1}{5}(3) - (13/5)(6)] = \frac{1}{6}(-6) = -1$$

$\Rightarrow \quad x = 1, y = 2, z = -1.$ **Ans.**

5.6 AUGMENTED MATRIX

Definition:

The matrix $A = \begin{bmatrix} a_{11} & a_{12} \dots a_{1n} \\ a_{21} & a_{22} \dots a_{2n} \\ \dots & \dots \quad \dots \quad \dots \\ a_{m1} & a_{m2} \dots a_{mn} \end{bmatrix}$

augmented by the matrix $K = \begin{bmatrix} k_1 \\ k_2 \\ \dots \\ k_m \end{bmatrix}$ is called the augmented matrix

of A and is written as A* = or [A, K] $\begin{bmatrix} a_{11} & a_{12} \dots a_{1n} & k_1 \\ a_{21} & a_{22} \dots a_{2n} & k_2 \\ \dots & \dots \quad \dots \quad \dots & \dots \\ a_{m1} & a_{m2} \dots a_{mn} & k_m \end{bmatrix}$

Also it is evident that the order of the matrix A* or [A, K] is $m \times (n + 1)$.

5.7 FUNDAMENTAL THEOREM

A system of m linear equations in n unknowns given by AX = K is consistent (i.e. has a solution) if and only if the matrix of coefficients A and the augmented matrix A of the system have the same rank.*

[If the above common rank is r then r of the unknowns can be expressed as linear combinations of the remaining n − r unknowns. When these n − r unknowns are assigned arbitrary values, the system has an infinite

number of solutions out of which (n – r + 1) are linearly independent whereas the rest are linear combinations of them.]

Proof:

Consider m non-homogeneous linear equations in n unknowns given by AX = K, where

$$A = \begin{bmatrix} a_{11} & a_{12} \ldots & \ldots a_{1n} \\ a_{21} & a_{22} \ldots & \ldots a_{2n} \\ \ldots & \ldots \ldots & \ldots \ldots \\ a_{m1} & a_{m2} \ldots & \ldots a_{mn} \end{bmatrix}, X = \begin{bmatrix} x_1 \\ x_2 \\ \ldots \\ x_n \end{bmatrix} \text{ and } K = \begin{bmatrix} k_1 \\ k_2 \\ \ldots \\ k_n \end{bmatrix}$$

Let r be the rank the matrix A and $C_1, C_2, \ldots, C_n$ be the column vectors of the matrix A, then $A = [C_1, C_2, \ldots, C_n]$ and so AX = K reduces to $[C_1, C_2, \ldots, C_n] \begin{bmatrix} x_1 \\ x_2 \\ \ldots \\ x_n \end{bmatrix} = k$

$$\Rightarrow \qquad C_1 x_1 + C_2 x_2 \ldots + C_n x_n = K \qquad \ldots\text{(i)}$$

Necessarrry Condition: Let the given system of equations posses a solution (*i.e.*, be consistent), then there must exist n scalars $b_1, b_2, \ldots, b_n$ which satisfy (i) *i.e.*, $C_1b_1 + C_2 b_2 + \ldots + C_n b_n = K$ ...(ii)

Since rank of A is r, so each n – r columns viz. $C_{r+1}, C_{r+2}, \ldots, C_n$ is a linear combination of $C_1, C_2, \ldots, C_n$.

∴ From (ii) we find that K is a linear combination of $C_1, C_2 \ldots, C_r$, since $C_{r+1}, C_{r+2}, \ldots, C_n$ in (ii) can be expressed in terms of $C_1, C_2, \ldots, C_n$.

∴ The maximum number of linearly independent columns of the augmented matrix [A, K] or A* is also r. Hence the rank of A* is r.

Thus A and A* are of the same rank r.

Sufficient Condition: Let the matrices A and A* be of the same rank rr. Then the number of linearly independent columns of the matrix A* is r. But the column vectors $C_1, C_2, \ldots, C_r$ of the matrix A* already form a linearly independent set and thus the matrix K can be expressed as a linear combination of the columns $C_1, C_2, \ldots, C_r$.

∴ There exist r scalarrs $b_1, b_2, \ldots, b_r$ such that

$$b_1C_1 + b_2C_2 + \ldots + b_rC_r + 0C_{r+1} + \ldots + 0C_n = K \qquad \ldots\text{(iii)}$$

From (i) and (iii) on comparing, we get

$x_1 = b_1, x_2 = b_2, ..., x_r = b_r, x_{r+1} = 0, ..., x_n = 0$ and these are the solutions of the system of equations given by AX = K.

Theorem:

If A be an n × n matrix, X and K be n × 1 matrices, then the system of equations AX = K possess a unique soloution if matrix A is non-singular.

Proof:

Let $A = [a_{ij}]$ and $|A| \neq 0$.

Then rank of A and augmented matrix [A, K] or A* are both n. We conclude that the system of equations AX = K is consistent.

From AX = K, we have

$A^{-1}(AX) = A^{-1}K$, premultiplying both sides by A^{-1}

$\Rightarrow \quad (A^{-1}A)X = A^{-1}K \Rightarrow IX = A^{-1}K \qquad \because A^{-1}A = I$

$\Rightarrow \quad X = A^{-1}K$ is the solution of the given system of equations.

Now let X_1 and X_2 be two sets of solutions of AX = K,

then $\quad AX_1 = K, AX_2 = K$

$\Rightarrow \quad AX_1 = AX_2$, as each is equal to K

$\Rightarrow \quad A^{-1}(AX_1) = A^{-1}(AX_2)$ premultiplying both sides by A^{-1}

$\Rightarrow \quad (A^{-1}A)X_1 = (A^{-1}A)X_2$

$\Rightarrow \quad IX_1 = IX_2, \qquad \because A^{-1}A = I$

$\Rightarrow \quad X_1 = X_2$

$\Rightarrow$ the solution is unique.

5.8 REDUCED ECHELON FORM OF A MATRIX

Definition: *If in an Echelon form matrix the first non-zero element in the ith row lies in jth column, and all other elements in the jth column are zero, then the matrix is said to be in reduced Echelon form.*

For example: $\begin{bmatrix} 1 & 0 & 2 & 3 \\ 0 & 1 & 3 & 2 \\ 0 & 0 & 0 & 0 \end{bmatrix}$ In the first non-zero element in the second row lies in the second column and all other elements in the second column are zero.

Example 1:

Show that the equations

$$x + 2y - z = 3,\ 3x - y + 2z = 1,\ 2x - 2y + 3z = 2$$

are consistent and solve them by the use of matrices.

Solution:

The given equations in the matrix form AX = K can be written as

$$\begin{bmatrix} 1 & 2 & -1 \\ 3 & -1 & 2 \\ 2 & -2 & 3 \end{bmatrix}\begin{bmatrix} x \\ y \\ z \end{bmatrix} = \begin{bmatrix} 3 \\ 1 \\ 2 \end{bmatrix}$$

∴ The augmented matrix $A^* = \begin{bmatrix} 1 & 2 & -1 & 3 \\ 3 & -1 & 2 & 1 \\ 2 & -2 & 3 & 2 \end{bmatrix}$.

⇒ $A^* \sim \begin{bmatrix} 1 & 2 & -1 & 3 \\ 0 & -7 & 5 & -8 \\ 0 & -6 & 5 & -4 \end{bmatrix}$, replacing R_2 and R_3 by $R_2 - 3R_1$ and $R_3 - 2R_1$ respectively.

$\sim \begin{bmatrix} 1 & 2 & -1 & 3 \\ 0 & -7 & 5 & -8 \\ 0 & 1 & 0 & 4 \end{bmatrix}$, replacing R_3 by $R_3 - R_2$

$\sim \begin{bmatrix} 1 & 0 & -1 & -5 \\ 0 & 0 & 5 & 20 \\ 0 & 1 & 0 & 4 \end{bmatrix}$, replacing R_1 R_1 by $R_1 - 2R_3$, $R_2 + 7R_3$ respectively ...(1)

This is a matrix in the reduced Echelon form having three non-zero orws hence its rank is 3.

Simultaneously we get reduced Echelon form of A viz.

$\begin{bmatrix} 1 & 0 & -1 \\ 0 & 0 & 5 \\ 0 & 1 & 0 \end{bmatrix}$ having three non-zero rows and hence the rank of A is also 3.

Thus, we observe that A and A* have the same rank and as such the given equations are consistent.

Example 2:

Are the following equations consistent?

$$x + y + 2z + w = 5$$
$$2x + 3y - z - 2w = 2$$
$$4x + 5y + 3z = 7$$

Solution:

The given equations in the matrix form AX = K can be written as

$$\begin{bmatrix} 1 & 1 & 2 & 1 \\ 2 & 3 & -1 & -2 \\ 4 & 5 & 3 & 0 \end{bmatrix} \begin{bmatrix} x \\ y \\ z \\ w \end{bmatrix} = \begin{bmatrix} 5 \\ 2 \\ 7 \end{bmatrix}$$ **(Note)**

$\therefore$ The augmented matrix $A^* = \begin{bmatrix} 1 & 1 & 2 & 1 & 5 \\ 2 & 3 & -1 & -2 & 2 \\ 4 & 5 & 3 & 0 & 7 \end{bmatrix}$

$\Rightarrow$ $A^* \sim \begin{bmatrix} 1 & 1 & 2 & 1 & 5 \\ 0 & 1 & -5 & -4 & -8 \\ 0 & 1 & -5 & -4 & -13 \end{bmatrix}$, replacing R_2 and R_3 by $R_2 - 2R_1$ and $R_3 - 4R_1$ respectively.

$\sim \begin{bmatrix} 1 & 0 & 7 & 5 & 13 \\ 0 & 1 & -5 & -4 & -8 \\ 0 & 0 & 0 & 0 & -5 \end{bmatrix}$, replacing R_1, R_3 by $R_1 - R_2$, $R_3 - R_2$ respectively

This is a matrix in the reduced Echelon form having three non-zero rows, hence its rank is 3.

Simultaneously we get the reduced Echelon form of A viz.

$\begin{bmatrix} 1 & 0 & 7 & 5 \\ 0 & 1 & -5 & -4 \\ 0 & 0 & 0 & 0 \end{bmatrix}$ which has two non-zero rows hence its rank is 2.

Thus we find that the ranks of A and A* are not the same, hence the given equations are not consistent *i.e.,* they cannot have any solutions.

Example 3:

Show that the following equations are consistent and find their solutions by matrix method:

$x_1 + x_2 + x_3 = 2,\ 4x_1 - x_2 + 2x_3 = -6,\ 3x_1 + x_2 + x_3 = -18.$

Solution:

In the matrix form AX = K, the given equations can be written as

$$\begin{bmatrix} 1 & 1 & 1 \\ 4 & -1 & 2 \\ 3 & 1 & 1 \end{bmatrix} \begin{bmatrix} x_1 \\ x_2 \\ x_3 \end{bmatrix} = \begin{bmatrix} 2 \\ -6 \\ -18 \end{bmatrix}.$$

∴ The augmented matrix $A^* = \begin{bmatrix} 1 & 1 & 1 & 2 \\ 4 & -1 & 2 & -6 \\ 3 & 1 & 1 & -18 \end{bmatrix}$

⇒ $A^* \sim \begin{bmatrix} 1 & 1 & 1 & 2 \\ 5 & 0 & 3 & -4 \\ 2 & 0 & 0 & -20 \end{bmatrix}$, replacing R_2, R_3 by $R_2 + R_1$ and $R_3 - R_1$ respectively

$\sim \begin{bmatrix} 1 & 1 & 1 & 2 \\ 5 & 0 & 3 & -4 \\ -1 & 0 & 0 & -10 \end{bmatrix}$, replacing R_3 by $-\frac{1}{2}R_3$

$\sim \begin{bmatrix} 0 & 1 & 1 & 12 \\ 0 & 0 & 3 & 46 \\ -1 & 0 & 0 & 10 \end{bmatrix}$, replacing R_1, R_2 by $R_1 + R_3$ and $R_2 + 5R_2$ respectively

⇒ $A^* \sim \begin{bmatrix} -1 & 0 & 0 & 10 \\ 0 & 1 & 1 & 12 \\ 0 & 0 & 1 & (46/3) \end{bmatrix}$, int erchanging rows and replacing R_3 by $\frac{1}{2}R_3$

This is a matrix in the reduced Echelon form having three non-zero rows, hence its rank is 3.

Simultaneously, we get the reduced Echelon form of viz.

$\begin{bmatrix} -1 & 0 & 0 \\ 0 & 1 & 1 \\ 0 & 0 & 1 \end{bmatrix}$, which has also three non-zero rows and hence its ranks is also 3.

Thus, we find the ranks of A and A* are the same and as such the given equations are consistent, and the matrix form of the given equations reduce to

$$\begin{bmatrix} -1 & 0 & 0 \\ 0 & 1 & 1 \\ 0 & 0 & 1 \end{bmatrix} \begin{bmatrix} x_1 \\ x_2 \\ x_3 \end{bmatrix} = \begin{bmatrix} 10 \\ 12 \\ (46/3) \end{bmatrix}$$

which gives $-x_1 = 10;$

$x_2 + x_3 = 12,\ x_3 = (46/3)$

$\Rightarrow \quad x_1 = -10;$

$x_2 = -(10/3),$

$x_3 = (46/3).$ **Ans.**

Example 4:

For what values of λ, *the equations* $x + y + z = 1$, $x + 2y + 4z = \lambda$, $x + 4y + 10z = \lambda^2$ *have a solution and solve completely in each case.*

Solution:

The given equations in the matrix form AX = K can be written as

$$\begin{bmatrix} 1 & 1 & 1 \\ 1 & 2 & 4 \\ 1 & 4 & 10 \end{bmatrix} \begin{bmatrix} x \\ y \\ z \end{bmatrix} = \begin{bmatrix} 1 \\ \lambda \\ \lambda^2 \end{bmatrix}$$

$\therefore$ The augmented matrix $A^* = \begin{bmatrix} 1 & 1 & 1 & 1 \\ 1 & 2 & 4 & \lambda \\ 1 & 4 & 10 & \lambda^2 \end{bmatrix}$

$\Rightarrow \quad A^* \sim \begin{bmatrix} 1 & 1 & 1 & 1 \\ 0 & 1 & 3 & \lambda - 1 \\ 0 & 3 & 9 & \lambda^2 - 1 \end{bmatrix}$, replacing R_2 and R_3 by $R_2 - R_1$ and $R_3 - R_1$ respectively

$\sim \begin{bmatrix} 1 & 0 & -2 & 2 - \lambda \\ 0 & 3 & 9 & 3\lambda - 3 \\ 0 & 3 & 9 & \lambda^2 - 1 \end{bmatrix}$, replacing R_1 by $R_1 - R_2$ and then R_2 by $3R_2$

$\Rightarrow \quad A^* \sim \begin{bmatrix} 1 & 0 & -2 & 2 - \lambda \\ 0 & 1 & 3 & \lambda - 1 \\ 0 & 0 & 0 & \lambda^2 - 3\lambda + 2 \end{bmatrix}$, replacing R_3 by $R_3 - R_2$ and R_2 by $\frac{1}{3}R_2$ ...(1)

Simultaneously we get the reduced Echelon form of A viz.

$\begin{bmatrix} 1 & 0 & -2 \\ 0 & 1 & 3 \\ 0 & 0 & 0 \end{bmatrix}$ which has two non-zero rows hence its rank is 2. ...(2)

From (1), (2) we conclude that if the given equations have solution then the ranks of A and A* must be the same viz. 2 and from (1) if the rank of A* is 2, then it must have two non-zero rows *i.e.,* $\lambda^2 - 3\lambda + 2 = 0 \Rightarrow \lambda = 1, 2$. **Ans.**

The matrix from of the given equations reduces to

$$\begin{bmatrix} 1 & 0 & -2 \\ 0 & 1 & 3 \\ 0 & 0 & 0 \end{bmatrix} \begin{bmatrix} x \\ y \\ z \end{bmatrix} = \begin{bmatrix} 2-\lambda \\ \lambda-1 \\ \lambda^2 - 3\lambda + 2 \end{bmatrix}$$

which is equivalent to

$$\left.\begin{array}{l} 1.x + 0.y - 2.z = 2-\lambda,\ 0.x + 1.y + 3.z = \lambda - 1 \\ 0.x + 0.y + 0.z = \lambda^2 - 3\lambda + 2 \end{array}\right\}$$

and

If $\lambda = 1$, then these are $x - 2z = 1$, $y + 3z = 0$.

As the rank of A and A* is 2, so two of the unknowns viz. x and y are expressed as a linear function of the remaining unknown z viz. $x = 2z + 1$, $y = -3z$.

By assigning arbitrary values to z, an infinite number of corresponding values of x and y can be obtained. Hence the system of equations has infinite number of solutions.

Assigning to arbitrary values 0, 1 to z, we have two sets of solutions of the given equations as

x	1	3
y	0	–3
z	0	1

Let any other solution of the given equations be $x = -1$, $y = 3$, $z = -1$ corresponding to the value -1 of z.

If this third solution is a linear combination of the first two solutions then a, b can be found as follows:

$$\left.\begin{array}{l} 1.a + 3.b = -1 \\ 0.a - 3.b = 3 \\ 0.a + 1.b = -1 \end{array}\right\}$$

$\Rightarrow a + 3b = -1,\ 3b = -3 \ \Rightarrow\ b = -1$ ...(4)

i.e., $b = -1, a = 2$. These values of a and b satisfy all the three equations given by (4). Hence the third solution is a linear solution of the first two solutions.

We can similarly solve for $\lambda = 2$ also.

Example 5:

Examine if the system of equations $x + y + 4z = 6$, $3x + 2y - 2z = 9$, $5x + y + 2z = 13$ is consistent? Find also the solution if it is consistent.

Solution:

The given equations are

$$x + y + 4z = 6$$
$$3x + 2y - 2z = 9$$
$$5x + y + 2z = 13$$

In the matrix form AX = K, these can be written as

$$\begin{bmatrix} 1 & 1 & 4 \\ 3 & 2 & -2 \\ 5 & 1 & 2 \end{bmatrix} \begin{bmatrix} x \\ y \\ z \end{bmatrix} = \begin{bmatrix} 6 \\ 9 \\ 13 \end{bmatrix}$$

$\therefore$ Then aughmented matrix $A^* = \begin{bmatrix} 1 & 1 & 4 & 6 \\ 3 & 2 & -2 & 9 \\ 5 & 1 & 2 & 13 \end{bmatrix}$

$\Rightarrow \quad A^* \sim \begin{bmatrix} 1 & 0 & 0 & 2 \\ 3 & -1 & -14 & -3 \\ 5 & -4 & -18 & -7 \end{bmatrix}$, replacing C_2, C_3 and C_4 by $C_2 - C_1, C_4 - 4C_1$ and $C_4 - 4C_1$ respectively.

$\Rightarrow \quad A^* \sim \begin{bmatrix} 1 & 0 & 0 & 2 \\ 0 & -1 & -14 & -9 \\ 0 & -4 & -18 & -17 \end{bmatrix}$, replacing R_2, R_3 by $R_2 - 3R_1$ and $R_3 - 5R_1$ respectively.

$\sim \begin{bmatrix} 1 & 0 & 0 & 2 \\ 0 & 1 & 14 & 9 \\ 0 & 4 & 18 & 17 \end{bmatrix}$, replacing R_2 by $-R_2$ and R_3 by $-R_3$

$$\sim \begin{bmatrix} 1 & 0 & 0 & 2 \\ 0 & 1 & 14 & 9 \\ 0 & 1 & -38 & -19 \end{bmatrix}, \text{ replacing } R_3 \text{ by } R_3 - 4R_2$$

$$\sim \begin{bmatrix} 1 & 0 & 0 & 2 \\ 0 & 1 & 14 & 9 \\ 0 & 0 & 1 & \frac{1}{2} \end{bmatrix}, \text{ replacing } R_3 \text{ by } -(1/38)\, R_3$$

This is a matrix having three non-zero rows and in the reduced Echelon form, hence its rank is 3.

Simultaneously we get the reduced Echelon form of A viz.

$\begin{bmatrix} 1 & 0 & 0 \\ 0 & 1 & 14 \\ 0 & 0 & 1 \end{bmatrix}$, which also has three non-zero rows and hence its rank is also 3.

Thus we find that the ranks of A and A* are the same and as such the given equations are consistent.

Now the matrix form of the given equations reduces to

$$\begin{bmatrix} 1 & 0 & 0 \\ 0 & 1 & 14 \\ 0 & 0 & 1 \end{bmatrix} \begin{bmatrix} x \\ y \\ z \end{bmatrix} = \begin{bmatrix} 2 \\ 9 \\ \frac{1}{2} \end{bmatrix}$$

which is equivalent to

$$1.x + 0.\, y + 0z = 2;\ 0.x + 1.y + 14.z = 9;\ 0\, x + 0.y + 1.z = \frac{1}{2}$$

$$\Rightarrow \quad x = 2;\ y + 14z = 9;\ z = \frac{1}{2}$$

$$\Rightarrow \quad x = 2;\ y = 9 - 14z;\ z = \frac{1}{2}$$

$$\Rightarrow \quad x = 2,\ y = 9 - 7 = 2,\ z = \frac{1}{2}.$$ **Ans.**

Example 6:

Solve the equations with the help of matrices considering specially the case when $\lambda = 2$:

$$\lambda x + 2y - 2z = 1;\ 4x + 2\lambda y - z = 2,\ 6x + 6y + \lambda z = 3.$$

Solution:

The given equation in the matrix form AX = K can be written as

$$\begin{bmatrix} \lambda & 2 & -2 \\ 4 & 2\lambda & -1 \\ 6 & 6 & \lambda \end{bmatrix}\begin{bmatrix} x \\ y \\ z \end{bmatrix} = \begin{bmatrix} 1 \\ 2 \\ 3 \end{bmatrix}$$

$\therefore$ The augmented matrix $A^* = \begin{bmatrix} \lambda & 2 & -2 & 1 \\ 4 & 2\lambda & -1 & 2 \\ 6 & 6 & \lambda & 3 \end{bmatrix}$

$\Rightarrow$ $A^* \sim \begin{bmatrix} 6\lambda & 12 & -12 & 6 \\ 12\lambda & 6\lambda^2 & -3\lambda & 6\lambda \\ 6\lambda & 6\lambda & \lambda^2 & 3\lambda \end{bmatrix}$, replacing R_1, R_2, R_3 by $5R_1, \lambda R_2, R_3$ respectively.

$\sim \begin{bmatrix} 6\lambda & 12 & -12 & 6 \\ 0 & 6\lambda^2 - 24 & 24 - 3\lambda & 6\lambda - 12 \\ 0 & 6\lambda - 12 & \lambda^2 + 12 & 3\lambda - 6 \end{bmatrix}$, replacing R_2, R_3, by $R_2 - 2R_1, R_3 - R_1$ respectively.

This is a matrix having three non-zero row[illegible] form. Hence the rank of A* is 3.

Simultaneously we get the reduced

$\begin{bmatrix} 6\lambda & 12 & -12 \\ 0 & 6\lambda^2 - 24 & 24 - 3\lambda \\ 0 & 6\lambda - 12 & \lambda^2 + 12 \end{bmatrix}$ which als[illegible]s and so its rank is also 3.

Thus, the ranks of A* and A are the same and as such the given equations have solutions.

The matrix form of the given equations then reduce to

$$\begin{bmatrix} 6\lambda & 12 & -12 \\ 0 & 6\lambda^2 - 24 & 24 - 3\lambda \\ 0 & 6\lambda - 12 & \lambda^2 + 12 \end{bmatrix}\begin{bmatrix} x \\ y \\ z \end{bmatrix} = \begin{bmatrix} 6 \\ 6\lambda - 12 \\ 3\lambda - 6 \end{bmatrix}$$

This gives $6\lambda x + 12y - 12z = 6;$

$(6\lambda^2 = 24)\ y + (24 - 3\lambda)\ z = 6\lambda - 12$

and $(6\lambda - 12)\, y + (\lambda^2 + 12) = 3\lambda - 6$

If $\lambda = 2$, then these equations reduce to

$$12x + 12y - 12z = 6;\ 18z = 0$$

which give $z = 0$ and $2x + 2y = 1$

i.e., $x = -y + \frac{1}{2},\ z = 0.y$ **(Note)**

By assigning arbitrary values to y, an infinite number of corresponding values of x and z can be obtained. Hence the given system of equations has an infinite number of solutions. (This can be ascertained from the ranks of A* and A also in the case when $\lambda = 2$)

Assigning two arbitrary values 0, 1 to y we have two sets of solutions of the given equations as

x	$\frac{1}{2}$	$-\frac{1}{2}$
y	0	1
z	0	0

Let any other solution of the given equation be $x = \frac{3}{2},\ y = -1,\ z = 0$ corresponding to the value -1 of y.

If this third solution is a linear combination of the first two solutions then a and b can be found as follows:

$$\frac{1}{2}a - \frac{1}{2}b = (3/2);$$

$$0.a + 1.b = -1,\ 0.a + 0.b = 0.$$

Solving the first two of these we get a = 1, b = – 1 2 which satisfy the third also.

Hence, this solution is a linear combination of the first two solutions. In this way we can get two linearly independent solutions of the given set of equations for $\lambda = 2$.

Example 7:

Examine, if the system of following equations is consistent. If consistent find the solution:

$$x + y + z = 6,\ 2x + 3y - 2z = 2,\ 5x + y + 2z = 13.$$

Solution:

The given equations in the matrix form AX = K can be written as

$$\begin{bmatrix} 1 & 1 & 1 \\ 2 & 3 & -2 \\ 5 & 1 & 2 \end{bmatrix} \begin{bmatrix} x \\ y \\ z \end{bmatrix} = \begin{bmatrix} 6 \\ 2 \\ 13 \end{bmatrix}$$

Then augmented matrix $A^* = \begin{bmatrix} 1 & 1 & 1 & 6 \\ 2 & 3 & -2 & 2 \\ 5 & 1 & 2 & 13 \end{bmatrix}$

$\Rightarrow \quad A^* \sim \begin{bmatrix} 1 & 1 & 1 & 6 \\ 0 & 1 & -4 & -10 \\ 4 & 0 & 1 & 7 \end{bmatrix}$ replacing R_2, R_3 by $R_2 - R_1$, $R_3 - R_1$ respectively

$\sim \begin{bmatrix} 1 & 0 & 5 & 16 \\ 0 & 1 & -4 & -10 \\ 4 & 0 & 5 & 17 \end{bmatrix}$, replacing R_1 by $R_1 - R_2$

$\sim \begin{bmatrix} 1 & 0 & 5 & 16 \\ 0 & 1 & -4 & -10 \\ 0 & 0 & -19 & -57 \end{bmatrix}$, replacing R_3 by $R_3 - 4R_1$

$\sim \begin{bmatrix} 1 & 0 & 5 & 16 \\ 0 & 1 & -4 & -10 \\ 0 & 0 & 1 & 3 \end{bmatrix}$, replacing R_3 by $-(1/19)\,R_3$

$\sim \begin{bmatrix} 1 & 0 & 0 & 1 \\ 0 & 1 & 0 & 2 \\ 0 & 0 & 1 & 3 \end{bmatrix}$, replacing R_1, R_2 by $R_1 - 5R_3$, $R_2 + 4R_3$ respectively

This is a matrix in the reduced Echelon form having three non-zero rows and hence the rank of A* is 3.

Simultaneously, we get the reduced Echelon form of A viz.

$\begin{bmatrix} 1 & 0 & 0 \\ 0 & 1 & 0 \\ 0 & 0 & 1 \end{bmatrix}$, which is equal to $\mathbf{I}_3$ and as such the rank of **A** is also 3.

Thus we observe that the rank of A and A* are the same and as such the given equations are consistent *i.e.,* have solutions which can be obtained as follows:

The matrix equation is

$$\begin{bmatrix}1 & 0 & 0\\0 & 1 & 0\\0 & 0 & 1\end{bmatrix}\begin{bmatrix}x\\y\\z\end{bmatrix}=\begin{bmatrix}1\\2\\3\end{bmatrix}$$

$\Rightarrow$ x = 1, y = 2, z = 3. **Ans.**

Example 8:

Solve the system of equations:

$$x + 2y - 3z - 4w = 6$$

$$x + 3y + z - 2w = 4$$

$$2x + 5y - 2z - 5w = 10.$$

Solution:

The given equations in the matrix form AX = K is

$$\begin{bmatrix}1 & 2 & -3 & -4\\1 & 3 & 1 & -2\\2 & 5 & -2 & -5\end{bmatrix}\begin{bmatrix}x\\y\\z\\w\end{bmatrix}=\begin{bmatrix}6\\4\\10\end{bmatrix}$$

The augmented matrix $A^* = \begin{bmatrix}1 & 2 & -3 & -4 & 6\\1 & 3 & 1 & -2 & 4\\2 & 5 & -2 & -5 & 10\end{bmatrix}$

$\Rightarrow$ $A^* \sim \begin{bmatrix}1 & 2 & -3 & -4 & 6\\0 & 1 & 4 & 2 & -2\\0 & 1 & 4 & 3 & -2\end{bmatrix}$, replacing R_2 and R_3 by $R_2 - R_1$ and $R_3 - 2R_1$ respectively

$\sim \begin{bmatrix}1 & 0 & -11 & -8 & 10\\0 & 1 & 4 & 2 & -2\\0 & 0 & 0 & 1 & -0\end{bmatrix}$, replacing R_1 R_3 by $R_1 - 2R_2$, $R_3 - R_2$ respectively

This is a matrix in the reduced Echelon form having three non-zero rows and hence the rank of A* is 3.

Simultaneously we get the reduced Echelon form of A viz.

$$\begin{bmatrix} 1 & 0 & -11 & -8 \\ 0 & 1 & 4 & 2 \\ 0 & 0 & 0 & 1 \end{bmatrix}$$ having three non – zero rows and hence the rank of A is also 3.

Thus we observe that A and A* have the same rank and as such the given equations have solutions which can be obtained as follows:

The matrix equation is

$$\begin{bmatrix} 1 & 0 & -11 & -8 \\ 0 & 1 & 4 & 2 \\ 0 & 0 & 0 & 1 \end{bmatrix} \begin{bmatrix} x \\ y \\ z \\ w \end{bmatrix} = \begin{bmatrix} 10 \\ -2 \\ 0 \end{bmatrix}$$

$\Rightarrow \quad x + 0y - 11z - 8w = 10;$

$y + 4y + 2w = -2$ and $w = 0$ **(Note)**

$\Rightarrow \quad y = -4z - 2,\ w = 0z\ ;\ x = 11z + 10$ **(Note)**

Thus, we find that as the rank of A and A* is 3, so three of the unknowns viz. x, y and w are expressed as a linear function of the remaining 4 – 3 *i.e.,* one unknown viz.

By assigning arbitrary values to z, an infinite number of corresponding values of x, y and w can be obtained. Hence the system of equations has infinite number of solutions.

Now, we can show that the system has only n – r + 1 *i.e.,* 4 – 3 + 1 *i.e.,* 2 linearly independent solutions.

Assigning two arbitrary values 0, 1 to z, we have two sets of solutions of the given equations as

x	10	21
y	–2	–6
z	0	1
w	0	0

Let any other solution of the given equations be

$$x = -1,\ y = 2,\ z = -1,\ w = 0,$$

corresponding to the value –1 of z.

If this solution is linear combination of the first two solutions then a, b can be found as follows:

$$\left.\begin{array}{r} 10a + 21b = -1 \\ -2a - 6b = 2 \\ 0.a + 1.b = -1 \\ 0.a + 0.b = 0 \end{array}\right\} \quad \textbf{(Note)}$$

$\Rightarrow$ $$10a + 21b = -1 \quad ...(1)$$

$$-2a + 6b = 2 \quad ...(2)$$

and $$b = -1 \quad ...(3)$$

Solving (1) and (3) we get a = 2, b = – 1.

These values of a and b satisfy (2) also. Hence the third solution is a linear combination of the first two solutions.

Example 9:

Can the following equations have solutions?

$$x + 2y + 3z = 4$$

$$2x + 3y + 8z = 7$$

$$x - y + 9z = 1$$

Solution:

The given equation in the matrix form AX = K can be written as

$$\begin{bmatrix} 1 & 2 & 3 \\ 2 & 3 & 8 \\ 1 & -1 & 9 \end{bmatrix} \begin{bmatrix} x \\ y \\ z \end{bmatrix} = \begin{bmatrix} 4 \\ 7 \\ 1 \end{bmatrix}$$

The augmented matrix $A^* = \begin{bmatrix} 1 & 2 & 3 & 4 \\ 2 & 3 & 8 & 7 \\ 1 & -1 & 9 & 1 \end{bmatrix}$

$\Rightarrow$ $A^* \sim \begin{bmatrix} 1 & 0 & 0 & 4 \\ 0 & -1 & 2 & -1 \\ 0 & -3 & 6 & -3 \end{bmatrix}$, replacing R_2, R_3 by $R_2 - 2R_1$, $R_3 - 3R_1$ respectively.

$\sim \begin{bmatrix} 1 & 0 & 0 & 4 \\ 0 & 1 & -2 & 1 \\ 0 & 0 & 0 & 0 \end{bmatrix}$, replacing R_2, R_3 by $-R_2$, $R_3 - 3R_2$ respectively.

This is a matrix in the reduced Echelon form having two non-zero rows and hence the rank of A* is 2.

Simultaneously we get the reduced Echelon form of A viz. $\begin{bmatrix} 1 & 0 & 0 \\ 0 & 1 & -2 \\ 0 & 0 & 0 \end{bmatrix}$,

which has two non-zero rows, hence its rank is 2.

Hence the ranks of A and A* are same and as such the given equations can have solutions.

Example 10:

Apply rank test to examine if the following system of equations is consistent, solve them $2x + 4y - z = 9,\ 3x - y + 5z = 5,\ 8x + 2y + 9z = 12.$

Solution:

The given equations in the matrix form AX = K can be written as

$$\begin{bmatrix} 2 & 4 & -1 \\ 3 & -1 & 5 \\ 8 & 2 & 9 \end{bmatrix} \begin{bmatrix} x \\ y \\ z \end{bmatrix} = \begin{bmatrix} 9 \\ 5 \\ 19 \end{bmatrix}$$

The augmented matrix $A^* = \begin{bmatrix} 2 & 4 & -1 & 9 \\ 3 & -1 & 5 & 5 \\ 8 & 2 & 9 & 19 \end{bmatrix}$

$$\Rightarrow \quad A^* \sim \begin{bmatrix} 2 & 4 & -1 & 9 \\ 3 & -1 & 5 & 5 \\ 2 & 4 & -1 & 9 \end{bmatrix}, \text{ replacing } R_3 \text{ by } R_3 - 2R_2$$

$$\sim \begin{bmatrix} 2 & 4 & -1 & 9 \\ 3 & -5 & 6 & -4 \\ 0 & 0 & 0 & 0 \end{bmatrix}, \text{ replacing } R_2\ R_3 \text{ by } R_2 - R_1 \text{ and } R_3 - R_1 \text{ respectively}$$

$$\Rightarrow \quad A^* \sim \begin{bmatrix} 2 & 4 & -1 & 9 \\ 0 & -7 & \frac{15}{2} & -\frac{17}{2} \\ 0 & 0 & 0 & 0 \end{bmatrix}, \text{ replacing } R_2 \text{ by } R_2 - \frac{1}{2}R_1$$

$$\sim \begin{bmatrix} 2 & 0 & \frac{19}{7} & \frac{29}{7} \\ 0 & 1 & \frac{13}{14} & -\frac{17}{14} \\ 0 & 0 & 0 & 0 \end{bmatrix}, \text{ replacing } R_2 \text{ by } -\frac{1}{7}R_2 \text{ and then } R_1 \text{ by } R_1 - 4R_2$$

$$\sim \begin{bmatrix} 1 & 0 & \frac{19}{14} & \frac{29}{14} \\ 0 & 1 & -\frac{13}{14} & \frac{17}{14} \\ 0 & 0 & 0 & 0 \end{bmatrix}, \text{ replacing } R_1 \text{ by } \frac{1}{2} R_1$$

This is a matrix in the reduced Echelon form having two non-zero rows and hence the rank of A* is 2.

Simultaneously, we get the reduced Echelon form of A viz.

$\begin{bmatrix} 1 & 0 & (19/14) \\ 0 & 1 & -(13/14) \\ 0 & 0 & 0 \end{bmatrix}$ which also has two non-zero rows and as such the rank of A is also 2.

Thus, we observe that the ranks of A and A* are the same and as such the given equations are consistent *i.e.,* have solutions which can be obtained as follows:

The matrix equation is

$$\begin{bmatrix} 1 & 0 & (19/14) \\ 0 & 1 & -(13/14) \\ 0 & 0 & 0 \end{bmatrix} \begin{bmatrix} x \\ y \\ z \end{bmatrix} = \begin{bmatrix} (29/14) \\ (17/14) \\ 0 \end{bmatrix}$$

$\Rightarrow \quad x + (19/14)\,z = (29/14) : y - (13/14)\,x = (17/14)$

$\Rightarrow \quad y = \frac{13}{14} z + \frac{17}{14};\ x = \frac{29}{14} - \frac{19}{14} z$

$\Rightarrow \quad x = -\frac{19}{14} z + \frac{29}{14};\ y = \frac{13}{14} z + \frac{17}{14}.$

Thus, we find that as the rank of A* and A is 2, so two of the unknowns viz. x and y are expressed as a linear function of the remaining 3 – 2 *i.e.,* one unknown viz. z.

By assigning arbitrary values to z, an infinite number of corresponding values of x and y can be obtained. hence the given system of equations has an infinite number of solutions.

Now, we can show that the system has only n – r + 1 *i.e.,* 3 – 2 + 1 *i.e.,* 2 linearly independent solutions

Assigning two arbitrary values 0, 1 to z, we have two sets of solutions of the given equations as

x	$\frac{29}{14}$	$\frac{10}{14}$
y	$\frac{17}{14}$	$\frac{30}{14}$
z	0	1

Let ay other solution of the given equation be x = (24/7), y (2/7), z = – 1, corresponding to the value –1 of z.

If this third solution is a linear combination of the first two solutions then a and b can be found as follows:

$$\left.\begin{aligned}(29/14)\,a + (10/14)\,b &= (24/7)\\ (17/14)\,a + (30/14)\,b &= (2/7)\\ 0.a + 1.b &= -1\end{aligned}\right\} \qquad \textbf{(Note)}$$

$$\Rightarrow \quad 29a + 10b = 48 \qquad ...(1)$$

$$17a + 30b = 4 \qquad ...(2)$$

$$b = -1 \qquad ...(3)$$

Solving (1) and (3) we get a = 2, b = –1, which satisfy (2) also. Hence the third solution is a linear combination of the first two solutions.

Example 11:

Solve the system of linear equations

2x – 3y + 4z = 3, x – 3z = – 2.

Solution:

The given equations in the matrix form AX = K can be written as

$$\begin{bmatrix} 2 & -3 & 4 \\ 1 & 0 & -3 \end{bmatrix} \begin{bmatrix} x \\ y \\ z \end{bmatrix} = \begin{bmatrix} 3 \\ -2 \end{bmatrix}$$

$\therefore$ The augmented matrix $A^* = \begin{bmatrix} 2 & -3 & 4 & 3 \\ 1 & 0 & -3 & -2 \end{bmatrix}$

$$\Rightarrow \quad A^* \sim \begin{bmatrix} 0 & -3 & 10 & 7 \\ 1 & 0 & -3 & -2 \end{bmatrix}, \text{ replacing } R_1 \text{ by } R_1 - 2R_2$$

$$\sim \begin{bmatrix} 1 & 0 & -3 & -2 \\ 0 & 1 & (10/3) & -(7/3) \end{bmatrix}, \text{ interchanging } R_1 \text{ and } R_2 \text{ and replacing } R_2 \text{ by } \frac{1}{3}R_2$$

This a matrix having two non-zero rows, hence its rank is 2.

Simultaneously we get $A \sim \begin{bmatrix} 1 & 0 & -3 \\ 0 & 1 & -(10/3) \end{bmatrix}$, which is also in the reduced Echelon form having two non-zero rows. Hence its rank is 2

Thus, we find that A and A* have the same rank 2 and thus two of the unknowns can be expressed as a linear function of the remaining unknown

Now the matrix form of the given equations reduce to

$$\begin{bmatrix} 1 & 0 & -3 \\ 0 & 1 & -(10/3) \end{bmatrix} \begin{bmatrix} x \\ y \\ z \end{bmatrix} = \begin{bmatrix} -2 \\ -(7/3) \end{bmatrix}$$

which is equivalent to $1.x + 0.y - 3z = -2;$

$$0.x + 1.y - \frac{10}{3} z = -\frac{7}{3}$$

i.e., $x - 3z = -2, -3y + 10z = 7$ *i.e.,* $x = 3z - 2,\ y - \frac{10}{3} z = -\frac{7}{3}$, for all z

i.e., $x = 3k - 2,\ y = (10/3)\,k - (7/3),\ z = k$, for all k. **Ans.**

Definition: *A linear equation of the type*

$$a_1x_1 + a_2x_2 + a_3x_3 + a_4x_4 + \ldots + a_nx_n = 0$$

is called a homogeneous linear equation.

Definition: *A system of homogeneous linear equations is given in the matrix form by **AX = O**, where **A** and **X** are the same notations*

Note 1: $x_1 = 0\ x_2 = \ldots = x_n$ is always a solution and is called the *trivial solution.*

An Important Theorem (Without Proof)

A homogeneous system of n linear equations in n unknowns, whose determinant of coefficients does not vanish, has only the trivial solution.

Note 2: The matrix of coefficients A and the augmented matrix A* being the same have equal ranks and thus the system is always consistent.

Note 3: Here **K** is zero matrix.

Theorem:

A system of m homogeneous equation in n unknowns $x_1, x_2, \ldots, x_n$ has a solution other then the trivial solution viz. $x_1 = 0 = x_2 = \ldots = x_n$ if and only if the rank r of the matrix of coefficients A is less then n, the number of unknowns.

If r = n, then n of the equations can be solved by Cramer's Rule for the unique solution $x_1 = 0 = x_2 = ... = x_n$ and the given system has no no-trivial solutions.

If r is the rank of A, the r of the n unknowns $x_1, x_2, ..., x_n$ can be expressed as linear combination of the remaining n – r unknowns to which arbitrary values may be assigned. Hence the system will have an infinite number of solutions out of which n – r are linearly independent and the remaining can be expressed as linear combination of these n – r.

The above theorem is illustrated in the following examples:

SOLVED EXAMPLES

Example 1:

Show by considering rank of an appropriate matrix, that the following system of equations, possesses no solution other than the trivial solutions $x = 0 = y = z$:

$$3x - y + z = 0,\ -15x + 6y - 5z = 0,\ 5x - 2y + 2z = 0.$$

Solution:

The given equations in the matrix form AX = O is given by

$$\begin{bmatrix} 3 & -1 & 1 \\ -15 & 6 & -5 \\ 5 & -2 & 2 \end{bmatrix} \begin{bmatrix} x \\ y \\ z \end{bmatrix} = \begin{bmatrix} 0 \\ 0 \\ 0 \end{bmatrix}$$

The matrix A of coefficients

$$= \begin{bmatrix} 3 & -1 & 1 \\ -15 & 6 & -5 \\ 5 & -2 & 2 \end{bmatrix}$$

$$\sim \begin{bmatrix} 0 & 0 & 1 \\ 0 & 1 & -5 \\ -1 & 0 & 2 \end{bmatrix},$$ replacing C_1, C_2 by $C_1 - 3C_3, C_2 + C_1$ respectively

$$\sim \begin{bmatrix} 0 & 0 & 1 \\ 0 & 1 & 0 \\ 1 & 0 & -2 \end{bmatrix},$$ replacing R_2, R_3 by $R_2 + 5R_1, -R_3$ respectively

$$\sim \begin{bmatrix} 0 & 0 & 1 \\ 0 & 1 & 0 \\ 1 & 0 & 0 \end{bmatrix}, \text{ replacing } R_3 \text{ by } R_3 + 2R_1$$

$$\Rightarrow \quad A \sim \begin{bmatrix} 1 & 0 & 0 \\ 0 & 1 & 0 \\ 0 & 0 & 1 \end{bmatrix}, \text{ interchanging } R_1 \text{ and } R_3$$

$$= I_3.$$

The rank of A is 3 and is equal to the number of unknowns viz x, y,z. Hence x = 0 = y = z and the given system has no non-trivial solutions.

Example 2:

Solve $\quad x + y + z + w = 0$

$$x + 3y + 2z + 4w = 0$$

$$2x + z - w = 0$$

Solution:

The given equation in the matrix form AX = O is given by

$$\begin{bmatrix} 1 & 1 & 1 & 1 \\ 1 & 3 & 2 & 4 \\ 2 & 0 & 1 & -1 \end{bmatrix} \begin{bmatrix} x \\ y \\ z \\ w \end{bmatrix} = \begin{bmatrix} 0 \\ 0 \\ 0 \end{bmatrix}$$

The matrix A of coefficients $= \begin{bmatrix} 1 & 1 & 1 & 1 \\ 1 & 3 & 2 & 4 \\ 2 & 0 & 1 & -1 \end{bmatrix}$

$$\Rightarrow \quad A \sim \begin{bmatrix} 1 & 1 & 1 & 1 \\ 0 & 2 & 1 & 3 \\ 2 & -2 & -1 & -3 \end{bmatrix}, \text{ replacing } R_2 \text{ and } R_3 \text{ by } R_2 - R_1 \text{ and } R_3\ 2R_1 \text{ respectively}$$

$$\sim \begin{bmatrix} 1 & 1 & 1 & 1 \\ 0 & 2 & 1 & 3 \\ 0 & 0 & 0 & 0 \end{bmatrix}, \text{ replacing } R_3 \text{ by } R_3 + R_2$$

$$\sim \begin{bmatrix} 1 & 1 & 1 & 1 \\ 0 & 1 & \frac{1}{2} & (3/2) \\ 0 & 0 & 0 & 0 \end{bmatrix}, \text{ replacing } C_2 \text{ by } C_2 - C_1$$

$$\sim \begin{bmatrix} 0 & 1 & 1 & 1 \\ 0 & 1 & \frac{1}{2} & (3/2) \\ 0 & 0 & 0 & 0 \end{bmatrix}, \text{ replacing } R_2 \text{ by } \frac{1}{2}R_2$$

This is a matrix in the reduced Echelon form having two non-zero rows, hence the rank of A is 2 and is less then the number of unknowns x, y, z and w *i.e.,* 4.

∴ Then matrix form of the given equations reduces to

$$\begin{bmatrix} 1 & 0 & 1 & 1 \\ 0 & 1 & \frac{1}{2} & (3/2) \\ 0 & 0 & 0 & 0 \end{bmatrix} \begin{bmatrix} x \\ y \\ z \\ w \end{bmatrix} = \begin{bmatrix} 0 \\ 0 \\ 0 \end{bmatrix}$$

⇒ $$1.x + 0.y + 1.\ x + 1.\ w = 0;$$

$$0.x + 1.\ y + \frac{1}{2}z + \frac{3}{2}w = 0$$

and $$0.x + 0.y + 0.z + 0.w = 0$$

From the first two equations we have

$$\left.\begin{aligned} y &= -z - w \\ y &= -\frac{1}{2}z - (3/2)w \end{aligned}\right\} \qquad \text{...(1)}$$

i.e., two of the unknowns viz. x and y have been expressed as linear combinations of the remaining two unknowns viz. z and w.

An infinite number of solutions of the given equations can be obtained by assigning arbitrary values to z and w.

We know that the system has n – r *i.e.,* 4 = 2 = 2 linearly independent solutions.

Take any two solutions of the system by assigning the following arbitrary values to z and w

$$z = 2, 4$$

$$w = 0, 2.$$

Then the solutions are given in the tabular form as

(Note: The corresponding values of x and y are calculated from the equations (1) above)

x	–2	–6
y	–1	–5
z	2	4
w	0	2

Let any other solution be

$$x = -2,\ y = -3,\ z = 0,\ w = 2, \qquad \text{...(2)}$$

If this solution is a linear combination of the first two solutions (given in the above table) then we can always find two constants λ and μ such that

$$\left.\begin{aligned} -2\lambda - 6\mu &= -2 \text{ ...(3)} \\ -\lambda - 5\mu &= -3 \text{ ... (4)} \\ 2\lambda + 4\mu &= 0 \text{ ... (5)} \\ 0.\lambda + 2\mu &= 2 \text{ ... (6)} \end{aligned}\right\} \qquad \textbf{(Note)}$$

From (6) we get $\mu = 1$

From (5) we get $\lambda = -2$.

These values of ł and m satisfy (3) and (4) also.

Hence the third solution [given by (2) above] is a linear combination of the first two solutions (given in the tabular form above).

Example 3:

Solve completely the system of equations:

$$x - 2y + z - w = 0$$

$$x + y - 2z + 3w = 0$$

$$4x + y - 5z + 8w = 0$$

$$5x - 7y + 2z - w = 0$$

Solution:

The given equation in the matrix form Ax = O is given by

$$\begin{bmatrix} 1 & -2 & 1 & -1 \\ 1 & 1 & -2 & 3 \\ 4 & 1 & -5 & 8 \\ 5 & -7 & 2 & -1 \end{bmatrix} \begin{bmatrix} x \\ y \\ z \\ w \end{bmatrix} = \begin{bmatrix} 0 \\ 0 \\ 0 \\ 0 \end{bmatrix}$$

The matrix A of coefficients

$$= \begin{bmatrix} 1 & -2 & 1 & -1 \\ 1 & 1 & -2 & 3 \\ 4 & 1 & -5 & 8 \\ 5 & -7 & 2 & -1 \end{bmatrix}$$

$$\sim \begin{bmatrix} 1 & -2 & 1 & -1 \\ 0 & 3 & -3 & 4 \\ 0 & 9 & -9 & 12 \\ 0 & 3 & -3 & 4 \end{bmatrix},$$ replacing R_2, R_3 and R_4 by $R_2 - R_1$, $R_3 - 4R_1$ and $R_4 - 5R_1$ respectively.

$$\Rightarrow A \sim \begin{bmatrix} 1 & 0 & -1 & \frac{5}{3} \\ 0 & 1 & -1 & \frac{4}{3} \\ 0 & 0 & 0 & 0 \\ 0 & 0 & 0 & 0 \end{bmatrix},$$ replacing R_3 and R_4 and $R_3 - 3R_2$ and $R_2 - R_2$ respectively, then R_2 by $\frac{1}{2} R_2$ and finally R_1 by $R_1 + R_2$

This is a matrix in the reduced Echelon form having two non-zero rows, hence the rank of A is 2 and is less than the number of unknowns viz. 4.

$\therefore$ The matrix form of the given equations reduces to

$$\begin{bmatrix} 1 & 0 & -1 & (5/3) \\ 0 & 1 & -1 & (4/3) \\ 0 & 0 & 0 & 0 \\ 0 & 0 & 0 & 0 \end{bmatrix} \begin{bmatrix} x \\ y \\ z \\ w \end{bmatrix} = \begin{bmatrix} 0 \\ 0 \\ 0 \\ 0 \end{bmatrix}$$

which is equivalent to

$$x - z + (5/3)w = 0, \; y - z + (4/3)w = 0$$

i.e., $\quad y = z - (4/3)w, \; x = z - (5/3)w$

$$\therefore X = \begin{bmatrix} x \\ y \\ z \\ w \end{bmatrix} = \begin{bmatrix} z-(5/3) \\ z-(4/3) \\ z \\ w \end{bmatrix}$$

By multiplication we find that

$$AX = \begin{bmatrix} 1 & 0 & -1 & (5/3) \\ 0 & 1 & -1 & (4/3) \\ 0 & 0 & 0 & 0 \\ 0 & 0 & 0 & 0 \end{bmatrix} \begin{bmatrix} z-(5/3)w \\ z-(4/3) \\ z \\ w \end{bmatrix}$$

$$= \begin{bmatrix} z-(5/3)w-z+(5/3) \\ 0+z-(4/3)w-z+(4/3) \\ 0+0+0+0 \\ 0+0+0+0 \end{bmatrix} = \begin{bmatrix} 0 \\ 0 \\ 0 \\ 0 \end{bmatrix} = O,$$

whatever the values of z and w may be.

$\therefore$ We have $x = \lambda - (5/3)\mu$, $y = \lambda - (4/3)\mu$, $z = \lambda$, $w = \mu$, where λ and μ can take any values, as the complete solution of the given system of equations. **Ans.**

Example 4:

Investigate for what value of λ, μ the simultaneous equations:

$$x + 2y + z = 8;\ 2x + y + 3z = 13,\ 3x + 4y - \lambda z = \mu$$

have (i) no solution, (ii) a unique solution and (iii) infinitely many solutions.

Solution:

The given equations in the matrix form AX = K can be written as

$$\begin{bmatrix} 1 & 2 & 1 \\ 2 & 1 & 3 \\ 3 & 4 & -\lambda \end{bmatrix} \begin{bmatrix} x \\ y \\ z \end{bmatrix} = \begin{bmatrix} 8 \\ 13 \\ \mu \end{bmatrix}$$

$\therefore$ The augmented matrix $A^* = \begin{bmatrix} 1 & 2 & 1 & 8 \\ 2 & 1 & 3 & 13 \\ 3 & 4 & -\lambda & \mu \end{bmatrix}$

$\Rightarrow \quad A^* \sim \begin{bmatrix} 1 & 2 & 1 & 8 \\ 0 & -3 & 1 & -3 \\ 0 & -2 & -\lambda-3 & \mu-24 \end{bmatrix}$, replacing R_2, R_3 by $R_2 - 2R_1$, $R_3 - 3R_1$ respectively.

$\sim \begin{bmatrix} 1 & 5 & 0 & 11 \\ 0 & -6 & 2 & -6 \\ 0 & -6 & -3\lambda-9 & 3\mu-72 \end{bmatrix}$, replacing R_1, R_2, R_3 by $R_1 - R_2$, $2R_2$ and $3R_3$ respectively.

$\sim \begin{bmatrix} 1 & 5 & 0 & 11 \\ 0 & 1 & -1/3 & 1 \\ 0 & 0 & -3\lambda-11 & 3\mu-65 \end{bmatrix}$, replacing R_3 and R_2 by $R_3 - R_2$ and $-\frac{1}{6}R_2$ respectively ...(i)

Case I: If $3\lambda + 11 \neq 0$, $3\mu - 66 \neq 0$ *i.e.*, $\lambda \neq -\frac{11}{3}$ $m \neq 22$, then from (i) matrix A* has three non-zero rows and is in the reduced Echelon form. Thus the matrix A* is of rank 3. Also then the matrix A = $\begin{bmatrix} 1 & 5 & 0 \\ 0 & 1 & -1/3 \\ 0 & 0 & -3\lambda - 1 \end{bmatrix}$

is also in reduced Echelon form having three non-zero rows and thus its rank is also 3. Also there are three unknown quantities x, y, z. Hence in the case $\lambda \neq -\frac{11}{3}$, $\mu \neq 22$, the given equations have a *unique solution.*

Case II: If $3\lambda + 11 = 0$, $3\mu - 66 \neq 0$ *i.e.*, $\lambda = \frac{11}{3}$, $\mu \neq 22$, then the rank of the matrix A* is 3 but that of A is 2, since in this case both A* and A are in the reduced Echelon form but A* has three non-zero rows, whereas A has two non-zero rows. Thus the ranks of A* and A are not the same and so there is **no solution** of the given equations.

Case III: If $3\lambda + 11 = 0$, $3\mu - 66 = 0$ *i.e.*, $\lambda = -\frac{11}{3}$, $m = 22$, then the ranks of A as well as A* are the same and each is 2 *i.e.*, less than the number of unknowns viz. x, y and z. Hence in this case two unknowns will be expressed in terms of the third and thus we shall have an *infinite number of solutions.*

Example 5:

Prove that if the system of equations: $x = ay + z$, $y = z + ax$, $z = x + y$ is consistent (having non-zero solution) then $a + 1 = 0$.

Solution:

The given equations can be rewritten as

$$x - ay - z = 0$$
$$ax - y + z = 0$$
$$x + y - z = 0$$

These equations in the matrix from AX – K can be rewritten as

$$\begin{bmatrix} 1 & -a & -1 \\ a & -1 & 1 \\ 1 & 1 & -1 \end{bmatrix} \begin{bmatrix} x \\ y \\ z \end{bmatrix} = \begin{bmatrix} 0 \\ 0 \\ 0 \end{bmatrix}$$

If the given system of equations has a non-zero solution then the matrix A have a rank < the number of unknown quantities x, y, z *i.e.*, 3 and $|A| \neq 0$. Here we have

$$A = \begin{bmatrix} 1 & -a & -1 \\ a & -1 & 1 \\ 1 & 1 & -1 \end{bmatrix}$$

$$\Rightarrow \quad A \sim \begin{bmatrix} 0 & -a-1 & -1 \\ a+1 & 0 & 1 \\ 0 & 0 & -1 \end{bmatrix}, \text{ replacing } C_1, C_2 \text{ by } C_1 + C_3, C_2 + C_3 \text{ respectively}$$

$$\sim \begin{bmatrix} 0 & -(a+1) & 0 \\ a+1 & 0 & 0 \\ 0 & 0 & -1 \end{bmatrix}, \text{ replacing } R_1, R_2 \text{ by } R_1 - R_3, R_2 + R_3 \text{ respectively} \quad \text{...(i)}$$

$$\sim \begin{bmatrix} 0 & 1 & 0 \\ 1 & 0 & 0 \\ 0 & 0 & 1 \end{bmatrix}, \text{ replacing } C_1, C_2, C_3, \text{ by } C_1/(a+1), -C_2/(a+1), -C_3 \text{ respecitvely}$$

$$\sim \begin{bmatrix} 1 & 0 & 0 \\ 0 & 1 & 0 \\ 0 & 0 & 1 \end{bmatrix}, \text{ interchanging } R_1 \text{ and } R_2$$

$$= I_2$$

i.e., the rank of A is 3 *i.e.*, equal to the number of unknowns viz. x, y, z.

But if $a + 1 = 0$, then from (i) we get

$$A \sim \begin{bmatrix} 0 & 0 & 0 \\ 0 & 0 & 0 \\ 0 & 0 & -1 \end{bmatrix},$$ which has one non-zero row and so the rank of A is 1 *i.e.*, < 3, the number of unknonws x, y, z.

$$\text{Also } |A| = \begin{vmatrix} 0 & -a & -1 \\ a & -1 & 1 \\ 1 & 1 & -1 \end{vmatrix}$$

$$= \begin{vmatrix} 0 & -a-a & -1 \\ a+1 & 0 & 1 \\ 0 & 0 & -1 \end{vmatrix}, \text{ adding } C_3 \text{ to } C_1 \text{ and } C_2$$

$$= \begin{vmatrix} 0 & 0 & -1 \\ 0 & 0 & 1 \\ 0 & 0 & -1 \end{vmatrix}, \text{ if } a+1=0$$

$= 0$, two rows being identical.

Hence the given equations are consistent if $a + 1 = 0$.

Example 6:

Show that the only real value of λ for which the following equations have non-zero solution is 6.

$$x + 2y + 3z = \lambda x,\ 3x + y + 2z = \lambda y,\ 2x + 2y + z = \lambda z.$$

Solution:

The given equations can be rewritten as

$$(1-\lambda)\,x + 2y + 3z = 0;$$

$$3x + (1-\lambda)y + 2z = 0$$

and $$2x + 3y + (1-y)\,z = 0$$

The equations in the matrix from AX = K can be rewritten as

$$\begin{bmatrix} 1-\lambda & 2 & 2 \\ 3 & 1-\lambda & 2 \\ 2 & 3 & 1-\lambda \end{bmatrix} \begin{bmatrix} x \\ y \\ z \end{bmatrix} = \begin{bmatrix} 0 \\ 0 \\ 0 \end{bmatrix}$$

If the given system of equations has a non-zero solution then the matrix A must have a rank < the number of unknown quantities x, y, z *i.e.*, 3 and $|A| = 0$.

$$\text{i.e.} \begin{bmatrix} 1-\lambda & 2 & 3 \\ 3 & 1-\lambda & 2 \\ 2 & 3 & 1-\lambda \end{bmatrix} = 0$$

$$\Rightarrow \quad \begin{bmatrix} 6-\lambda & 6-\lambda & 6-\lambda \\ 3 & 1-\lambda & 2 \\ 2 & 3 & 1-\lambda \end{bmatrix} = 0, \text{ replacing } R_1 \text{ by } R_1 + R_2 + R_3$$

$$\Rightarrow (6-\lambda)\begin{vmatrix} 1 & 1 & 1 \\ 3 & 1-\lambda & 2 \\ 2 & 3 & 1-\lambda \end{vmatrix} = 0, \text{ taking out } (6-\lambda) \text{ common from } R_1$$

$$\Rightarrow (6-\lambda)\begin{vmatrix} 1 & 0 & 0 \\ 3 & -2-\lambda & -1 \\ 2 & 1 & -1-\lambda \end{vmatrix} = 0, \text{ applying } C_2 - C_1 \text{ and } C_3 - C_1$$

$$\Rightarrow (6-\lambda)\begin{vmatrix} -1-\lambda & -1 \\ 1 & -1-\lambda \end{vmatrix} = 0$$

$\Rightarrow \quad (6-\lambda) \mid (2+\lambda)(1+\lambda) + 1] = 0$

$\Rightarrow \quad (6-\lambda)[\lambda^2 + 3\lambda + 3] = 0$

$\Rightarrow \quad \lambda = 6, \frac{1}{2}[-3 \pm \sqrt{(9-12)}]$

$\Rightarrow$ $\lambda - 6$ (the other roots being imaginary) is the only real value of l for which the given system of equations has a non-zero solution.

Example 7:

Investigate for what values of λ and μ, the simultaneous equations: $x + y + z = 6$, $x + 2y + 3z = 10$ and $x + 2y + \lambda y + \lambda z = \mu$ have (1) no solution, (2) unique solution, (3) infinite solutions.

Solution:

The given equations in the matrix form AX = K can be written as

$$\begin{bmatrix} 1 & 1 & 1 \\ 1 & 2 & 3 \\ 1 & 2 & \lambda \end{bmatrix}\begin{bmatrix} x \\ y \\ z \end{bmatrix} = \begin{bmatrix} 6 \\ 10 \\ \mu \end{bmatrix}$$

$$\therefore \text{ Then augmented matrix } A^* = \begin{bmatrix} 1 & 1 & 1 & 6 \\ 1 & 2 & 3 & 10 \\ 1 & 2 & \lambda & \mu \end{bmatrix}$$

$$\Rightarrow \quad A^* = \begin{bmatrix} 1 & 0 & -1 & 2 \\ 0 & 1 & 2 & 4 \\ 0 & 0 & \lambda-3 & \mu-0 \end{bmatrix}, \text{ replacing } R_2, R_3 \text{ by } R_2 - R_1, R_3\ R_2 \text{ respectively and then } R_1 \text{ by } R_1 - R_2. \quad ...(1)$$

Now following cases arise;

Case I: If $\lambda - 3 = \mu - 10 \neq 0$ *i.e.*, $\lambda \neq 3$, $\mu \neq 10$, then form (1) A* is in the reduced Echelen form having three non-zero rows and A is in the reduced Echelon form having two non-zero rows.

$\therefore$ The ranks of A* and A are 3 and 2 respectively which being different, the given equations have **no solution.**

Case II: If $\lambda - 3 = 0$, $\mu - 10 = 0$ *i.e.*, $\lambda = 3$, $\mu = 10$, then from (1) we find that both A* and A are in the reduced Echelon form having three non-zero rows and hence the ranks of A* and A are each 3 and these being the same the given equations are consistent. Also there are there are three unknowns viz x, y, z so the solution is *unique* in this case

Case III: If $\lambda - 3 = \mu - 10 \neq 0$ *i.e.*, $\lambda = 3$, $\mu \neq 10$, then form (1) we find that the matrices A* and A are in the reduced Echelon form having two non-zero rows each. Hence, the ranks of A* and A are the same each being 2, which is less than the number of unknowns x, y, z. Therefore in this case two unknowns will be expressed in terms of the third and thus we shall have an infinite number of solutions.

EXERCISES

1. Solve the following equations:

   $x_1 + x_2 + x_3 = 0$, $x_1 + 2x_2 - x_3 = 0$, $2x_1 + x_2 + 3x_3 = 0$.

2. Find the rank of the coefficient matrix for the following system of homogeneous equations over the field of real numbers and compute all the solutions:

   $x_1 + 2x_2 - 3x_3 + 4x_4 = 0$, $x_1 + 3x_2 - x_3 = 0$, $6x_1 + x_2 + 3x_3 = 0$.

3. Solve completely the following equations with the help of matrices:

   (i) $x - y - z + t = 0$, $x - y + 2z - t = 0$, $3x + y + t = 0$.

   (ii) $2w + 3x - y - z = 0$, $4w - 6x - 2y + 2z = 0$,

   $-6w + 12x + 3y - 4z = 0$.

4. Solve by matrix method:

   $x + 2y - z = 3$, $3x - y + 2z = 1$, $2x - 2y + 3z = 2$.

   Ans. $x = -1$, $y = 4$ $z = 4$

5. Solve by matrix method:

   $x - 2y + 3z = 2$, $2z - 3z = 3$, $x + y + z = 0$.

6. $x + 2y - z = 1$, $3x + y + z = 2$, $x - y + z = 0$.

   Ans. $x = 0$, $y = 1$, $z = 1$

7. $x - y + 2z = 3,\ 2x + z = 1,\ 3x + 2t + z = 4.$

Ans. $x = -1,\ y = 2,\ z = 3$

8. $2x + 3y + z = 9,\ x - 2y + 3z = 6,\ 3x + y + 2z = 8.$

Ans. $x = -\frac{5}{14};\ y = \frac{20}{14};\ z = \frac{7}{2}$

9. Find the inverse of the coefficient matrix of the following system of equations:

$x + y + z = 1,\ x + 2y + 2z = 1,\ x + 2y + 3z = 0$

and hence solve them.

Ans. $\begin{bmatrix} 2 & -1 & 0 \\ -1 & 2 & -1 \\ 0 & -1 & 1 \end{bmatrix}$; $x = 1,\ y = 1,\ z = -1.$

10. $x_1 - 3x_2 + x_3 + x_3 = 2,\ 2x_1 + x_2 + 3x_3 = 3,\ x_1 + 5x_2 + 5x_3 = 2.$

Ans. $x_1 = 1,\ x_2 = -(1/5),\ x_3 = (3/5).$

11. $x_1 - x_2 + x_3 = 2,\ 2x_1 - x_2 + 2x_3 = -6,\ 3x_1 + x_2 + x_3 = -18.$

12. Show that the equations

$2x + 6y + 11 = 0,\ 6x + 20y - 6z + 3 = 0,\ 6y - 13z + 1 = 0$ are not consistent.

Show that if in the following problems the given equations are consistent, then solve them.

6

EIGEN VALUES AND EIGEN VECTORS

6.1 ZERO DIVISORS

'If A and **B** are two singular matrices, it is possible to obtainthe result **AB = O,** where neither **A = O** nor **B = O, O** being the null matrix.' In such a case **A** and **B** are called *proper divisors of zero.*

If however, **A** and **B** two square matrices of order n such that

$$\mathbf{AB} = \mathbf{O}, \qquad \text{...(i)}$$

then if A is non-singular, $\mathbf{A} \neq \mathbf{O}$, $\mathbf{A}^{-1}$ exists and $\mathbf{A}^{-1} \neq \mathbf{O}$.

Pre-multiplying (i) with $\mathbf{A}^{-1}$, we get

$\mathbf{A}^{-1}\,\mathbf{AB} = \mathbf{A}^{-1}\,\mathbf{O}$ or $1\,\mathbf{B} = \mathbf{O}$, $\because\ \mathbf{A}^{-1}.\mathbf{O} = \mathbf{O}$ and $\mathbf{A}^{-1}\,\mathbf{A} = \mathbf{I}$

$\Rightarrow \qquad \mathbf{B} = \mathbf{O}.$

Hence we conclude that

if $\qquad \mathbf{A} \neq \mathbf{O}$, then $\mathbf{AB} = \mathbf{O} \Rightarrow \mathbf{B} = \mathbf{O}$.

Similarly. if $\mathbf{B} \neq \mathbf{O}$, then $\mathbf{AB} = \mathbf{O} \Rightarrow \mathbf{A} = \mathbf{O}$.

Hence non-singular matrices are not proper divisor of zero.

Note. If $\mathbf{AB} = \mathbf{O}$ and $\mathbf{B} \neq \mathbf{O}$, then **A** is called a *left zero divisor* and if $\mathbf{AB} = \mathbf{O}$ and $\mathbf{A} \neq \mathbf{O}$, then **B** is called a *right zero divisor.*

Matrix Polynomial

An expression of the form:

$$F(\lambda) = A_0 + A_1 + A_1\lambda + + A_2\lambda^2 + \ldots, + A_{m-1}\lambda^{m-1} + A_m\lambda^m$$

where $A_0, A_1, A_2, \ldots A_m$ are all square matrix of the same order is called a matrix polynomial of degree m provided A_m is not a null matrix. The symbol λ is called inderminate. It the order of each of the matrix coefficients $A_0, A_1, A_2 \ldots, A_m$ is n. then we say that the matrix polynomial is n rowed according to this definition of a matrix polynomial. Each square matrix can be expressed as matrix polynomial with zero degree.

Equality of Polynomial

Two matrix polynomials are equal iff the coefficient of like power of λ are the same.

6.2 CHARACTERISTIC EQUATION AND ROOTS OF A MATRIX

Let $\mathbf{A} = [a_{ij}]$ be an $n \times n$ matrix.

(i) **Characteristic Matrix of A:** The matrix $\mathbf{A} - \lambda\mathbf{I}$ is called *the characteristic matrix of A,* where **I** is the identity matrix.

(ii) **Characteristic Polynomial of A:** The determinant $|\mathbf{A} - \lambda\mathbf{I}|$ is called the *characteristic polynomial of A.*

(iii) **Characteristic Equation of A:** The equation $|\mathbf{A} - \lambda\mathbf{I}| = 0$ is know as the characteristic equation of **A** and is roots are called the *characteristic* or *latent roots* or *eigenvalues* or *characteristic values* or *latent values* or *proppervalues of A.*

(iv) **Spectrum of A:** The set of all eigenvalues of the matrix **A** is called the *spectrum of A.*

(v) **Eigen Value Problem:** The problem of finding the eigenvalues of a matrix is know as as *eigen-value problem.*

Example 1:

Find the characteristic roots of the matrix

$$\begin{bmatrix} 1 & 2 & 3 \\ 0 & 2 & 3 \\ 0 & 0 & 2 \end{bmatrix}.$$

Solution:

Here $\mathbf{A} = \begin{bmatrix} 1 & 2 & 3 \\ 0 & 2 & 3 \\ 0 & 0 & 2 \end{bmatrix}$ and $\mathbf{I} = \begin{bmatrix} 1 & 0 & 0 \\ 0 & 1 & 0 \\ 0 & 0 & 1 \end{bmatrix}$

$$\therefore |\mathbf{A} - \lambda\mathbf{I}| = \begin{vmatrix} 1-\lambda & 2-0 & 3-0 \\ 0-0 & 2-\lambda & 3-0 \\ 0-0 & 0-0 & 2-\lambda \end{vmatrix} = \begin{vmatrix} 1-\lambda & 2 & 3 \\ 0 & 2-\lambda & 3 \\ 0 & 0 & 2-\lambda \end{vmatrix}$$

$$= (1 - \lambda) \begin{vmatrix} 2-\lambda & 3 \\ 0 & 2-\lambda \end{vmatrix}, \text{ expanding with respect to } C_1$$

$$= (1 - \lambda)(2 - \lambda)^2$$

$\therefore$ The characteristic equation of the matrix **A** is

$$(1 - \lambda)(2 - \lambda)^2 = 0$$

and its roots (*i.e.*, characteristic roots) are 1, 2, 2. **Ans.**

Example 2:

Find the eigen values of of the matrix

$$A = \begin{bmatrix} -2 & 2 & -3 \\ 2 & 1 & -6 \\ -1 & -2 & 0 \end{bmatrix}.$$

Solution:

Here $\mathbf{A} = \begin{bmatrix} -2 & 2 & -3 \\ 2 & 1 & -6 \\ -1 & -2 & 0 \end{bmatrix}$ and $\mathbf{I} = \begin{bmatrix} 1 & 0 & 0 \\ 0 & 1 & 0 \\ 0 & 0 & 1 \end{bmatrix}$

$$\therefore |\mathbf{A} - \lambda\mathbf{I}| = \begin{bmatrix} -2-\lambda & 2-0 & -3-0 \\ 2-0 & 1-\lambda & -6-0 \\ -1-0 & -2-0 & 0-\lambda \end{bmatrix}$$

$$\Rightarrow |\mathbf{A} - \lambda\mathbf{I}| = \begin{bmatrix} -2-\lambda & 2 & -3 \\ 2 & 1-\lambda & -6 \\ -1 & -2 & -\lambda \end{bmatrix}$$

$$= \begin{bmatrix} -3-\lambda & 0 & -3-\lambda \\ 0 & -3-\lambda & -6-2\lambda \\ -1 & -2 & -\lambda \end{bmatrix}, \text{ applying } R_1 + R_3,\ R_2 + 2R_3$$

$$= \begin{vmatrix} -3-\lambda & 0 & 0 \\ 0 & -3-\lambda & -6-2\lambda \\ -1 & -2 & -\lambda+1 \end{vmatrix}, \text{ applying } C_3 - C_1$$

$$= (3 + \lambda) \begin{vmatrix} 3+\lambda & -2(3+\lambda) \\ 2 & -(\lambda-1) \end{vmatrix}, \text{ expanding with respect to } R_1$$

$$= (3 + \lambda) \begin{vmatrix} 3+\lambda & 0 \\ 2 & -\lambda+5 \end{vmatrix}, \text{ applying } C_2 + 2C_1$$

$= (3 + \lambda)^2 (5 - \lambda).$

$\therefore$ The characteristic equation of the matrix **A** is

$$(3 + \lambda)^2 (5 - \lambda) = 0$$

and its roots (*i.e.,* eigen values of **A**) are 5, – 3, – 3. **Ans.**

Example 3:

Find the characteristic roots of the matrix

$$A = \begin{bmatrix} 1 & -3 & 3 \\ 3 & -5 & 3 \\ 6 & -6 & 4 \end{bmatrix}.$$

Solution:

Here $\mathbf{A} = \begin{bmatrix} 1 & -3 & 3 \\ 3 & -5 & 3 \\ 6 & -6 & 4 \end{bmatrix}$ and $\mathbf{I} = \begin{bmatrix} 1 & 0 & 0 \\ 0 & 1 & 0 \\ 0 & 0 & 1 \end{bmatrix}$

$$\therefore |\mathbf{A} - \lambda\mathbf{I}| = \begin{vmatrix} 1-\lambda & -3-0 & 3-0 \\ 3-0 & -5-\lambda & 3-0 \\ 6-0 & -6-0 & 4-\lambda \end{vmatrix} = \begin{vmatrix} 1-\lambda & -3 & 3 \\ 3 & -5-\lambda & 3 \\ 6 & -6 & 4-\lambda \end{vmatrix}$$

$$= \begin{vmatrix} 1-\lambda & 0 & 3 \\ 3 & -2-\lambda & 3 \\ & & \end{vmatrix}, \text{ applying } R_3 - R_2$$

$$= \begin{vmatrix} 1-\lambda & 0 & 3 \\ 3 & -2-\lambda & 3 \\ 3 & 0 & 1-\lambda \end{vmatrix}, \text{ applying } R_3 - R_2$$

$$= (-2 - \lambda) \begin{vmatrix} 1-\lambda & 3 \\ 3 & 1-\lambda \end{vmatrix}, \text{ expanding w.r. to } C_2$$

$$= (-2 - \lambda) [(1 - \lambda)^2 - 9] = (\lambda + 2) [9 - (1 - \lambda)^2]$$

$$= (\lambda + 2) [9 - (1 - 2\lambda + \lambda^2)] = (\lambda + 2) (8 + 2\lambda - \lambda^2)$$

$\therefore$ The characteristic equation of the matrix **A** is

$$(\lambda + 2) (\lambda^2 - 2\lambda - 8) = 0$$

and its roots are $-2, \frac{1}{2}\left[2 \pm \sqrt{(4+32)}\right]$ *i.e.,*, – 2, 4, – 2. **Ans.**

Example 4:

Find the characteristic roots of the matrix

$$A = \begin{bmatrix} 2 & 2 & 1 \\ 1 & 3 & 1 \\ 1 & 2 & 2 \end{bmatrix}.$$

Solution:

Here

$$\mathbf{A} = \begin{bmatrix} 2 & 2 & 1 \\ 1 & 3 & 1 \\ 1 & 2 & 2 \end{bmatrix} \text{ and } \mathbf{I} = \begin{bmatrix} 1 & 0 & 0 \\ 0 & 1 & 0 \\ 0 & 0 & 1 \end{bmatrix}$$

$$\therefore |\mathbf{A} - \lambda \mathbf{I}| = \begin{vmatrix} 2-\lambda & 2-0 & 1-0 \\ 1-0 & 3-\lambda & 1-0 \\ 1-0 & 2-0 & 2-\lambda \end{vmatrix} = \begin{vmatrix} 2-\lambda & 2 & 1 \\ 1 & 3-\lambda & 1 \\ 1 & 2 & 2-\lambda \end{vmatrix}$$

$$= \begin{vmatrix} -\lambda & 0 & 1 \\ -1 & 1-\lambda & 1 \\ -3+2\lambda & -2+2\lambda & 2-\lambda \end{vmatrix}, \text{ applying } C_1 - 2C_3, \ C_2 - 2C_3$$

$$= \begin{vmatrix} 0 & 0 & 1 \\ -1 & 1-\lambda & 1 \\ -3+4\lambda-\lambda^2 & -2+2\lambda & 2-\lambda \end{vmatrix}, \text{ applying } C_1 + \lambda C_3$$

$$= \begin{vmatrix} \lambda - 1 & 1-\lambda \\ (\lambda-1)(3-\lambda) & -2(1-\lambda) \end{vmatrix} = (l-1)^2 \begin{vmatrix} 1 & -1 \\ 3-\lambda & 2 \end{vmatrix}$$

$$= (\lambda - 1)^2 (2 + 3 - \lambda) = (\lambda - 1)^2 (5 - \lambda).$$

$\therefore$ The characteristic equation of the matrix **A** is

$$(\lambda - 1)^2 (5 - \lambda) = 0$$

and its roots (or characteristic roots of **A**) are 1, 5. **Ans.**

Example 5:

Find the characteristic roots of the matrix

$$A = \begin{bmatrix} 1 & 3 & 0 \\ 3 & -2 & -1 \\ 0 & -1 & 1 \end{bmatrix}.$$

Solution:

Here we have

$$|\mathbf{A} - \lambda\mathbf{I}| = \begin{vmatrix} 1-\lambda & 3-0 & 0-0 \\ 3-0 & -2-\lambda & -1-0 \\ 0-0 & -1-0 & 1-\lambda \end{vmatrix} \qquad \textbf{(Note)}$$

$$= \begin{vmatrix} 1-\lambda & 3 & 0 \\ 3 & -2-\lambda & -1 \\ 0 & -1 & 1-\lambda \end{vmatrix}$$

$$= -(1-\lambda) \begin{vmatrix} -2-\lambda & -1 \\ -1 & 1-\lambda \end{vmatrix} - 3 \begin{vmatrix} 3 & -1 \\ 0 & 1-\lambda \end{vmatrix}$$

$$= (1-\lambda)\,[(2+\lambda)(\lambda-1) - 1] - 3\,[3\,(1-\lambda)]$$

$$= (1-\lambda)\,[(2\lambda - 2 + \lambda^2 - \lambda - \lambda) - 9]$$

$$= (1-\lambda)(\lambda^2 + \lambda - 12) = (1-\lambda)(\lambda+4)(\lambda-3).$$

∴ The characteristic equation of the matrix **A** is

$$(1-\lambda)(\lambda-3)(\lambda+4) = 0.$$

Hence the characteristic roots of **A** are 1, 3, – 4. **Ans.**

Example 6:

Deterimine the characteristic equation and roots of the matrix

$$A = \begin{bmatrix} 1 & -1 & 4 \\ 0 & 3 & 7 \\ 0 & 0 & 5 \end{bmatrix}.$$

Solution:

Here

$$|\mathbf{A} - \lambda\mathbf{I}| = \begin{vmatrix} 1-\lambda & -1-0 & 4-0 \\ 0-0 & 3-\lambda & 7-0 \\ 0-0 & 0-0 & 5-\lambda \end{vmatrix} \qquad \textbf{(Note)}$$

$$= \begin{vmatrix} 1-\lambda & -1 & 4 \\ 0 & 3-\lambda & 7 \\ 0 & 0 & 5-\lambda \end{vmatrix} = (1-1) \begin{vmatrix} 3-\lambda & 7 \\ 0 & 5-\lambda \end{vmatrix}$$

$$= (1-\lambda)(3-\lambda)(5-\lambda)$$

∴ The characteristic equation of the matrix **A** is

$$(1 - \lambda)(3 - \lambda)(5 - \lambda) = 0.$$

Hence the characterstic roots of **A** are 1, 3, 5. **Ans.**

Example 7:

Obtain the characteristic roots of the matrix

$$A = \begin{bmatrix} b & c & a \\ c & a & b \\ a & b & c \end{bmatrix}.$$

Solution:

Here

$$|\mathbf{A} - \lambda\mathbf{I}| = \begin{vmatrix} b-\lambda & c-0 & a-0 \\ c-0 & a-\lambda & b-0 \\ a-0 & b-0 & c-\lambda \end{vmatrix} = \begin{vmatrix} b-\lambda & c & a \\ c & a-\lambda & b \\ a & b & c-\lambda \end{vmatrix}$$

$$= \begin{vmatrix} b-\lambda+c+a & c & a \\ c+a-\lambda+b & a-\lambda & b \\ a+b+c-\lambda & b & c-\lambda \end{vmatrix}, \text{ applying } C_1 + C_2 + C_3$$

$$= (a + b + c - \lambda)\begin{vmatrix} 1 & c & a \\ 1 & a-\lambda & -b \\ 1 & b & c-\lambda \end{vmatrix}$$

$$= (a + b + c - \lambda)\begin{vmatrix} 1 & c & a \\ 0 & a-\lambda-c & b-a \\ 0 & b-c & c-\lambda-a \end{vmatrix},$$

applying $R_2 - R_1$, $R_3 - R_1$

$$= (a + b + c - \lambda)[(a - c - \lambda)(c - a - \lambda) - (b - a)(b - c)]$$

$$= (a + b + c - \lambda)[(a - c)(c - a) - \lambda\ \{(c - a) + (a - c)\} + \lambda^2 - (b - a)(b - c)]$$

$$= (a + b + c - \lambda)[\lambda^2 + 2ac - a^2 - c^2 - b^2 + bc + ab - ac]$$

$$= (a + b + c - \lambda)(\lambda^2 - a^2 - b^2 - c^2 + ab + bc + ca).$$

$\therefore$ The characteristic equation of **A** is $|\mathbf{A} - \lambda\mathbf{I}| = 0$

i.e., $(a + b + c - \lambda)(\lambda^2 - a^2 - b^2 - c^2 + ab + bc + ca) = 0$

and the characteristic roots of **A** are

$$a + b + c,\ \pm\sqrt{(a^2 + b^2 + c^2 - ab - bc - ca)}.$$

Example 8:

Find the eigenvalues of the matrix

$$A = \begin{bmatrix} -3 & 2 & 2 \\ -6 & 5 & 2 \\ -7 & 4 & 4 \end{bmatrix}.$$

Solution:

Here we have

$$|\mathbf{A} - \lambda\mathbf{I}| = \begin{vmatrix} -3-\lambda & 2 & 2 \\ -6 & 5-\lambda & 2 \\ -7 & 4 & 4-\lambda \end{vmatrix}$$

$$= \begin{vmatrix} -3-\lambda & 0 & 2 \\ -6 & 3-\lambda & 2 \\ -7 & \lambda & 4-\lambda \end{vmatrix}, \text{ applying } C_2 - C_3$$

$$= -(3+\lambda)\begin{vmatrix} 3-\lambda & 2 \\ \lambda & 4-\lambda \end{vmatrix} + 2\begin{vmatrix} -6 & 3-\lambda \\ -7 & \lambda \end{vmatrix}$$

$$= -(3+\lambda)[(3-\lambda)(4-\lambda) - 2\lambda] + 2[-6\lambda + 7(3-\lambda)]$$

$$= -(3+\lambda)(\lambda^2 - 9\lambda + 12) + 2(21 - 13\lambda)$$

$$= 6 - 11\lambda + 6\lambda^2 - \lambda^3 = (1-\lambda)(2-\lambda)(3-\lambda)$$

$\therefore$ The characterstic equation of **A** is $|\mathbf{A} - \lambda\mathbf{I}| = 0$

$\Rightarrow (1-\lambda)(2-\lambda)(3-\lambda) = 0 \Rightarrow \lambda = 1, 2, 3.$

$\therefore$ The required eigenvalues of **A** are 1, 2, 3. **Ans.**

Example 9:

Find the eigenvalues of $A = \begin{bmatrix} 3 & 1 & 4 \\ 0 & 2 & 6 \\ 0 & 0 & 5 \end{bmatrix}$.

Solution:

Here $|\mathbf{A} - \lambda\mathbf{I}| = \begin{vmatrix} 3-\lambda & 1 & 4 \\ 0 & 2-\lambda & 6 \\ 0 & 0 & 5-\lambda \end{vmatrix}$

$$= (3-\lambda)\begin{vmatrix} 2-\lambda & 6 \\ 0 & 5-\lambda \end{vmatrix} = (3-\lambda)(2-\lambda)(5-\lambda)$$

$\therefore$ The characteristic equation of A is $|\mathbf{A} - \lambda\mathbf{I}| = 0$

$\Rightarrow (3-\lambda)(2-\lambda)(5-\lambda) = 0 \Rightarrow \lambda = 2, 3, 5.$

$\therefore$ The required eigenvalues of **A** are 2, 3, 5. **Ans.**

Example 10:

Find the eigen values (or latent roots) of

$$A = \begin{bmatrix} 8 & -6 & 2 \\ -6 & 7 & -4 \\ 2 & -4 & 3 \end{bmatrix}.$$

Solution:

Here $|\mathbf{A} - \lambda\mathbf{I}| = \begin{vmatrix} 8-\lambda & -6 & 2 \\ -6 & 7-\lambda & -4 \\ 2 & -4 & 3-\lambda \end{vmatrix}$

$= (8 - \lambda) \{(7 - \lambda) (3 - \lambda) - 16\} + 6 \{- 6 (3 - \lambda) + 8\} + 2 \{24 - 2 (7 - \lambda)\}$

$= (8 - \lambda) (5 - 10\lambda + \lambda^2) + 6 (6\lambda - 10) + 2 (10 + 2\lambda)$

$= 45\lambda + 18\lambda^2 - \lambda^3.$

$\therefore$ The characteristic equation of **A** is $- 45\lambda + 18\lambda^2 - \lambda^3 = 0$ which gives $\lambda (\lambda^2 - 18\lambda + 45) = 0 \Rightarrow \lambda (\lambda - 3) (\lambda - 15) = 0 \Rightarrow \lambda = 0, 3, 15.$

$\therefore$ The required latent roots or eigenvalues of **A** are 0, 3 and 15.

Example 11:

If $a_1, a_2, a_3..., a_n$ are the characteristic roots of the n-square matrix **A** *and m is a scalar, then show that the characteristic roots of* **A** $- \mu$**I** *are* $a_1 - \mu, ..., a_n - \mu.$

Solution:

Since $a_1, a_2, ..., a_n$ are the characteristic roots of the matrix **A**, so from 10.2 we have

$$|\mathbf{A} - \lambda\mathbf{I}| = (\lambda - a_1) (\lambda - a_2) ... (\lambda - a_n) \qquad ...(i)$$

Now the characteristic function of $\mathbf{A} - \mu\mathbf{I}$

$= |(\mathbf{A} - \mu\mathbf{I}) - \lambda\mathbf{I}| = |\mathbf{A} - (\mu + \lambda) \mathbf{I}|$

$= \{(\mu + \lambda) - a_1\} \{(\mu + \lambda) - a_2\} \{(\mu + \lambda) - a_3\}...\{(\mu + \lambda) - a_n\}$, from (i)

$= \{\lambda - (a_1 - \mu)\} \{\lambda - (a_2 - \mu)\}...\{\lambda - (a_n - \mu)\},$

rearranging the terms in each bracket.

$\therefore$ The characteristic roots of $\mathbf{A} - \mu\mathbf{I}$ are $(a_1 - \mu), (a_2 - \mu)$ etc.

Example 12:

Find the characteristic roots of the matrix

$$A=\begin{bmatrix} \cos\theta & -\sin\theta \\ -\sin\theta & -\cos\theta \end{bmatrix}.$$

Solution:

Here $A=\begin{bmatrix} \cos\theta & -\sin\theta \\ -\sin\theta & -\cos\theta \end{bmatrix}$ and $I=\begin{bmatrix} 1 & 0 \\ 0 & 1 \end{bmatrix}$

$$\therefore |A - \lambda I| = \begin{vmatrix} \cos\theta-\lambda & -\sin\theta-0 \\ -\sin\theta-0 & -\cos\theta-\lambda \end{vmatrix}$$

$$= (\cos\theta - \lambda)(-\cos\theta - \lambda) - (\sin^2\theta)$$

$$= -(\cos^2\theta - \lambda^2) - \sin^2\theta = \lambda^2 - 1.$$

$\therefore$ The characteristic equation of the matrix **A** is $\lambda^2 - 1 = 0$ and its roots *i.e.,* characteristic roots are ± 1. **Ans.**

6.3 CAYLEY-HAMILTON THEOREM

Statement: *Every square matrix satisfies its characteristic equation or if* $|A - \lambda I| = (-1)^n [\lambda^n + a_1\lambda^{n-1} + a_2\lambda^{a-2} + \dots + a_n]$ *be the characterstic polynomial of an* $n \times n$ *matrix* $A = [a_{ij}]$, *then the matrix equation*

$X_n + a_1X^{a-1} + \dots + a_n I = O$ *is satisfied by* $X = A$

i.e. $\quad A^n + a_1 A^{n-1} + \dots + a_n 1 + O.$

Proof. $\because$ the elements of (A – lI) are at the most of first degree in l.

$\therefore$ The elements of Adj $(A - \lambda I)$ are at the most of degree $(n - 1)$ in λ and the coefficients of various powers of λ being polynomials in the a_{ij}.

$\therefore$ Adj $(A - \lambda I)$ can be written as

$$B = B_0 \lambda^{n-2} + B_1\lambda^{n-1} + \dots + B_{n-1},$$

where $B_0, B_i, \dots, B_{n-1}$ are $n \times n$ matrices, their elements being polynomials in a_{ij}.

Also from we know if $A = [a_{ij}]$ be an $n \times n$ matrix, then $A.(\text{Adj } A) = (\text{Adj } A).A - |A|.I$, where **I** is an $n \times n$ identity matrix.

Therefore $(A - \lambda I).\ \text{Adj}\ (A - \lambda I) = |A - \lambda I|.I$

$\Rightarrow \qquad (A - \lambda I).B\ |A - \lambda I|.I, \ \because\ B\ \text{Adj}\ (A - \lambda I)$

$\Rightarrow (A - \lambda I)(B_0\lambda^{n-1} + B_1\lambda^{n-2} + \dots + B_{n-1})$

$$= (-1)^n [\lambda^n + a_1\lambda^{n-1} + a_2\lambda^{n-2} + \dots + a_n] I.$$

Comparing coefficients of like powers of λ on both sides, we get

$$-IB_0 = (-1)^n I;$$

$$AB_0 - IB_1 = (-1)^n a_1 I;$$

$$\mathbf{AB_1 - IB_1 = (-1)^n a_2 I;}$$

... ...

... ...

$$\mathbf{AB_{n-1} = (-1)^n a_n I.}$$

Now, pre-multiplying these equations by $\mathbf{A^n, A^{n-1}, ..., A, I}$ respectively and adding the results so obtained we get

$$A^n(-IB_0) + A^{n-1}(AB_0 - IB_1) + A^{n-2}(AB_1 - IB_2) + ... + I(AB_{n-1})$$

$$= (-1)^n [IA^n + a_1 IA^{n-1} + a_2 IA^{n-2} + ... a_n I.I]$$

$$\Rightarrow O = (-1)^n [A^n + a_1 A^{n-1} + a_2 A^{n-2} + ... + a_n I]$$

where **O** is the null matrix.

Hence $\mathbf{A^n + a_1A^{n-1} + a_2A^{n-2} + ... + a_nI = O}$.

Cor. I: Multiplying the result of 10.3 above by $\mathbf{A^{m-n}}$, where $m \geq n$ and m is a positive integer, we get

$$\mathbf{A^m + a_1A^{m-1} + a_2A^{m-2} + ... + a_nA^{m-n} = O}$$

i.e., any positive integral power $\mathbf{A^m}$ of **A** can be linearly expressed in terms of **I.A., ...,** $\mathbf{A^{n-1}}$.

Cor. II: In 10.3 above we have proved that

$$A^n + a_1A^{n-1} + a_2A^{n-2} + ... + a_n I = O \quad ...(i)$$

$$\Rightarrow -a_nI = A[A^{n-1} + a_1A^{n-2} + a_2A^{n-3} + ... + a_{n-1} I] \quad \text{(Note)}$$

$$\Rightarrow -a_nA^{-1} I = A^{n-1} + a_1A^{n-2} + ... + a_{n-1} I \quad ...(ii)$$

$$\Rightarrow (-1)\, AB_{n-1} A^{-1} = -A^{n-1} - a_1 A^{n-2} - ... - a_{n-4} I,$$

$$\because AB_{n-1} = (-1)^n a_nI$$

$$\Rightarrow B_{n-1} = (-1)^n [-A^{n-1} - a_1A^{n-2} - ... - a_{n-1} I] = \text{Adj } A \quad ...(iii)$$

Cor. III: From result (ii) of Cor. II above we have

$$\mathbf{A^{-1} = \frac{1}{a^n} [A^{n-1} + a_1A^{n-2} + ... + a_{n-1} I]} \quad ...(iv)$$

which shows that $\mathbf{A^{-1}}$ can be expressed linearly in terms of $\mathbf{A^{n-1}}$, $\mathbf{A^{n-2}}$, **..., I.**

6.4 CHARACTERISTIC VECTORS (OR EIGENVECTORS)

Let us consider the linear transformation

$$\mathbf{K = AX} \quad ...(i)$$

which transforms a column vector **X** by means of a square matrix **A** into another column vector **K**.

If **X** be a vector which transforms to its multiple $\mu\mathbf{X}$ by the above transformation (i), then we have $\mu\mathbf{X} = \mathbf{AX}$ ...(ii)

$$\Rightarrow \qquad \mathbf{AX} - \mu\mathbf{IX} = \mathbf{O} \Rightarrow (\mathbf{A} - \mu\mathbf{I})\,\mathbf{X} = \mathbf{O}. \qquad \text{...(iii)}$$

This equation (ii) when written in full given n homogeneous equations in $x_1, x_2, ..., x_n$, which are n unknowns. Then n equations will have a non-zero solution only if $|\mathbf{A} - \mu\mathbf{I}| = 0$ *i.e.,* the coefficient matrix is singular.

This equation is called the characteristic equation of the transformation and is the same as the characteristic equation of the matrix **A**. This equation has n roots and corresponding to each root, the equation (ii) has a non-zero solution

$$\mathbf{X} = \begin{bmatrix} x_1 \\ x_2 \\ \cdots \\ \cdots\cdot \\ x_n \end{bmatrix}$$

which is defined as *characteristic vector* or *Eigen vector* or *latent vector* or *invariant vector.*

6.5 THEOREMS ON LATENT ROOTS (OR CHARACTERISTIC ROOTS)

Theorem 1:

If square matrix A of order n has latent roots $\lambda_1, \lambda_2, ..., \lambda_n$, then A' has also the same latent roots.

Proof:

$$\text{Let } \mathbf{A} = \begin{bmatrix} a_{11} & a_{12} & a_{13}\cdots\cdots & a_{1n} \\ a_{21} & a_{22} & a_{23}\cdots\cdots & a_{2n} \\ \cdots & \cdots & \cdots\cdots\cdots & \cdots \\ a_{n1} & a_{n2} & a_{n3}\cdots\cdots & a_{nn} \end{bmatrix}$$

$\therefore$ The characteristic equation of A is

$$|\mathbf{A} - \lambda\mathbf{I}| = \begin{vmatrix} a_{11}-\lambda & a_{12} & \cdots & a_{1n} \\ a_{21} & a_{22}-\lambda & \cdots & a_{2n} \\ \cdots & \cdots & \cdots & \cdots \\ a_{n1} & a_{n2} & \cdots & a_{nn}-\lambda \end{vmatrix} = 0$$

Also $\mathbf{A}' = \begin{bmatrix} a_{11} & a_{21} & a_{31} & \cdots\cdots & a_{n1} \\ a_{12} & a_{12} & a_{32} & \cdots\cdots & a_{n2} \\ \cdots & \cdots & \cdots & \cdots\cdots & \cdots \\ a_{1n} & a_{2n} & a_{3n} & \cdots\cdots & a_{nn} \end{bmatrix}$

$\therefore$ The characteristic equation of A' is

$$|\mathbf{A}' - \lambda \mathbf{I}| = \begin{vmatrix} a_{11} - \lambda & a_{21} & \cdots\cdots & a_{n1} \\ a_{12} & a_{22} - \lambda & \cdots\cdots & a_{n2} \\ \cdots & \cdots & \cdots\cdots & \cdots \\ a_{1n} & a_{2n} & \cdots\cdots & a_{nn} - \lambda \end{vmatrix} = 0 \qquad \text{...(ii)}$$

Also we know that the value of a determinant remains unaltered if rows are changed into columns and thus we find that the determinants given by (i) and (ii) are the same, the diagonal elements being the same.

Hence, from (i) and (ii) we conclude that the characteristic equations of **A** and **A'** are the same. Consequently the latent roots of **A** and **A'** are the same.

Theorem 2:

If **A** *is an* $n \times n$ *triangular matrix, then the elements of the principal diagonal are the characteristic roots of* **A.**

Proof:

Let $\mathbf{A} = \begin{bmatrix} a_{11} & a_{12} & a_{13} & \cdots\cdots & a_{1n} \\ 0 & a_{22} & a_{23} & \cdots\cdots & a_{2n} \\ 0 & 0 & a_{33} & \cdots\cdots & a_{3n} \\ \cdots & \cdots & \cdots & \cdots\cdots & \cdots \\ 0 & 0 & 0 & \cdots\cdots & a_{nn} \end{bmatrix}$

(Here we have taken upper triangular matrix).

$\therefore$ The characteristic equation of **A** is

$$|\mathbf{A} - \lambda \mathbf{I}| = \begin{vmatrix} a_{11} - \lambda & a_{12} & a_{13} & \cdots & a_{1n} \\ 0 & a_{22} - \lambda & a_{23} & \cdots & a_{2n} \\ 0 & 0 & a_{33} - \lambda & \cdots & a_{3n} \\ \cdots & \cdots & \cdots & \cdots & \cdots \\ 0 & 0 & 0 & \cdots & a_{nn} - \lambda \end{vmatrix} = 0$$

$$\Rightarrow (a_{11}-\lambda)\begin{vmatrix} a_{22}-\lambda & a_{23} & \cdots & a_{2n} \\ 0 & a_{33}-\lambda & \cdots & a_{3n} \\ \cdots & \cdots & \cdots & \cdots \\ 0 & 0 & \cdots & a_{nm}-\lambda \end{vmatrix}$$

$$\Rightarrow (a_{1,}-\lambda)(a_{22}-\lambda)\begin{vmatrix} a_{33}-\lambda & a_{34} & \cdots & a_{3n} \\ 0 & a_{44}-\lambda & \cdots & a_{4n} \\ \cdots & \cdots & \cdots & \cdots \\ 0 & 0 & \cdots & a_{nn}-\lambda \end{vmatrix} = 0,$$ expanding with respect to C_2

$\Rightarrow (a_{11}-\lambda)(a_{22}-\lambda)(a_{33}-\lambda)\ldots(a_{nn}-\lambda) = 0$, proceeding in this way

$$\lambda = a_{11}, a_{22}, a_{33} \ldots a_{nn}$$

i.e., the latent roots (or characteristic roots) of **A** are the elements of the principal diagonal of **A**. **Hence proved.**

Theorem 3:

The characteristic roots of a hermitian matrix are all real.

Proof:

Let **A** be the hermitian matrix. Then, we know that

$$\mathbf{AX} = \lambda\mathbf{X}. \qquad \ldots(i)$$

where λ is a characteristic root of **A** and **X** the corresponding characteristic vector.

From (i) we get $\mathbf{X}^\theta \mathbf{AX} = \mathbf{X}^\theta \lambda\mathbf{X}$, premultiplying both sides by $\mathbf{X}^\theta$

$$\Rightarrow \quad \mathbf{X}^\theta \mathbf{AX} = \lambda\mathbf{X}^\theta\mathbf{X}. \qquad \ldots(ii)$$

where $\mathbf{X}^\theta$ is the transposed conjugate of X

Also if **A** is a hermitian matrix, then by definition we have

$$\mathbf{A}^\theta = \mathbf{A} \qquad \ldots(iii)$$

Now taking tranposed conjugate of both sides of (ii) we get

$$\mathbf{X}^\theta\mathbf{AX} = \bar{\lambda}\,\mathbf{X}^\theta\mathbf{X}, \text{ using (iii) also.} \qquad \ldots(iv)$$

$\therefore$ From (ii) and (iv) we get $\lambda\mathbf{X}^\theta\mathbf{X} = \bar{\lambda}\,\mathbf{X}^\theta\mathbf{X}$

$\Rightarrow (\lambda - \bar{\lambda})\,\mathbf{X}^\theta\mathbf{X} = \mathbf{O} \Rightarrow \quad \lambda - \lambda = 0, \quad \because \mathbf{X}^\theta\mathbf{X} \neq \mathbf{O}$

$\Rightarrow \lambda = \bar{\lambda} \Rightarrow \lambda$ is real. **Hence proved.**

Theorem 4:

The characteristic roots of a real symmetric matrix are all real.

Proof:

Do as Theorem 3 above. Here all the elements of **A** are real and as such it is particular case of Theorem 3 above.

Theorem 5:

The characteristic roots of a skew-hermitian matrix are either purely imaginaries or zero.

Proof:

Let **A** be a skew-hermitian matrix, then we know that **iA** is hermitian.

If λ be a characteristic of **A**, then $|\mathbf{A} - \lambda\mathbf{I}| = \mathbf{0}$

$\Rightarrow(i\lambda)$ is real, since the characteristic roots of a hermitian matrix are all real.

$\Rightarrow\lambda$ is either purely imaginary or zero.

i.e., the characteristic roots of a skew hermitian matrix **A** are either purely imaginary of zero. **Hence proved.**

Theorem 6:

The characteristic roots of real skew-symmetric matrix are purely imaginaries or zero.

Proof:

Do as Theorem **5** above. Here all the elements of **A** re real and as such it is a particular case of Theorem **V** above.

Theorem 7:

The characteristic roots of a unitary matrix are of unit modulus.

Proof:

Let **A** be a unitary matrix. Let λ be a characteristic root of **A** and **X** the corresponding characteristic vector.

Then $\mathbf{AX} = \lambda\mathbf{X}$ ...(i)

Taking transposed conjugate of both sides of (i), we get

$$(\mathbf{AX})^{\theta} = (\lambda\mathbf{X})^{\theta} \Rightarrow \mathbf{X}^{\theta}\,\mathbf{A}^{\theta} = \bar{\lambda}\,\mathbf{X}^{\theta} \qquad \text{...(ii)}$$

$\therefore$ From (i) and (ii) we get $\mathbf{X^\theta A^\theta AX} = \bar{\lambda}\ \mathbf{X^q}\ \lambda\mathbf{X}$ **(Note)**

$\Rightarrow \mathbf{X^\theta (A^\theta A) X} = \bar{\lambda}\ \lambda\ \mathbf{X^\theta X}$

$\Rightarrow \mathbf{X^\theta (I) X} = \bar{\lambda}\ \lambda\ \mathbf{X^\theta X}$, $\quad\because$ A is unitary

$\Rightarrow \mathbf{X^\theta X} (1 - \bar{\lambda}\ \bar{\lambda}) = O$

$\Rightarrow 1 - \lambda\bar{\lambda} = 0$, $\quad\because \mathbf{X^\theta X \neq O}$

$\lambda\lambda = 1 \Rightarrow |\bar{\lambda}|^2 = \lambda\bar{\lambda} = 1$

i.e., the characteristic roots of **A** are of unit modulus. **Hence proved.**

Theorem 8:

The characteristic roots of an orthogonal matrix are of unit modulus.

Proof:

Do as Theorem 7 above remembering that if all the elements of the unitary matrix **A** are real then **A** is an orthogonal matrix.

6.6 AN IMPORTANT THEOREM

The scalar λ is a characteristic root of the matrix **A** *if and only if the matrix* **A** – **λI** *is singular.*

Proof:

Let $\mathbf{A} = [a_{ij}]_{n\times n}$, $\mathbf{X} = [x_1\ x_2\ \ldots\ x_n]'$

then $\quad \mathbf{AX} = \mathbf{lX}$

reduces to

$$\begin{bmatrix} a_{11} & a_{12} & \cdots & a_{1n} \\ a_{21} & a_{22} & \cdots & a_{2n} \\ \cdots & \cdots & \cdots & \cdots \\ a_{n1} & a_{n2} & \cdots & a_{nn} \end{bmatrix} \begin{bmatrix} x_1 \\ x_2 \\ \cdots \\ x_n \end{bmatrix} = \lambda \begin{bmatrix} x_1 \\ x_2 \\ \cdots \\ x_n \end{bmatrix}$$

$$= \lambda \begin{bmatrix} 1 & 0 & 0 & \cdots & 0 \\ 0 & 1 & 0 & \cdots & 0 \\ \cdots & \cdots & \cdots & \cdots & \cdots \\ \cdots & \cdots & \cdots & \cdots & \cdots \\ 0 & 0 & 0 & \cdots & 1 \end{bmatrix} \begin{bmatrix} x_1 \\ x_2 \\ \cdots \\ \cdots \\ x_n \end{bmatrix}$$

$$= \begin{bmatrix} \lambda & 0 & \dots & 0 \\ 0 & \lambda & \dots & 0 \\ \dots & \dots & \dots & \dots \\ 0 & 0 & \dots & \lambda \end{bmatrix} \begin{bmatrix} x_1 \\ x_2 \\ \dots \\ x_n \end{bmatrix}$$

$$\Rightarrow \quad \left\{ \begin{bmatrix} a_{11} & a_{12} & \dots & a_{1n} \\ a_{21} & a_{22} & \dots & a_{2n} \\ \dots & \dots & \dots & \dots \\ a_{n1} & a_{n2} & \dots & a_{nn} \end{bmatrix} - \begin{bmatrix} \lambda & 0 & \dots & 0 \\ 0 & \lambda & \dots & 0 \\ \dots & \dots & \dots & \dots \\ 0 & 0 & \dots & \dots \end{bmatrix} \right\} \begin{bmatrix} x_1 \\ x_2 \\ \dots \\ x_n \end{bmatrix} = \mathbf{O}$$

$$\Rightarrow \quad \begin{bmatrix} a_{11}-\lambda & a_{12} & \dots & a_{1n} \\ a_{21} & a_{22}-\lambda & \dots & a_{2n} \\ \dots & \dots & \dots & \dots \\ a_{n1} & a_{n2} & \dots & a_{nn}-\lambda \end{bmatrix} \begin{bmatrix} x_1 \\ x_2 \\ \dots \\ x_n \end{bmatrix} = \mathbf{O}$$

$\Rightarrow \quad \mathbf{(A - \lambda I)\ X = O}$

i.e., $\quad \mathbf{AX = \lambda X \Rightarrow (A - \lambda I)\ X = O,}$

which is a homogeneous system of linear equations whose coefficient is $\mathbf{A - \lambda I}$.

Now as we require a vector $\mathbf{X \neq O}$, so we must have

$$|\mathbf{A - \lambda I}| = \mathbf{0}$$

i.e., the matrix $\mathbf{A - \lambda I}$ must be singular.

Example 1:

Verify Cayley-Hamilton's Theorem for the matrix

$$A = \begin{bmatrix} 0 & 0 & 1 \\ 3 & 1 & 0 \\ -2 & 1 & 4 \end{bmatrix}. \textit{ Hence compute } A^{-1}.$$

Solution:

Here $\mathbf{I} = \begin{bmatrix} 1 & 0 & 0 \\ 0 & 1 & 0 \\ 0 & 0 & 1 \end{bmatrix}$ and $\mathbf{A} = \begin{bmatrix} 0 & 0 & 1 \\ 3 & 1 & 0 \\ -2 & 1 & 14 \end{bmatrix}$

$$\therefore |\mathbf{A - \lambda I}| = \begin{vmatrix} 0-\lambda & 0 & 1 \\ 3 & 1-\lambda & 0 \\ -2 & 1 & 4-\lambda \end{vmatrix}$$

$$= -\lambda\ \{(1-\lambda)(4-\lambda)\} + 1\ \{3 + 2\ (1-\lambda))$$

$$= -\lambda\,(4 - 5\lambda + \lambda^2) + 5 - 2\lambda$$
$$= 5 - 6\lambda + 5\lambda^2 - \lambda^2$$

$\therefore$ The characteristic equation of **A** is

$$\lambda^3 - 5\lambda^2 + 6\lambda - 5 = 0 \qquad \text{...(i)}$$

Also we have

$$\mathbf{A}^2 = \begin{bmatrix} 0 & 0 & 1 \\ 3 & 1 & 0 \\ -2 & 1 & 4 \end{bmatrix}\begin{bmatrix} 0 & 0 & 1 \\ 3 & 1 & 0 \\ -2 & 1 & 4 \end{bmatrix} = \begin{bmatrix} -2 & 1 & 4 \\ 3 & 1 & 3 \\ -5 & 5 & 14 \end{bmatrix}$$

and $\mathbf{A}^3 = \mathbf{A}^2.\mathbf{A} = \begin{bmatrix} -2 & 1 & 4 \\ 3 & 1 & 3 \\ -5 & 5 & 14 \end{bmatrix}\begin{bmatrix} 0 & 0 & 1 \\ 3 & 1 & 0 \\ -2 & 1 & 4 \end{bmatrix}$

$$= \begin{bmatrix} -5 & 5 & 14 \\ -3 & 4 & 15 \\ -13 & 19 & 51 \end{bmatrix}$$

$\therefore \qquad \mathbf{A^3 - 5\,A^2 + 6A - 5I}$

$$= \begin{bmatrix} -5 & 5 & 14 \\ -3 & 4 & 15 \\ -13 & 19 & 51 \end{bmatrix} - 5\begin{bmatrix} -2 & 1 & 4 \\ 3 & 1 & 3 \\ -5 & 5 & 14 \end{bmatrix} + 6\begin{bmatrix} 0 & 0 & 1 \\ 3 & 1 & 0 \\ -2 & 1 & 4 \end{bmatrix} - 5\begin{bmatrix} 1 & 0 & 0 \\ 0 & 1 & 0 \\ 0 & 0 & 1 \end{bmatrix}$$

$$= \begin{bmatrix} -5 & 5 & 14 \\ -3 & 4 & 15 \\ -13 & 19 & 51 \end{bmatrix} - \begin{bmatrix} -10 & 5 & 20 \\ 15 & 5 & 15 \\ -25 & 25 & 70 \end{bmatrix} + \begin{bmatrix} 0 & 0 & 6 \\ 18 & 6 & 0 \\ -12 & 16 & 424 \end{bmatrix} - \begin{bmatrix} 5 & 0 & 0 \\ 0 & 5 & 0 \\ 0 & 0 & 5 \end{bmatrix}$$

$$= \begin{bmatrix} -5+10+0-5 & 5-5+0-0 & 14-20+6-0 \\ -3-15+18-0 & 4-5+6-5 & 15-15+0-0 \\ -13+25-12-0 & 19-25+6-0 & 51-70+24-5 \end{bmatrix}$$

$$= \begin{bmatrix} 0 & 0 & 0 \\ 0 & 0 & 0 \\ 0 & 0 & 0 \end{bmatrix} = \mathbf{O}, \text{ where } \mathbf{O} \text{ is the null matrix.}$$

Hence the matrix **A** satisfies its characteristic equation given by (i). Hence Cayley-Hamilton's Theorem is satisfied by the matrix **A**.

Again $\quad \mathbf{A^3 - 5\,A^2 + 6\,A - 5\,I = O}$

$\Rightarrow \qquad \mathbf{5\,I = A^3 - 5\,A^2 + 6\,A}$

Multiplying both sides by $\mathbf{A}^{-1}$, we get

$$\mathbf{5A^{-1} = A^2 - 5\,A + 6\,I}, \qquad \because \qquad \mathbf{AA^{-1} = I}$$

$$= \begin{bmatrix} -2 & 1 & 4 \\ 3 & 1 & 3 \\ -5 & 5 & 14 \end{bmatrix} - 5\begin{bmatrix} 0 & 0 & 1 \\ 3 & 1 & 0 \\ -2 & 1 & 4 \end{bmatrix} + 6\begin{bmatrix} 1 & 0 & 0 \\ 0 & 1 & 0 \\ 0 & 0 & 1 \end{bmatrix}, \text{ from (ii)}$$

$$= \begin{bmatrix} -2 & 1 & 4 \\ 3 & 1 & 3 \\ -5 & 5 & 14 \end{bmatrix} + \begin{bmatrix} 0 & 0 & -5 \\ -15 & -5 & 0 \\ 10 & -5 & -20 \end{bmatrix} + \begin{bmatrix} 6 & 0 & 0 \\ 0 & 6 & 0 \\ 0 & 0 & 6 \end{bmatrix}$$

$$= \begin{bmatrix} -2+0+6 & 1+0+0 & 4-5+0 \\ 3-15+0 & 1-5+6 & 3+0+0 \\ -5+10+0 & 5-5+0 & 14-20+6 \end{bmatrix} = \begin{bmatrix} 4 & 1 & -1 \\ -12 & 2 & 3 \\ 5 & 0 & 0 \end{bmatrix}$$

$$\Rightarrow A^{-1} = \frac{1}{5}\begin{bmatrix} 4 & 1 & -1 \\ -12 & 2 & 3 \\ 5 & 0 & 0 \end{bmatrix}.$$

Example 2:

Show that the matrix $A = \begin{bmatrix} 1 & 2 & 1 \\ 0 & 1 & -1 \\ 3 & -1 & 1 \end{bmatrix}$ *satisfies.*

Cayley-Hamilton Theorem.

Solution:

Here we have

$$|A - \lambda I| = \begin{vmatrix} 1-\lambda & 2 & 1 \\ 0 & 1-\lambda & -1 \\ 3 & -1 & 1-\lambda \end{vmatrix}$$

$$= (1-\lambda)\{(1-\lambda)(1-\lambda) - 1\} - 2\{3\} + 1\{-3(1-\lambda)\}$$

$$= (1-\lambda)^3 - (1-\lambda) - 6 - 3(1-\lambda) = -9 + \lambda + 3\lambda^2 - \lambda^3$$

$\therefore$ The characteristic equation of **A** is $\lambda^3 - 3\lambda^2 - \lambda + 9 = 0$ ...(i)

Also we have

$$A^2 = \begin{bmatrix} 1 & 2 & 1 \\ 0 & 1 & -1 \\ 3 & -1 & 1 \end{bmatrix} \times \begin{bmatrix} 1 & 2 & 1 \\ 0 & 1 & -1 \\ 3 & -1 & 1 \end{bmatrix} = \begin{bmatrix} 4 & 3 & 0 \\ -3 & 2 & -2 \\ 6 & 4 & 5 \end{bmatrix}$$

and $A^3 = A^2.A = \begin{bmatrix} 4 & 3 & 0 \\ -3 & 2 & -2 \\ 6 & 4 & 1 \end{bmatrix} \times \begin{bmatrix} 1 & 2 & 1 \\ 0 & 1 & -1 \\ 3 & -1 & 1 \end{bmatrix}$

$$= \begin{bmatrix} 4 & 11 & 1 \\ -9 & -2 & -7 \\ 21 & 11 & 7 \end{bmatrix}$$

$\therefore$ $\mathbf{A^3 - 3A^2 - A + 9I}$

$$= \begin{bmatrix} 4 & 11 & 1 \\ -9 & -2 & -7 \\ 21 & 11 & 7 \end{bmatrix} - 3\begin{bmatrix} 4 & 3 & 0 \\ -3 & 2 & -2 \\ 6 & 4 & 5 \end{bmatrix}$$

$$- \begin{bmatrix} 1 & 2 & 1 \\ 0 & 1 & -1 \\ 3 & -1 & 1 \end{bmatrix} + 9\begin{bmatrix} 1 & 0 & 0 \\ 0 & 1 & 0 \\ 0 & 0 & 1 \end{bmatrix}$$

$$= \begin{bmatrix} 4 & 11 & 1 \\ -9 & -2 & -7 \\ 21 & 11 & 7 \end{bmatrix} + \begin{bmatrix} -12 & -9 & 0 \\ 9 & -6 & 6 \\ -18 & -12 & -15 \end{bmatrix}$$

$$+ \begin{bmatrix} -1 & -2 & -1 \\ 0 & -1 & 1 \\ -3 & 1 & -1 \end{bmatrix} + \begin{bmatrix} 9 & 0 & 0 \\ 0 & 9 & 0 \\ 0 & 0 & 0 \end{bmatrix}$$

$$= \begin{bmatrix} 0 & 0 & 0 \\ 0 & 0 & 0 \\ 0 & 0 & 0 \end{bmatrix} = \mathbf{O}$$, where **O** is the null matrix.

Hence the matrix satisfies its characteristic equation given by (i). Hence Cayley Hamilton theorem is verified for the given matrix **A**.

Example 3:

Verify that matrix $\mathbf{A} = \begin{bmatrix} 2 & 2 & 1 \\ 1 & 3 & 1 \\ 1 & 2 & 2 \end{bmatrix}$ *satisfies its characteristic equation.*

Solution:

We can prove that the characteristic equation of the matrix **A** is $(1 - \lambda)^2 (5 - \lambda) = 0$

$$\Rightarrow \quad -(\lambda^3 - 7\lambda^2 - 11\lambda - 5) = 0$$

$$\Rightarrow \quad \lambda^3 - 7\lambda^2 + 11\lambda - 5 = 0 \qquad \text{...(i)}$$

Also $\mathbf{A}^2 = \begin{bmatrix} 2 & 2 & 1 \\ 1 & 3 & 1 \\ 1 & 2 & 2 \end{bmatrix} \times \begin{bmatrix} 2 & 2 & 1 \\ 1 & 3 & 1 \\ 1 & 2 & 2 \end{bmatrix}$

$$= \begin{bmatrix} 4+2+1 & 4+6+2 & 2+2+2 \\ 2+3+1 & 2+9+2 & 1+3+2 \\ 2+2+2 & 2+6+4 & 1+2+4 \end{bmatrix} = \begin{bmatrix} 7 & 12 & 6 \\ 6 & 13 & 6 \\ 6 & 12 & 7 \end{bmatrix}$$

$\mathbf{A^3 = A^2.A}$

$$= \begin{bmatrix} 7 & 12 & 6 \\ 6 & 13 & 6 \\ 6 & 12 & 7 \end{bmatrix} \cdot \begin{bmatrix} 2 & 2 & 1 \\ 1 & 3 & 1 \\ 1 & 2 & 2 \end{bmatrix}$$

$$= \begin{bmatrix} 14+12+6 & 14+36+12 & 7+12+12 \\ 12+13+6 & 12+39+12 & 6+13+12 \\ 12+12+7 & 12+36+14 & 6+12+14 \end{bmatrix}$$

$$= \begin{bmatrix} 32 & 62 & 31 \\ 31 & 63 & 31 \\ 31 & 62 & 32 \end{bmatrix}$$

$\mathbf{A^3 - 7\,A^2 + 11A - 5I}$

$$= \begin{bmatrix} 32 & 62 & 31 \\ 31 & 63 & 31 \\ 31 & 62 & 32 \end{bmatrix} - 7 \begin{bmatrix} 7 & 12 & 6 \\ 6 & 13 & 6 \\ 6 & 12 & 7 \end{bmatrix} + 11 \begin{bmatrix} 2 & 2 & 1 \\ 1 & 3 & 1 \\ 1 & 2 & 2 \end{bmatrix} - 5 \begin{bmatrix} 1 & 0 & 0 \\ 0 & 1 & 0 \\ 0 & 0 & 1 \end{bmatrix}$$

$$= \begin{bmatrix} 32 & 62 & 31 \\ 31 & 63 & 31 \\ 31 & 62 & 32 \end{bmatrix} + \begin{bmatrix} -49 & -84 & -42 \\ -42 & -91 & -42 \\ -42 & -84 & -49 \end{bmatrix} + \begin{bmatrix} 22 & 22 & 11 \\ 11 & 33 & 11 \\ 11 & 22 & 22 \end{bmatrix} + \begin{bmatrix} -5 & 0 & 0 \\ 0 & -5 & 0 \\ 0 & 0 & -5 \end{bmatrix}$$

$$= \begin{bmatrix} 32-49+22-5 & 62-84+22+0 & 31-42+11+0 \\ 31-42+11+0 & 63-91+33-5 & 31-42+11+0 \\ 31-42+11+0 & 62-84+22+0 & 32-49+22-5 \end{bmatrix}$$

$$= \begin{bmatrix} 0 & 0 & 0 \\ 0 & 0 & 0 \\ 0 & 0 & 0 \end{bmatrix}$$

= **O**, where **O** is the null matrix.

Hence the matrix **A** satisfies its characteristic equation given by (i)

Example 4:

Find the characteristic vectors of the matrix A given in above.

Solution:

As in above we can find that the characteristic equation **A** is $(1 - \lambda)^2 (5 - \lambda) = 0$ and so the characteristic roots of **A** are 1, 1, 5.

The equation $(\mathbf{A} - \lambda\mathbf{I})\ \mathbf{X} = \mathbf{O}$ for the matrix **A** is

$$\begin{bmatrix} 2-\lambda & 2 & 1 \\ 1 & 3-\lambda & 1 \\ 1 & 1 & 2-\lambda \end{bmatrix} \begin{bmatrix} x_1 \\ x_2 \\ x_3 \end{bmatrix} = \begin{bmatrix} 0 \\ 0 \\ 0 \end{bmatrix} \quad \text{...(i)}$$

Putting $\lambda = 1$ in the above equation, we get

$$\begin{bmatrix} 1 & 2 & 1 \\ 1 & 2 & 1 \\ 1 & 2 & 1 \end{bmatrix} \begin{bmatrix} x_1 \\ x_2 \\ x_3 \end{bmatrix} = \begin{bmatrix} 0 \\ 0 \\ 0 \end{bmatrix}$$

The corresponding characteristic vector is given by the equation

$$x_1 + 2x_2 + x_3 = 0$$

$\therefore$ The characteristic vector corresponding to $\lambda = 1$ may be taken as $(1, -1, 1)$. **Ans.**

Putting $\lambda = 5$ in (i), we get

$$\begin{bmatrix} 2-5 & 2 & 1 \\ 1 & 3-5 & 1 \\ 1 & 2 & 2-5 \end{bmatrix} \begin{bmatrix} x_1 \\ x_2 \\ x_3 \end{bmatrix} = \begin{bmatrix} 0 \\ 0 \\ 0 \end{bmatrix}$$

$$\Rightarrow \begin{bmatrix} -3 & 2 & 1 \\ 1 & -2 & 1 \\ 1 & 2 & -3 \end{bmatrix} \begin{bmatrix} x_1 \\ x_2 \\ x_3 \end{bmatrix} = \begin{bmatrix} 0 \\ 0 \\ 0 \end{bmatrix}$$

The corresponding characteristic vector is given by the equations $-3x_1 + 2x_2 + x_3 = 0$, $x_1 - 2x_2 + x_3 =$ and $x_1 + 2x_2 - 3x_3 = 0$.

Solving these we get $\dfrac{x_1}{1} = \dfrac{x_2}{1} = \dfrac{x_3}{1}$

$\therefore$ The characteristic vector corresponding to $\lambda = 5$ may be taken as $(1,1,1)$. **Ans.**

Example 5:

Verify Cayley Hamilton's Theorem for the matrix

$$A = \begin{bmatrix} 2 & -1 & 1 \\ -1 & 2 & -1 \\ 1 & -1 & 2 \end{bmatrix}.$$

Find the characteristic equation of the matrix $\mathbf{A} = \begin{bmatrix} 2 & -1 & 1 \\ -1 & 2 & -1 \\ 1 & -1 & 2 \end{bmatrix}$ *and verify that it is satisfied by A.*

Solution:

Here $|\mathbf{A} - \lambda\mathbf{I}| = \begin{vmatrix} 2-\lambda & -1 & 1 \\ -1 & 2-\lambda & -1 \\ 1 & -1 & 2-\lambda \end{vmatrix}$

$$= \begin{vmatrix} 2-\lambda & 0 & 1 \\ -1 & 1-\lambda & -1 \\ 1 & \lambda-1 & 2-\lambda \end{vmatrix}, \text{ replacing } C_2 \text{ by } C_2 + C_3$$

$$= (1-\lambda) \begin{vmatrix} 2-\lambda & 0 & 0 \\ -1 & 1 & -1 \\ 1 & 1 & 2-\lambda \end{vmatrix}$$

$$= (1-\lambda) \left\{ (2-\lambda) \begin{vmatrix} 1 & -1 \\ 1 & 2-\lambda \end{vmatrix} + \begin{vmatrix} -1 & 1 \\ 1 & 1 \end{vmatrix} \right\}$$

$$= (1-\lambda)\{(2-\lambda)(3-\lambda) + (-1-1)\}$$

$$= (1-\lambda)\{4 - 5\lambda + \lambda^2\} = 4 - 5\lambda + \lambda^2 - 4\lambda + 5\lambda^2 - \lambda^3$$

$$= -\lambda^3 + 6\lambda^2 - 9\lambda + 4.$$

∴ The characteristic equation of the matrix A is

$$\lambda^3 - 6\lambda^2 + 9\lambda - 4 = 0 \qquad \text{...(i)}$$

Now $\mathbf{A}^2 = \begin{bmatrix} 2 & -1 & 1 \\ -1 & 2 & -1 \\ 1 & -1 & 2 \end{bmatrix} \begin{bmatrix} 2 & -1 & 1 \\ -1 & 2 & -1 \\ 1 & -1 & 2 \end{bmatrix}$

$$\Rightarrow \mathbf{A}^2 = \begin{bmatrix} 6 & -5 & 5 \\ -5 & 6 & -5 \\ 5 & -5 & 6 \end{bmatrix}$$

And $\mathbf{A}^3 = \mathbf{A}^2.\mathbf{A} = \begin{bmatrix} 6 & -5 & 5 \\ -5 & 6 & -5 \\ 5 & -5 & 6 \end{bmatrix} \begin{bmatrix} 2 & -1 & 1 \\ -1 & 2 & -1 \\ 1 & -1 & 2 \end{bmatrix}$

$$= \begin{bmatrix} 22 & -21 & 21 \\ -21 & 22 & -21 \\ 21 & -21 & 22 \end{bmatrix}$$

$\therefore$ $\mathbf{A^3 - 6A^2 + 9A - 4I}$

$$= \begin{bmatrix} 22 & -21 & 21 \\ -21 & 22 & -21 \\ 21 & -21 & 22 \end{bmatrix} - 6\begin{bmatrix} 6 & -5 & 5 \\ -5 & 6 & -5 \\ 5 & -5 & 6 \end{bmatrix} + 9\begin{bmatrix} 2 & -1 & 1 \\ -1 & 2 & -1 \\ 1 & -1 & 2 \end{bmatrix} - 4\begin{bmatrix} 1 & 0 & 0 \\ 0 & 1 & 0 \\ 0 & 0 & 1 \end{bmatrix}$$

$$= \begin{bmatrix} 22 & -21 & 21 \\ -21 & 22 & -21 \\ 21 & -21 & 22 \end{bmatrix} + \begin{bmatrix} -36 & 30 & -30 \\ 30 & -36 & 30 \\ -30 & 30 & -36 \end{bmatrix} + \begin{bmatrix} 18 & -9 & 9 \\ -9 & 18 & -9 \\ 9 & -9 & 18 \end{bmatrix}$$

$$+ \begin{bmatrix} -4 & 0 & 0 \\ 0 & -4 & 0 \\ 0 & 0 & -4 \end{bmatrix}$$

$$= \begin{bmatrix} 22-36+18-4 & -21+30-9+0 & 21-30+9+0 \\ -21+30-9+0 & 22-36+18-4 & -21+30-9+0 \\ 21-30+9+0 & -21+30-9+0 & 22-36+18-4 \end{bmatrix}$$

$$= \begin{bmatrix} 0 & 0 & 0 \\ 0 & 0 & 0 \\ 0 & 0 & 0 \end{bmatrix} = O, \text{ where } O \text{ is the null matrix}$$

Hence **A** satisfies is characteristic equation given by (i).

Hence Cayley Hamilton Theorem is satisfied by the matrix **A**.

SOLVED EXAMPLES

Example 1:

Find the eigen-values and eigen-vectors of the matrix

$$A = \begin{bmatrix} 2 & 1 & 1 \\ -11 & 4 & 5 \\ -1 & 1 & 0 \end{bmatrix}.$$

Solution:

Here $|\mathbf{A} - \lambda\mathbf{I}|$

$$= \begin{vmatrix} 2-\lambda & 1 & 1 \\ -11 & 4-\lambda & 5 \\ -1 & 1 & -\lambda \end{vmatrix} = \begin{vmatrix} 2-\lambda & 9 & 1 \\ -11 & -1-\lambda & 5 \\ -1 & 1+\lambda & -\lambda \end{vmatrix}, \text{ applying } C_2 - C_3$$

$$= (2-\lambda)\begin{vmatrix} -(1+\lambda) & 5 \\ (1+\lambda) & -\lambda \end{vmatrix} + \begin{vmatrix} -11 & -(1+\lambda) \\ -1 & (1+\lambda) \end{vmatrix}$$

$= (2 - \lambda)\,[\lambda\,(1 + \lambda) - 5\,(1 + \lambda)] + [-\,11\,(1 + \lambda) - (1 + \lambda)]$

$= (\lambda + 1)\,[(2 - \lambda)\,(\lambda - 5) - 12] = (\lambda + 1)\,(-\,\lambda^2 + 7\lambda - 22)$

$= -\,(\lambda + 1)\,(\lambda^2 - 7\lambda + 22)$

$\therefore$ The characteristic equation of **A** is $(\lambda + 1)(\lambda^2 - 7\lambda + 22) = 0$ its roots *i.e.*, required eigen-values of **A** are

$$-1,\ \frac{1}{2}\left[7 \pm \sqrt{(49-88)}\right] \Rightarrow -1,\ \frac{1}{2}\left[7 \pm \sqrt{(-39)}\right].$$ **Ans.**

Now the equation **(A – λI) X = O**, for the matrix **A** is

$$\begin{bmatrix} 2-\lambda & 1 & 5 \\ -11 & 4-\lambda & 5 \\ -1 & 1 & -\lambda \end{bmatrix}\begin{bmatrix} x_1 \\ x_2 \\ x_3 \end{bmatrix} = \mathbf{O}$$

Putting $\lambda = -\,1$ in (i) we get

$$\begin{bmatrix} 3 & 1 & 1 \\ -11 & 5 & 5 \\ -1 & 1 & 1 \end{bmatrix}\begin{bmatrix} x_1 \\ x_2 \\ x_3 \end{bmatrix} = \mathbf{O}$$

The corresponding eigen-vector is given by the equations

$$3x_1 + x_2 + x_3 = 0,\ -\,11x_1 + 5x_2 + 5x_3 = 0$$

and $\quad -\,x_1 + x_2 + x_3 = 0$, which give $x_1 = 0 = x_2 = x_3$

These being all zero the eigen vector corresponding to $\lambda = -\,1$ cannot be evaluated. **Ans.**

Similarly, corresponding to $\lambda = \frac{1}{2}\left[7 \pm \sqrt{(-39)}\right]$ we do not get any eigen vectors.

Example 2:

Find the eigen-vectors of

$$A = \begin{bmatrix} -2 & 2 & -3 \\ 2 & 1 & -6 \\ -1 & -2 & 0 \end{bmatrix}.$$

Solution:

We have calculated

$$|\mathbf{A} - \lambda\mathbf{I}| = \begin{vmatrix} -2-\lambda & 2 & -3 \\ 2 & 1-\lambda & -6 \\ -1 & -2 & -\lambda \end{vmatrix}$$

and the eigen values are given by $\lambda = 5, -3, -3$.

Now the equation **$(A - \lambda I)\ X = O$,** for the matrix A is

$$\begin{bmatrix} -2-\lambda & 2 & -3 \\ 2 & 1-\lambda & -6 \\ -1 & -2 & -\lambda \end{bmatrix} \begin{bmatrix} x_1 \\ x_2 \\ x_3 \end{bmatrix} = \mathbf{O}$$

Putting $\lambda = 5$, in (i) we get

$$\begin{bmatrix} -7 & 2 & -3 \\ 2 & -4 & -6 \\ -1 & -2 & -5 \end{bmatrix} \begin{bmatrix} x_1 \\ x_2 \\ x_3 \end{bmatrix} = O$$

The corresponding eigen-vector is given by the equations

$-7x_1 + 2x_2 - 3x_3 = 0,\ 2x_1 - 4x_2 - 6x_3 = 0,\ -x_1 - 2x_2 - 5x_3 = 0$

These give $\dfrac{x_1}{1} = \dfrac{x_2}{2} = \dfrac{x_3}{-1}$

$\therefore$ The eigen-vector corresponding to $\lambda = 5$ may be taken as $(1, 2, -1)$.

Putting $\lambda = -3$ in (i) we get

$$\begin{bmatrix} 1 & 2 & -3 \\ 2 & 4 & -6 \\ -1 & -2 & 3 \end{bmatrix} \begin{bmatrix} x_1 \\ x_2 \\ x_3 \end{bmatrix} = \mathbf{O}$$

The corresponding eigen-vector is given by the equations

$x_1 + 2x_2 - 3x_3 = 0,\ 2x_1 + 4x_2 - 6x_3 = 0,\ -x_1 - 2x_2 + 3x_3 = 0$

These reduce to $x_1 + 2x_2 - 3x_2 = 0$ only and so non-zero solution of this cannot be found, hence non eigen-vector can be derived for $\lambda = -3$.

Example 3:

Find the characteristic vector of

$$A = \begin{bmatrix} 1 & 2 & 3 \\ 0 & 2 & 3 \\ 0 & 0 & 2 \end{bmatrix}.$$

Solution:

We have calculated

$$|\mathbf{A} - \lambda\mathbf{I}| = \begin{vmatrix} 1-\lambda & 2 & 3 \\ 0 & 2-\lambda & 3 \\ 0 & 0 & 2-\lambda \end{vmatrix}$$

and the eigen-values are given by $\lambda = 1, 2, 2$.

Now the equation $(\mathbf{A} - \lambda\mathbf{I})\, \mathbf{X} = \mathbf{O}$, for the matrix **A** is

$$\begin{vmatrix} 1-\lambda & 2 & 3 \\ 0 & 2-\lambda & 3 \\ 0 & 0 & 2-\lambda \end{vmatrix} \begin{bmatrix} x_1 \\ x_2 \\ x_3 \end{bmatrix} = \mathbf{O} \qquad \text{...(i)}$$

Putting $\lambda =$ in (i) we get

$$\begin{bmatrix} 0 & 2 & 3 \\ 0 & 1 & 3 \\ 0 & 0 & 1 \end{bmatrix} \begin{bmatrix} x_1 \\ x_2 \\ x_3 \end{bmatrix} = \mathbf{O}$$

The corresponding characteristic vector is given by the equations $2x_2 + x_3 = 0$, $x_2 + x_3 = 0$, $x_3 = 0$ which do not give non-zero solution of these equations and hence no characteristic vector can be derived for $\lambda = 1$.

Putting $\lambda = 2$ in (i) we get

$$\begin{bmatrix} -1 & 2 & 3 \\ 0 & 0 & 3 \\ 0 & 0 & 0 \end{bmatrix} \begin{bmatrix} x_1 \\ x_2 \\ x_3 \end{bmatrix} = \mathbf{O}$$

The corresponding characteristic vector is given by the equations

$$-x_1 + 2x_2 - 3x_3 = 0,\ 3x_3 = 0$$

These gives $\dfrac{x_1}{2} = \dfrac{x_2}{1} = \dfrac{x_3}{0}$.

$\therefore$ The characteristic vector corresponding to $\lambda = 2$ may be takne as (2, 1, 0). **Ans.**

Example 4:

Find the eigen-vectors of the matrix

$$A = \begin{bmatrix} 3 & 1 & 4 \\ 0 & 2 & 6 \\ 0 & 0 & 5 \end{bmatrix}.$$

Solution:

We have calculated

$$|\mathbf{A} - \lambda \mathbf{I}| = \begin{vmatrix} 3-\lambda & 1 & 4 \\ 0 & 2-\lambda & 6 \\ 0 & 0 & 5-\lambda \end{vmatrix}$$

and the characteristic roots (or eigenvalues) are given by $\lambda = 2, 3, 5$.

Now the equation $(\mathbf{A} - \lambda\mathbf{I})\,\mathbf{X} = \mathbf{O}$, for the matrix **A** is

$$\begin{vmatrix} 3-\lambda & 1 & 4 \\ 0 & 2-\lambda & 6 \\ 0 & 0 & 5-\lambda \end{vmatrix} \begin{bmatrix} x_1 \\ x_2 \\ x_3 \end{bmatrix} = \mathbf{O} \qquad \text{...(i)}$$

Putting $\lambda = 5$ in (i) we get

$$\begin{bmatrix} -2 & 1 & 4 \\ 0 & -3 & 6 \\ 0 & 0 & 0 \end{bmatrix} \begin{bmatrix} x_1 \\ x_2 \\ x_3 \end{bmatrix} = \mathbf{O}$$

The corresponding eigen-vector is given by the equations

$$-2x_1 + x_2 + 4x_3 = 0, \; -3x_2 + 6x_3 = 0,$$

which give $\quad 3x_2 = 6x_3 \quad \Rightarrow \dfrac{x_2}{1} = \dfrac{x_3}{1}$ ...(ii)

Now $\dfrac{x_2}{2} = \dfrac{x_3}{1} = k$ (say), then from $-2x_1 + x_2\ 4x_3 = 0$ we get

$$2x_1 = x_2 + 4x_3 = 2k + 4k = 6k \Rightarrow x_1 = 3k$$

$$\Rightarrow \quad \frac{x_1}{3} = k. \text{ So we get } \frac{x_1}{3} = \frac{x_2}{2} = \frac{x_3}{1}$$

∴ The characteristic vector corresponding to $\lambda = 5$ may be taken (3, 2, 1).

For $\lambda = 2, 3$ we find that $|\mathbf{A} - \lambda\mathbf{I}| = \mathbf{0}$ and so non-zero solutions of (i) cannot be evaluatged in these cases *i.e.*, eigenvectors cannot be calculated.

Example 5:

Determine the characteristic roots and obtain the characteristic equation of the matrix $A = \begin{bmatrix} 1 & 0 & 2 \\ 0 & 1 & 2 \\ 1 & 2 & 0 \end{bmatrix}$.

Solution:

Here

$$|\mathbf{A} - \lambda\mathbf{I}| = \begin{vmatrix} 1-\lambda & 0-0 & 2-0 \\ 0-0 & 1-\lambda & 0-0 \\ 1-0 & 2-0 & 0-\lambda \end{vmatrix} = \begin{vmatrix} 1-\lambda & 0 & 2 \\ 0 & 1-\lambda & 2 \\ 1 & 2 & -\lambda \end{vmatrix}$$

$$= (1-\lambda)\begin{vmatrix} 1-\lambda & 2 \\ 2 & -\lambda \end{vmatrix} + \begin{vmatrix} 0 & 2 \\ 1-\lambda & 2 \end{vmatrix}$$, expanding with respect to C_1

$$= (1-\lambda)\,[-\lambda(1-\lambda) - 4] - 2(1-\lambda)$$

$$= (1-\lambda)\,[-\lambda + \lambda^2 - 4 - 2]$$

$$= (1-\lambda)\,[\lambda^2 - \lambda - 6] = -(\lambda - 1)(\lambda + 2)(\lambda - 3)$$

∴ The characteristic equation of the matrix **A** is

$$(\lambda - 1)(\lambda + 2)(\lambda - 3) = 0.$$

Hence the characteristic roots of A are 1, – 2, 3. **Ans.**

Example 6:

If $a + b + c = 0$, *find the characteristic roots of the matrix*

$$A = \begin{bmatrix} a & c & b \\ c & b & a \\ b & a & c \end{bmatrix}.$$

Solution:

Here

$$|\mathbf{A} - \lambda\mathbf{I}| = \begin{vmatrix} a-\lambda & c-0 & b-0 \\ c-0 & b-\lambda & a-0 \\ b-0 & a-0 & c-\lambda \end{vmatrix} = \begin{vmatrix} a-\lambda & c & b \\ c & b-\lambda & a \\ b & a & c-\lambda \end{vmatrix}$$

$$= \begin{vmatrix} a+b+c-\lambda & c & b \\ c+b+a-\lambda & b-\lambda & a \\ b+a+c-\lambda & a & c-\lambda \end{vmatrix}$$, replacing C_1 by $C_1 + C_2\ C_3$

$$= \begin{vmatrix} -\lambda & c & b \\ -\lambda & b-\lambda & a \\ -\lambda & a & c-\lambda \end{vmatrix}$$, since $a + b + c = 0$ (given)

$$= \begin{vmatrix} -\lambda & c & b \\ 0 & b-\lambda-c & a-b \\ 0 & a-c & c-\lambda-b \end{vmatrix}, \text{ replacing } R_2, R_3 \text{ by } R_2 - R_1 \text{ and } R_3 - R_1$$

$$= -\lambda \begin{vmatrix} b-c-\lambda & a-b \\ a-c & c-b-\lambda \end{vmatrix}, \text{ expanding with respect to } C_1$$

$$= -\lambda\,[(b-c-\lambda)(c-b-\lambda)-(a-b)(a-c)]$$

$$= -\lambda\,[bc - b^2 - b\lambda - c^2 + cb + c\lambda - \lambda c + b\lambda + \lambda^2 - a^2 + ac + ba - bc]$$

$$= \lambda\,[(a^2+b^2+c^2-ab-bc-ca)-\lambda^2] \qquad \text{...(i)}$$

Also $a+b+c=0 \Rightarrow (a+b+c)^2 = 0$

$$\Rightarrow a^2+b^2+c^2+2ab+2bc+2ca = 0$$

$$\Rightarrow 2\,(ab+bc+ca) = -(a^2+b^2+c^2) \qquad \textbf{(Note)}$$

$$\Rightarrow -(ab+bc+ca) = \frac{1}{2}(a^2+b^2+c^2)$$

∴ From (i) we get

$$|\mathbf{A}-\lambda\mathbf{I}| = \lambda\,[\{(a^2+b^2+c^2)+\frac{1}{2}(a^2+b^2+c^2)\}-\lambda^2] \qquad \textbf{(Note)}$$

$$= \lambda\,[(3/2)(a^2+b^2+c^2)-\lambda^2$$

∴ The characteristic equation of A is $\lambda\,[3/2\,(a^2+b^2+c^2)-\lambda^2] = 0$ which gives $\lambda = 0$ or $\lambda^2 = (3/2)(a^2+b^2+c^2)$.

∴ The required roots are $0, \pm\,(3/2)(a^2+b^2+c^2)]^{1/2}$. **Ans.**

Example 7:

Find latent roots and latent vectors of the matrix

$$A = \begin{bmatrix} a & h & g \\ 0 & b & 0 \\ 0 & 0 & c \end{bmatrix}.$$

Solution:

Here $|\mathbf{A}-\lambda\mathbf{I}| = \begin{vmatrix} a-\lambda & h & g \\ 0 & b-\lambda & 0 \\ 0 & 0 & c-\lambda \end{vmatrix}$

$$= (a-\lambda)(b-\lambda)(c-\lambda)$$

∴ The characteristic equation of the matrix **A** is

$$(\lambda - a)(\lambda - b)(\lambda - c) = 0$$

and the characteristic or latent roots of **A** are a, b, c. **Ans.**

Again the equation **(A – λI) X = O** for the matrix **A** is

$$\begin{vmatrix} a-\lambda & h & g \\ 0 & b-\lambda & 0 \\ 0 & 0 & c-\lambda \end{vmatrix}\begin{bmatrix} x_1 \\ x_2 \\ x_3 \end{bmatrix} = \begin{bmatrix} 0 \\ 0 \\ 0 \end{bmatrix} \quad \text{...(i)}$$

Putting $\lambda = a$ in the above equation we get

$$\begin{bmatrix} 0 & h & g \\ 0 & b-a & 0 \\ 0 & 0 & c-a \end{bmatrix}\begin{bmatrix} x_1 \\ x_2 \\ x_3 \end{bmatrix} = \begin{bmatrix} 0 \\ 0 \\ 0 \end{bmatrix}$$

The corresponding characteristic vector is given by the equations

$$0.x_1 + hx_2 + gx_3 = 0,\ (b-a)\,x_2 = 0,\ (c-a)\,x_3 = 0$$

∴ The characteristic vector corresponding to $\lambda = a$ may be taken as (a, 0, 0). **Ans.**

Similarly we can find the characteristic vector corresponding to $\lambda = b$ and $\lambda = c$ as (– h, a – b, 0) and (– h, 0, a – c). **Ans.**

Example 8:

Find the characteristic roots and vectors of the matrix $A = \begin{bmatrix} -2 & -1 \\ 5 & 4 \end{bmatrix}$.

Solution:

Here $|\mathbf{A} - \lambda\mathbf{I}| = \begin{vmatrix} -2-\lambda & -1-0 \\ 5-0 & 4-\lambda \end{vmatrix}$

$$= (-2-\lambda)(4-\lambda) - 5(-1) = -8 + 2\lambda + \lambda^2 + 5$$

$$= \lambda^2 - 2\lambda - 3 = (\lambda - 3)(\lambda + 1).$$

∴ The characteristic equation of the matrix **A** is $(\lambda - 3)(\lambda + 1) = 0$ and the characteristic roots of **A** therefore are – 1, 3. **Ans.**

The equation **(A – lI) X = O** for the matrix **A** is

$$\begin{bmatrix} -2-\lambda & -1 \\ 5 & 4-\lambda \end{bmatrix}\begin{bmatrix} x_1 \\ x_2 \end{bmatrix} = \begin{bmatrix} 0 \\ 0 \end{bmatrix} \quad \text{...(i)}$$

Putting l = – 1 in the above equation we get

$$\begin{bmatrix} -2+1 & -1 \\ 5 & 4+1 \end{bmatrix}\begin{bmatrix} x_1 \\ x_2 \end{bmatrix} = \begin{bmatrix} 0 \\ 0 \end{bmatrix} \Rightarrow \begin{bmatrix} -1 & -1 \\ 5 & 5 \end{bmatrix}\begin{bmatrix} x_1 \\ x_2 \end{bmatrix} = \begin{bmatrix} 0 \\ 0 \end{bmatrix}$$

The corresponding characteristic vector is given by the equation

$$-x_1 - x_2 = 0 \text{ or } x_1 + x_2 = 0 \text{ or } \frac{x_1}{1} = \frac{x_2}{-1}$$

∴ The characteristic vector corresponding to $\lambda = -1$ may be taken as (1, – 1).

Ans.

Putting $\lambda = 3$ in (i) we get

$$\begin{bmatrix} -2-3 & -1 \\ 5 & 4-3 \end{bmatrix}\begin{bmatrix} x_1 \\ x_2 \end{bmatrix} = \begin{bmatrix} 0 \\ 0 \end{bmatrix} \Rightarrow \begin{bmatrix} -5 & -1 \\ 5 & 1 \end{bmatrix}\begin{bmatrix} x_1 \\ x_2 \end{bmatrix} = \begin{bmatrix} 0 \\ 0 \end{bmatrix}$$

∴ The corresponding characteristic vector is given by the equation

$$-5x_1 - x_2 = 0 \text{ or } 5x_1 + x_2 = 0 \text{ or } \frac{x_1}{1} = \frac{x_2}{-5}.$$

∴ The characteristic vector corresponding to $\lambda = 3$ may be taken as (1 – 5).

Ans.

Example 9:

If $B = \begin{bmatrix} 2 & \sqrt{2} \\ \sqrt{2} & 1 \end{bmatrix}$ *then find the characteristic equation of B and verify that the matrix B satisfies this equation. Also find the characteristic roots and the corresponding characteristic vectors of B.*

Solution:

Here $\mathbf{B} = \begin{bmatrix} 2 & \sqrt{2} \\ \sqrt{2} & 1 \end{bmatrix}$ and $\mathbf{I} = \begin{bmatrix} 1 & 0 \\ 0 & 1 \end{bmatrix}$

$$\therefore |\mathbf{B} - \lambda\mathbf{I}| = \begin{vmatrix} 2-\lambda & \sqrt{2} \\ \sqrt{2} & 1-\lambda \end{vmatrix}$$

$$= [(2-\lambda)(1-\lambda) - \sqrt{2}\sqrt{2}] = 2 - 2\lambda + \lambda^2 - 2$$

$$= \lambda^2 - 3\lambda = \lambda(\lambda - 3)$$

∴ The characteristic equation of the matrix B is

$$\lambda(\lambda - 3) = 0.$$

Ans.

and the characteristic roots of **B** are 0 and 3. **Ans.**

Also $\mathbf{B}^2 = \begin{bmatrix} 2 & \sqrt{2} \\ \sqrt{2} & 1 \end{bmatrix} \times \begin{bmatrix} 2 & \sqrt{2} \\ \sqrt{2} & 1 \end{bmatrix}$

$$= \begin{bmatrix} 2.2+\sqrt{2}.\sqrt{2} & 2.\sqrt{2}+\sqrt{2}.1 \\ \sqrt{2}.2+1.\sqrt{2} & \sqrt{2}.\sqrt{2}+1.1 \end{bmatrix} = \begin{bmatrix} 6 & 3\sqrt{2} \\ 3\sqrt{2} & 3 \end{bmatrix}$$

$$\therefore \mathbf{B}^2 - 3\mathbf{B} = \begin{bmatrix} 6 & 3\sqrt{2} \\ 3\sqrt{2} & 3 \end{bmatrix} - 3\begin{bmatrix} 2 & \sqrt{2} \\ \sqrt{2} & 1 \end{bmatrix}$$

$$\Rightarrow \mathbf{B}^2 - 3\mathbf{B} = \begin{bmatrix} 6 & 3\sqrt{2} \\ 3\sqrt{2} & 3 \end{bmatrix} + \begin{bmatrix} -6 & -3\sqrt{2} \\ -3\sqrt{2} & -3 \end{bmatrix}$$

$$= \begin{bmatrix} 6-6 & 3\sqrt{2}-3\sqrt{2} \\ 3\sqrt{2}-3\sqrt{2} & 3-3 \end{bmatrix} = \begin{bmatrix} 0 & 0 \\ 0 & 0 \end{bmatrix}$$

$= \mathbf{O}$, where $\mathbf{O}$ is the null matrix.

Hence the matrix B satisfies its characteristic equation given by

$$\lambda(\lambda - 3) = 0 \Rightarrow \lambda^2 - 3\lambda = 0.$$

The equation $(\mathbf{B} - \lambda\mathbf{I})\,\mathbf{X} = \mathbf{O}$ for the matrix B is

$$\begin{bmatrix} 2-\lambda & \sqrt{2} \\ \sqrt{2} & 1-\lambda \end{bmatrix}\begin{bmatrix} x_1 \\ x_2 \end{bmatrix} = \mathbf{O} \qquad \text{...(i)}$$

Putting $\lambda = 0$ in (i) we get

$$\begin{bmatrix} 2 & \sqrt{2} \\ \sqrt{2} & 1 \end{bmatrix}\begin{bmatrix} x_1 \\ x_2 \end{bmatrix} = \mathbf{O}$$

The corresponding characteristic vector is given by the equations

$$2x_1 + \sqrt{2x_2} = 0 \text{ and } \sqrt{2x_1} + x_2 = 0. \qquad \textbf{(Note)}$$

Taking any one of them we get

$$\frac{x_1}{\sqrt{2}} = \frac{x_2}{-2} \Rightarrow \frac{x_1}{1} = \frac{x_2}{-\sqrt{2}}$$

$\therefore$ The characteristic vector corresponding to $\lambda = 0$ may be taken as $(1, -\sqrt{2})$.

Putting $\lambda = 3$ in (i) we get

$$\begin{bmatrix} -1 & \sqrt{2} \\ \sqrt{2} & -2 \end{bmatrix}\begin{bmatrix} x_1 \\ x_2 \end{bmatrix} = \mathbf{O}$$

The corresponding characteristic vector is given by the equations

$$-x_1 + \sqrt{2x_2} = 0 \text{ and } \sqrt{2x_1} - 2x_2 = 0.$$

Taking any one of them we get

$$\frac{x_1}{2} = \frac{x_2}{\sqrt{2}} \Rightarrow \frac{x_1}{\sqrt{2}} = \frac{x_2}{1}$$

$\therefore$ The characteristic vector corresponding to $\lambda = 3$ may be taken as $\left(\sqrt{2}, 1\right)$. **Ans.**

Example 10:

Show that the matrix $A = \begin{bmatrix} 0 & c & -b \\ -c & 0 & a \\ b & -a & 0 \end{bmatrix}$ *satisfies its characteristic equation. Also find* A^{-1}.

Solution:

Here $|\mathbf{A} - \lambda \mathbf{I}| = \begin{vmatrix} 0-\lambda & c-0 & -b-0 \\ -c-0 & 0-\lambda & a-0 \\ b-0 & -a-0 & a-\lambda \end{vmatrix}$

$$= -\lambda \begin{vmatrix} -\lambda & a \\ -a & -\lambda \end{vmatrix} - c \begin{vmatrix} -c & a \\ b & -\lambda \end{vmatrix} - b \begin{vmatrix} -c & -\lambda \\ b & -a \end{vmatrix}$$

$$= -\lambda (\lambda^2 + a^2) - c (c\lambda - ab) - b (ca + b\lambda)$$

$$= -\lambda^3 - \lambda (a^2 + b^2 + c^2)$$

$\therefore$ The characteristic equation of $\mathbf{A}$ is

$$\lambda^3 + \lambda (a^2 + b^2 + c^2) = 0 \qquad \text{...(i)}$$

Also we have

$$\mathbf{A}^2 = \begin{bmatrix} 0 & c & -b \\ -c & 0 & a \\ b & -a & 0 \end{bmatrix} \begin{bmatrix} 0 & c & -b \\ -c & 0 & a \\ b & -a & -0 \end{bmatrix}$$

$$= \begin{bmatrix} -(c^2 + b^2) & ab & ca \\ ab & -(c^2 + a^2) & bc \\ ac & bc & -(a^2 + b^2) \end{bmatrix}$$

$\therefore \mathbf{A}^3 = \mathbf{A}^2.\mathbf{A}$

$$= \begin{bmatrix} -(b^2 + c^2) & ab & ca \\ ab & -(c^2 + a^2) & bc \\ ac & bc & -(a^2 + b^2) \end{bmatrix} \begin{bmatrix} 0 & c & -b \\ -c & 0 & a \\ b & -a & 0 \end{bmatrix}$$

$$= \begin{bmatrix} 0 & -c(b^2+c^2+a^2) & b(b^2+c^2+a^2) \\ c(c^2+a^2+b^2) & 0 & -a(b^2+c^2+a^2) \\ -b(c^2+a^2+b^2) & a(c^2+a^2+b^2) & 0 \end{bmatrix}$$

$$= -(a^2+b^2+c^2)\begin{bmatrix} 0 & c & -b \\ -c & 0 & a \\ b & -a & 0 \end{bmatrix}$$

taking out $-(a^2 + b^2 + c^2)$ common **(Note)**

$\Rightarrow \mathbf{A^3 = -(a^2 + b^2 + c^2)\ A}$

$\Rightarrow \mathbf{A^3 + (a^2 + b^2 + c^2)\ A = O}$, which shows that the matrix A satisfies its characteristic equation given by (i).

Again multiplying both sides of (ii) by $\mathbf{A^{-2}}$, we get

$$\mathbf{A} = -(a^2 + b^2 - c^2)\ \mathbf{A^{-1}} \qquad \textbf{(Note)}$$

which gives $\quad \mathbf{A^{-1}} = -\dfrac{1}{(a^2+b^2+c^2)}\ \mathbf{A}$

$$= -\frac{1}{(a^2+b^2+c^2)}\begin{bmatrix} 0 & c & -b \\ -c & 0 & a \\ b & -a & 0 \end{bmatrix}. \qquad \textbf{Ans.}$$

Example 11:

Prove that the matrices A and $\boldsymbol{B^{-1}AB}$ *have the same latent roots.*

Solution:

We know that two matrices have the same latent roots (or characteristic roots) if their characteristic equations are the same.

Let $\mathbf{B^{-1}\ AB = C}$, then

$$\mathbf{C - \lambda I = B^{-1}\ AB - \lambda I.} \qquad \text{...(i)}$$

Also $\mathbf{B^{-1}\ \lambda IB = B^{-1}\ \lambda B = \lambda B^{-1}\ B = \lambda I}$

$\therefore$ From (i) we get $\mathbf{C - \lambda I = B^{-1}\ AB - B^{-1}\ \lambda IB}$

$$\mathbf{= B^{-1}\ (A - \lambda I)\ B}$$

$$\Rightarrow \mathbf{|\ C - \lambda I\ | = |\ B^{-1}\ |\ |\ A - \lambda I\ |\ |\ B\ |}$$

$$\mathbf{= |\ A - \lambda I\ |\ |\ B^{-1}\ |\ |\ B\ |}$$

$$\mathbf{= |\ A - \lambda I|\ |\ B^{-1}\ B\ |}$$

$$\mathbf{= |\ A - \lambda I\ |\ |\ I\ | = |\ A - \lambda I\ |}$$

$\therefore |\mathbf{C} - \lambda\mathbf{I}| = 0 \Rightarrow |\mathbf{A} - \lambda\mathbf{I}| = 0$

Hence the characteristic equation of **C** and **A** are the same *i.e.,* **C** and **A** or $\mathbf{B}^{-1}$ **AB** and **A** have the same latent roots. **Hence proved.**

Example 12:

Prove that the eigen values of a diagonal matrix are given by its diagonal elements.

Solution:

Let $\mathbf{A} = \begin{bmatrix} a_{11} & 0 & \dots & 0 \\ 0 & a_{22} & \dots & 0 \\ \dots & \dots & \dots & \dots \\ 0 & 0 & \dots & a_{nn} \end{bmatrix}$

be a diagonal matrix with $a_{11}, a_{22}, \dots, a_{nn}$ as diagonal elements.

Then the characteristic equation is

$$|\mathbf{A} - \lambda\mathbf{I}| = 0 \Rightarrow \begin{bmatrix} a_{11}-\lambda & 0 & \dots & 0 \\ 0 & a_{22}-\lambda & \dots & 0 \\ \dots & \dots & \dots & \dots \\ 0 & 0 & \dots & a_{nn}-\lambda \end{bmatrix} = 0$$

$\Rightarrow (a_{11} - \lambda)(a_{22} - \lambda) \dots (a_{nn} - \lambda) = 0$

$\Rightarrow \lambda = a_{11}, a_{22}, \dots, a_{nn}$ are the eigen values of the matrix **A** and are the diagonal elements of the diagonal matrix **A**. **Hence proved.**

Example 13:

Find the spectrum of the matrix

$$A = \begin{bmatrix} -2 & 2 & -3 \\ 2 & 1 & -6 \\ -1 & -2 & 0 \end{bmatrix}.$$

Solution:

We can find that the eigen values of the matrix **A** are 5, – 3, – 3.

Also we know spectrum of **A** is the set of eigen values of **A**.

$\therefore$ Required spectrum of **A** = {5, – 3, – 3,}

= {5, – 3} **Ans.**

Example 14:

The equation $AX = \lambda X$ has non-trivial solution X iff λ is a characteristic value of A.

Solution:

Let λ_1 be a characteristic value of **A** and $\mathbf{X}_1$ be the corresponding characteristic vector of **A**, then

$$\mathbf{AX}_1 = \lambda_1\mathbf{X}_1$$
$$= \lambda_1\ \mathbf{IX}_1 = (\lambda_1\mathbf{I})\ \mathbf{X}_1$$

$\Rightarrow \mathbf{AX}_1 - (\lambda_1\mathbf{I})\ \mathbf{X}_1 = \mathbf{O}$, where **O** is the null matrix

$\Rightarrow \quad (\mathbf{A} - \lambda_1\mathbf{I})\ \mathbf{X}_1 = \mathbf{O}$

$\Rightarrow \quad (\mathbf{A} - \lambda_1\mathbf{I}) = \mathbf{O}, \qquad \because \mathbf{X}_1 \neq \mathbf{O}$

$\Rightarrow |\ \mathbf{A} - \lambda_1\ \mathbf{I}\ | = 0$

Hence every characteristic value of λ of **A** is a root of its characteristic equation.

Conversely if λ_1 be any root of the characteristic equation $|\mathbf{A} - \lambda_1\ \mathbf{I}| = 0$, then the equation

$$(\mathbf{A} - \lambda_1\ \mathbf{I})\ \mathbf{X} = \mathbf{O}$$

must possess a non-zero vector $\mathbf{X}_1$, such that

$$\mathbf{AX}_1 = \lambda_1\ \mathbf{IX}_1 = \mathbf{l}_1\ \mathbf{X}_1.$$

Hence every root l of the characteristic equation of A is a characteristic value of A. **Hence proved.**

Example 15:

Find the characteristic roots of the matrix $\mathbf{A} = \begin{bmatrix} 1 & 4 \\ 2 & 3 \end{bmatrix}$ *and verify Cayley Hamilton theorem for the matrix.*

Solution:

Here $\mathbf{A} = \begin{bmatrix} 1 & 4 \\ 2 & 3 \end{bmatrix}$ and $\mathbf{I} = \begin{bmatrix} 1 & 0 \\ 0 & 1 \end{bmatrix}$

$\therefore$ We have $|\mathbf{A} - \lambda\mathbf{I}| = \begin{vmatrix} 1-\lambda & 4 \\ 2 & 3-\lambda \end{vmatrix}$

$$= (1 - \lambda)\ (3 - \lambda) - 2.4$$
$$= 3 - \lambda - 3\lambda + \lambda^2 - 8 = \lambda^2 - 4\lambda - 5$$

$\therefore$ The characteristic equation of A is $\lambda^2 - 4\lambda - 5 = 0$ and its roots *i.e.*, characteristic roots of A are 5, – 1. **Ans.**

Now $A^2 = \begin{bmatrix} 1 & 4 \\ 2 & 3 \end{bmatrix}\begin{bmatrix} 1 & 4 \\ 2 & 3 \end{bmatrix}$

$= \begin{bmatrix} 1.1+4.2 & 1.4+4.3 \\ 2.1+3.2 & 2.4+3.3 \end{bmatrix} = \begin{bmatrix} 9 & 16 \\ 8 & 17 \end{bmatrix}$

$\therefore$ $\mathbf{A^2 - 4A - 5I}$

$= \begin{bmatrix} 9 & 16 \\ 8 & 17 \end{bmatrix} - 4\begin{bmatrix} 1 & 4 \\ 2 & 3 \end{bmatrix} - 5\begin{bmatrix} 1 & 0 \\ 0 & 1 \end{bmatrix}$

$= \begin{bmatrix} 9 & 16 \\ 8 & 17 \end{bmatrix} + \begin{bmatrix} -4 & -16 \\ -8 & -12 \end{bmatrix} + \begin{bmatrix} -5 & 0 \\ 0 & -5 \end{bmatrix}$

$= \begin{bmatrix} 9-4-5 & 16-16+0 \\ -8+0 & 17-12-5 \end{bmatrix} = \begin{bmatrix} 0 & 0 \\ 0 & 0 \end{bmatrix}$

$=$ **O**, where **O** is the null matrix.

Hence **A** statisfies its characteristic equation $\lambda^2 - 4\lambda - 5 = 0$. Hence Cayley Hamilton Theorem is verified by the matrix **A**.

Example 16:

Find the characteristic roots and inverse of the matrix

$$A = \begin{bmatrix} 5 & 6 \\ 1 & 2 \end{bmatrix}.$$

Solution:

Here $|\mathbf{A} - \lambda\mathbf{I}| = \begin{vmatrix} 5-\lambda & 6 \\ 1 & 2-\lambda \end{vmatrix}$

$= (5 - \lambda)(2 - \lambda) - 6 = 4 - 7\lambda + \lambda^2$

$\therefore$ The characteristic equation of **A** is $\lambda^2 - 7\lambda + 4 = 0$. ...(i)

Now as A must satisfy Cayley Hamilton's Theorem, so we get

$\mathbf{A^2 - 7A + 4I = O}$, where **O** is the null matrix

$\Rightarrow \quad \mathbf{I} = -\frac{1}{4}\mathbf{A}^2 + \frac{7}{4}\mathbf{A}$

Multiplying both sides by $\mathbf{A}^{-1}$ we get

$\mathbf{A}^{-1} = -\frac{1}{4}\mathbf{A} + \frac{7}{4}\mathbf{I} \qquad \because \qquad \mathbf{AA^{-1} = I}$

$= -\frac{1}{4}\begin{bmatrix} 5 & 6 \\ 1 & 2 \end{bmatrix} + \frac{7}{4}\begin{bmatrix} 1 & 0 \\ 0 & 1 \end{bmatrix} = \begin{bmatrix} -\frac{5}{4} & -\frac{3}{2} \\ -\frac{1}{4} & -\frac{1}{2} \end{bmatrix} + \begin{bmatrix} \frac{7}{4} & 0 \\ 0 & \frac{7}{4} \end{bmatrix}$

$$= +\frac{7}{4}\begin{bmatrix} -\frac{5}{4} & -\frac{3}{2}+0 \\ -\frac{1}{4}+0 & -\frac{1}{2}+\frac{7}{4} \end{bmatrix} = \begin{bmatrix} \frac{1}{2} & -\frac{3}{2} \\ -\frac{1}{4} & \frac{5}{4} \end{bmatrix} = \frac{1}{4}\begin{bmatrix} 2 & -6 \\ -1 & 5 \end{bmatrix}.$$ **Ans.**

Also the characteristic roots of **A** are the roots of (i)

i.e., $\lambda = \frac{1}{2}\left[7 \pm \sqrt{(49-16)}\right] = \frac{1}{2}\left[7 \pm \sqrt{(33)}\right]$

Example 17:

Verify Cayley Hamilton's Theorem for the matrix $\mathbf{A} = \begin{bmatrix} 1 & 0 & 2 \\ 0 & 2 & 1 \\ 2 & 0 & 3 \end{bmatrix}$. *Hence or otherwise compute* $\mathbf{A}^{-1}$.

Solution:

Here $|\mathbf{A} - \lambda \mathbf{I}| = \begin{vmatrix} 1-\lambda & 0 & 2 \\ 0 & 2-\lambda & 1 \\ 2 & 0 & 3-\lambda \end{vmatrix}$

$= (1 - \lambda) \{(2 - \lambda) (3 - \lambda) - 0\} + \{0 - 2 (2 - \lambda)\}$

$= (1 - \lambda) (2 - \lambda) (3 - \lambda) - 4 (2 - \lambda)$

$= (2 - \lambda) [(1 - \lambda) (3 - \lambda) - 4] = (2 - \lambda) \{3 - 4) + \lambda^2 - 4\}$

$= (2 - \lambda) (\lambda^2 - 4\lambda - \lambda) = - \lambda^3 + 6\lambda^2 - 7\lambda - 2$

$\therefore$ The characteristic equation of **A** is

$$\boldsymbol{\lambda^3 - 6\lambda^2 + 7\lambda + 2 = 0.} \qquad ...(i)$$

Now $\mathbf{A}^2 = \begin{bmatrix} 1 & 0 & 2 \\ 0 & 2 & 1 \\ 2 & 0 & 3 \end{bmatrix}\begin{bmatrix} 1 & 0 & 2 \\ 0 & 2 & 1 \\ 2 & 0 & 3 \end{bmatrix} = \begin{bmatrix} 5 & 0 & 8 \\ 2 & 4 & 5 \\ 8 & 0 & 13 \end{bmatrix}$

and $\mathbf{A}^3 = \mathbf{A}^2.\mathbf{A} = \begin{bmatrix} 5 & 0 & 8 \\ 2 & 4 & 5 \\ 8 & 0 & 13 \end{bmatrix}\begin{bmatrix} 1 & 0 & 2 \\ 0 & 2 & 1 \\ 2 & 0 & 3 \end{bmatrix} = \begin{bmatrix} 21 & 0 & 34 \\ 12 & 8 & 23 \\ 34 & 0 & 55 \end{bmatrix}$

$\therefore \mathbf{A}^3 - 6\mathbf{A}^2 + 7\mathbf{A} + 2\mathbf{I}$

$= \begin{bmatrix} 21 & 0 & 34 \\ 12 & 8 & 23 \\ 34 & 0 & 55 \end{bmatrix} - 6\begin{bmatrix} 5 & 0 & 8 \\ 2 & 4 & 5 \\ 8 & 0 & 13 \end{bmatrix} + 7\begin{bmatrix} 1 & 0 & 2 \\ 0 & 2 & 1 \\ 2 & 0 & 3 \end{bmatrix} + \begin{bmatrix} 1 & 0 & 0 \\ 0 & 1 & 0 \\ 0 & 0 & 1 \end{bmatrix}$

$= \begin{bmatrix} 21 & 0 & 34 \\ 12 & 8 & 23 \\ 34 & 0 & 55 \end{bmatrix} + \begin{bmatrix} -30 & 0 & -48 \\ -12 & -24 & -30 \\ -48 & 0 & -78 \end{bmatrix} + \begin{bmatrix} 7 & 0 & 14 \\ 0 & 14 & 7 \\ 14 & 0 & 21 \end{bmatrix} + \begin{bmatrix} 2 & 0 & 0 \\ 0 & 2 & 0 \\ 0 & 0 & 2 \end{bmatrix}$

$$= \begin{bmatrix} 21-30+7+2 & 0+0+0+0 & 34-48-14+0 \\ 12-12+0+0 & 8-24+14+2 & 23-30+7+0 \\ 34-48+14+0 & 0+0+0+0 & 55-78+21+2 \end{bmatrix}$$

$$= \begin{bmatrix} 0 & 0 & 0 \\ 0 & 0 & 0 \\ 0 & 0 & 0 \end{bmatrix} = \mathbf{O}, \text{ where } \mathbf{O} \text{ is the null matrix.}$$

Hence A satisfies its characteristic equation given by (i).

Hence Cayley Hamilton's Theorem is satisfied by A *i.e.,*

$$\mathbf{A^3 - 6A^2 + 7A + 2I = O \Rightarrow 2I = - A^3 + 6A^2 - 7A.}$$

Multiplying both sides by $\mathbf{A^{-1}}$, we get

$\mathbf{2A^{-1} = - A^2 + 6A - 7I, \because AA^{-1} = I, IA^{-1} = A^{-1}}$

$$= - \begin{bmatrix} 5 & 0 & 8 \\ 2 & 4 & 5 \\ 8 & 0 & 13 \end{bmatrix} + 6 \begin{bmatrix} 1 & 0 & 2 \\ 0 & 2 & 1 \\ 2 & 0 & 3 \end{bmatrix} - 7 \begin{bmatrix} 1 & 0 & 0 \\ 0 & 1 & 0 \\ 0 & 0 & 1 \end{bmatrix}$$

$$= \begin{bmatrix} -5 & 0 & -8 \\ -2 & -4 & -5 \\ -8 & 0 & -13 \end{bmatrix} + \begin{bmatrix} 6 & 0 & 12 \\ 0 & 12 & 6 \\ 12 & 0 & 18 \end{bmatrix} + \begin{bmatrix} -7 & 0 & 0 \\ 0 & -7 & 0 \\ 0 & 0 & -7 \end{bmatrix}$$

$$= \begin{bmatrix} -5+6-7 & 0+0+0 & -8+12+0 \\ -2+0+0 & -4+12-7 & -5+6+0 \\ -8+12+0 & 0+0+0 & -13+18-7 \end{bmatrix} = \begin{bmatrix} -6 & 0 & 4 \\ -2 & 1 & 1 \\ 4 & 0 & -2 \end{bmatrix}$$

$$\Rightarrow \quad A^{-1} = \frac{1}{2} \begin{bmatrix} -6 & -0 & 4 \\ -2 & 1 & 1 \\ 4 & 0 & -2 \end{bmatrix} = \begin{bmatrix} -3 & 0 & 2 \\ -1 & \frac{1}{2} & \frac{1}{2} \\ 2 & 0 & -1 \end{bmatrix}$$ **Ans**

Example 18:

Find the characteristic equation of the matrix

$$A = \begin{bmatrix} 1 & 3 & 7 \\ 4 & 2 & 3 \\ 1 & 2 & 1 \end{bmatrix} \text{ and hence find } A^{-1}.$$

Also verify Cayley Hamilton's Theorem for **A**.

Solution:

Here $|\mathbf{A} - \lambda\mathbf{I}|$

$$= \begin{vmatrix} 1-\lambda & 3 & 7 \\ 4 & 2-\lambda & 3 \\ 1 & 2 & 1-\lambda \end{vmatrix} = \begin{vmatrix} \lambda & 1 & 6+\lambda \\ 0 & -6-\lambda & -1+4\lambda \\ 1 & 2 & 1-\lambda \end{vmatrix}, \text{ applying } R_1 - R_3,\ R_2 - 4R_3$$

$$= -\lambda \begin{vmatrix} -6-\lambda & -1+4\lambda \\ 2 & 1-\lambda \end{vmatrix} + \begin{vmatrix} 1 & 6+\lambda \\ -6-\lambda & -1+4\lambda \end{vmatrix}$$

$$= -\lambda\,[-(6+\lambda)(1-\lambda) - 2(4\lambda-\lambda)] + [(4\lambda-\lambda) + (6+\lambda)^2]$$

$$= -\lambda\,(-6 + 6\lambda - \lambda + \lambda^2 - 8\lambda + 2) + (\lambda^2 + 16\lambda + 35)$$

$$= -\lambda^3 + 4\lambda^2 + 20\lambda + 35.$$

∴ The characteristic equation of the matrix A is

$$\lambda^3 - 4\lambda^2 - 20\lambda - 35 = 0 \qquad \text{...(i)}$$

∴ By Cayley Hamilton's Theorem, we have

$\mathbf{A^3 - 4A^2 - 20A - 35I = O}$, where **O** is the null matrix

⇒ $\mathbf{35I = A^3 - 4A^2 - 20A.}$

Multiplying both sides by $\mathbf{A^{-1}}$, we get

$$\mathbf{35A^{-1} = A^2 - 4A - 20I}, \quad \because\ \mathbf{AA^{-1} = I} \qquad \text{...(ii)}$$

Now $\mathbf{A^2} = \begin{bmatrix} 1 & 3 & 7 \\ 4 & 2 & 3 \\ 1 & 2 & 1 \end{bmatrix} \times \begin{bmatrix} 1 & 3 & 7 \\ 4 & 2 & 3 \\ 1 & 2 & 1 \end{bmatrix}$

$$= \begin{bmatrix} 1+12+7 & 3+6+14 & 7+9+7 \\ 4+8+3 & 12+4+6 & 28+6+3 \\ 1+8+1 & 3+4+2 & 7+6+1 \end{bmatrix} = \begin{bmatrix} 20 & 23 & 23 \\ 15 & 22 & 37 \\ 10 & 9 & 14 \end{bmatrix}$$

∴ From (ii) we get

$$35\mathbf{A}^{-1} = \begin{bmatrix} 20 & 23 & 23 \\ 15 & 22 & 37 \\ 10 & 9 & 14 \end{bmatrix} - 4\begin{bmatrix} 1 & 3 & 7 \\ 4 & 2 & 3 \\ 1 & 2 & 1 \end{bmatrix} - 20\begin{bmatrix} 1 & 0 & 0 \\ 0 & 1 & 0 \\ 0 & 0 & 1 \end{bmatrix}$$

$$= \begin{bmatrix} 20-4-20 & 23-12-0 & 23-28-0 \\ 15-16-0 & 22-8-20 & 37-12-0 \\ 10-4-0 & 9-8-0 & 14-4-20 \end{bmatrix} = \begin{bmatrix} -4 & 11 & -5 \\ -1 & -6 & 25 \\ 6 & 1 & -10 \end{bmatrix}$$

⇒ $\mathbf{A}^{-1} = (1/35) \begin{bmatrix} -4 & 11 & -5 \\ -1 & -6 & 25 \\ 6 & 1 & -10 \end{bmatrix}$. **Ans.**

Verify Cayley Hamilton's Theorem for yourself.

Example 19:

Using the characteristic equation of the matrix

$$A = \begin{bmatrix} 1 & 0 & 1 \\ 0 & 1 & 0 \\ 0 & 0 & 1 \end{bmatrix}, \textit{ find } A^{-1}.$$

Solution:

Here $|\mathbf{A} - \lambda\mathbf{I}|$

$$= \begin{vmatrix} 1-\lambda & 0 & 1-\lambda \\ 0 & 1-\lambda & 0 \\ 0 & 0 & 1-\lambda \end{vmatrix}$$

$$= (1 - \lambda)^3 = -(\lambda^3 - 3\lambda^2 - 1)$$

$\therefore$ The characteristic equation of the matrix is

$$\lambda^2 - 3\lambda^2 + 3\lambda - \lambda = 0 \qquad \text{...(i)}$$

Also $\mathbf{A}^2 = \mathbf{A}.\mathbf{A} = \begin{bmatrix} 1 & 0 & 1 \\ 0 & 1 & 0 \\ 0 & 0 & 1 \end{bmatrix}\begin{bmatrix} 1 & 0 & 1 \\ 0 & 1 & 0 \\ 0 & 0 & 1 \end{bmatrix}$

$$= \begin{bmatrix} 1 & 0 & 2 \\ 0 & 1 & 0 \\ 0 & 0 & 1 \end{bmatrix}$$

and $\mathbf{A}^3 = \mathbf{A}^2.\mathbf{A} = \begin{bmatrix} 1 & 0 & 2 \\ 0 & 1 & 0 \\ 0 & 0 & 1 \end{bmatrix}\begin{bmatrix} 1 & 0 & 1 \\ 0 & 1 & 0 \\ 0 & 0 & 1 \end{bmatrix}$

$$= \begin{bmatrix} 1 & 0 & 3 \\ 0 & 1 & 0 \\ 0 & 0 & 1 \end{bmatrix}$$

$\therefore \mathbf{A}^3 - 3\mathbf{A}^2 + 3\mathbf{A} - \mathbf{I}$

$$= \begin{bmatrix} 1 & 0 & 3 \\ 0 & 1 & 0 \\ 0 & 0 & 1 \end{bmatrix} - 3\begin{bmatrix} 1 & 0 & 2 \\ 0 & 1 & 0 \\ 0 & 0 & 1 \end{bmatrix} + 3\begin{bmatrix} 1 & 0 & 1 \\ 0 & 1 & 0 \\ 0 & 0 & 1 \end{bmatrix} - \begin{bmatrix} 1 & 0 & 0 \\ 0 & 1 & 0 \\ 0 & 0 & 1 \end{bmatrix}$$

$$= \begin{bmatrix} 1-3+3-1 & 0+0+0+0 & 3-6+3-0 \\ 0+0+0+0 & 1-3+3-1 & 0+0+0+0 \\ 0+0+0+0 & 0+0+0+0 & 1-3+3-1 \end{bmatrix} = \begin{bmatrix} 0 & 0 & 0 \\ 0 & 0 & 0 \\ 0 & 0 & 0 \end{bmatrix}$$

$= \mathbf{O}$, where $\mathbf{O}$ is the null matrix

$\Rightarrow \mathbf{A}^3 - 3\mathbf{A}^2 + 3\mathbf{A} - \mathbf{I} = \mathbf{O} \Rightarrow \mathbf{I} = \mathbf{A}^3 - 3\mathbf{A}^2 + 3\mathbf{A}$

$\Rightarrow \mathbf{A^{-1} = A^2 + 3I}$, multiplying both sides by A^{-1}

$$= \begin{bmatrix} 1 & 0 & 2 \\ 0 & 1 & 0 \\ 0 & 0 & 1 \end{bmatrix} - 3\begin{bmatrix} 1 & 0 & 1 \\ 0 & 1 & 0 \\ 0 & 0 & 1 \end{bmatrix} + 3\begin{bmatrix} 1 & 0 & 0 \\ 0 & 1 & 0 \\ 0 & 0 & 1 \end{bmatrix}$$

$$= \begin{bmatrix} 1-3+3 & 0+0+0 & 2-3+0 \\ 0+0+0 & 1-3+0 & 0+0+0 \\ 0+0+0 & 0+0+0 & 1-3+3 \end{bmatrix} = \begin{bmatrix} 1 & 0 & -1 \\ 0 & 1 & 0 \\ 0 & 0 & 1 \end{bmatrix}$$ **Ans.**

Example 20:

If matrix $A = \begin{bmatrix} 2 & -1 & 1 \\ -1 & 2 & -1 \\ 1 & -1 & 2 \end{bmatrix}$, *find the characteristic roots of A. Verify the Cayley Hamilton's Theorem and hence compute* A^{-1}.

Solution:

We can calculate the characteristic equation of the matrix **A** as

$\lambda^3 - 6\lambda^2 + 9\lambda - 4 = 0$, which can be rewritten as

$(\lambda - 1)(\lambda^2 - 5\lambda + 4) = 0$

$\Rightarrow (\lambda - 1)(\lambda - 1)(\lambda - 4) = 0 \Rightarrow l = 1, 1, 4$

$\therefore$ The characteristic roots of **A** are 1, 1 and 4. **Ans.**

Also an we can prove that Cayley Hamilton's theorem is satisfied by the matrix **A** and we can

$$\mathbf{A^3 - 6A^2 + 9A - 4I = O \Rightarrow 4I = A^3 - 6A^2 + 9A}$$

Multiplying both sides by $\mathbf{A^{-1}}$, we get

$$\mathbf{4A^{-1} = A^2 - 6A + 9I}, \quad \because \mathbf{A^{-1}\,I = A^{-1},\ AA^{-1} = I}$$

$$= \begin{bmatrix} 6 & -5 & 5 \\ -5 & 6 & -5 \\ 5 & -5 & 6 \end{bmatrix} - 6\begin{bmatrix} 2 & -1 & 1 \\ -1 & 2 & -1 \\ 1 & -1 & 2 \end{bmatrix} + 9\begin{bmatrix} 1 & 0 & 0 \\ 0 & 1 & 0 \\ 0 & 0 & 1 \end{bmatrix}$$

$$= \begin{bmatrix} 6 & -5 & 5 \\ -5 & 6 & -5 \\ 5 & -5 & 6 \end{bmatrix} + \begin{bmatrix} -12 & 6 & -6 \\ 6 & -12 & 6 \\ -6 & 6 & -12 \end{bmatrix} + \begin{bmatrix} 9 & 0 & 0 \\ 0 & 9 & 0 \\ 0 & 0 & 9 \end{bmatrix}$$

$$= \begin{bmatrix} 6-12+9 & -5+6+0 & 5-6+0 \\ -5+6+0 & 6-12+9 & -5+6+0 \\ 5-6+0 & -5+6+0 & 6-12+9 \end{bmatrix} = \begin{bmatrix} 3 & 1 & -1 \\ 1 & 3 & 1 \\ -1 & 1 & 3 \end{bmatrix}$$

$$\Rightarrow \mathbf{A}^{-1} = \frac{1}{4}\begin{bmatrix} 3 & 1 & -1 \\ 1 & 3 & 1 \\ -1 & 1 & 3 \end{bmatrix} = \begin{bmatrix} -\frac{3}{4} & \frac{1}{4} & -\frac{1}{4} \\ \frac{1}{4} & \frac{3}{4} & \frac{1}{4} \\ -\frac{1}{4} & \frac{1}{4} & \frac{3}{4} \end{bmatrix}.$$ **Ans.**

Example 21:

Using Cayley Hamilton's Theorem, compute the inverse of the matrix

$$A = \begin{bmatrix} 1 & 2 & 1 \\ 0 & 3 & 2 \\ 1 & 0 & 1 \end{bmatrix}.$$

Solution:

Here $|\mathbf{A} - \lambda \mathbf{I}|$

$$= \begin{vmatrix} 1-\lambda & 2 & 1 \\ 0 & 3-\lambda & 2 \\ 1 & 0 & 1-\lambda \end{vmatrix}$$

$$= (1 - \lambda) \begin{vmatrix} 3-\lambda & 2 \\ 0 & 1-\lambda \end{vmatrix} + \begin{vmatrix} 2 & 1 \\ 3-\lambda & 2 \end{vmatrix}$$

$$= (1 - \lambda) \{(3 - \lambda)(1 - \lambda)\} + \{4 - (3 - \lambda)\}$$

$$= (1 - \lambda)(3 - 4\lambda + \lambda^2) + (1 + \lambda)$$

$$= -(\lambda^3 - 5\lambda^2 + 6\lambda - 4)$$

$\therefore$ The characteristic equation of the matrix is

$$\lambda^3 - 5\lambda^2 + 6\lambda - 4 = 0$$

$\therefore$ By Cayley Hamilton Theorem, we have

$\mathbf{A}^3 - 5\mathbf{A}^2 + 6\mathbf{A} - 4\mathbf{I} = \mathbf{O}$, where $\mathbf{O}$ is the null matrix

$\Rightarrow 4\mathbf{I} = \mathbf{A}^3 - 5\mathbf{A}^2 + 6\mathbf{A}$

$\Rightarrow 4\mathbf{A}^{-1} = \mathbf{A}^2 - 5\mathbf{A} + 6\mathbf{I}$, multiplying both sides by $\mathbf{A}^{-1}$

$$\Rightarrow \mathbf{A}^{-1} = \frac{1}{4}[A^2 - 5A + 6I]$$

Now $\mathbf{A}^2 = \mathbf{A}.\mathbf{A}. = \begin{bmatrix} 1 & 2 & 1 \\ 0 & 3 & 2 \\ 1 & 0 & 1 \end{bmatrix}\begin{bmatrix} 1 & 2 & 1 \\ 0 & 3 & 2 \\ 1 & 0 & 1 \end{bmatrix}$

$$= \begin{bmatrix} 2 & 8 & 6 \\ 2 & & 8 \\ 2 & 2 & 2 \end{bmatrix}$$

∴ From (i) we get

$$4A^{-1} = \begin{bmatrix} 2 & 8 & 6 \\ 2 & 9 & 8 \\ 2 & 2 & 2 \end{bmatrix} - 5\begin{bmatrix} 1 & 2 & 1 \\ 0 & 3 & 2 \\ 1 & 0 & 1 \end{bmatrix} + 6\begin{bmatrix} 1 & 0 & 0 \\ 0 & 1 & 0 \\ 0 & 0 & 1 \end{bmatrix}$$

$$= \begin{bmatrix} 2-5+6 & 8-10+0 & 6-5+0 \\ 2+0+0 & 9=15+6 & 8-10+0 \\ 2-5+0 & 2+0+0 & 2-5+6 \end{bmatrix} = \begin{bmatrix} 3 & -2 & 1 \\ 2 & 0 & -2 \\ -3 & 2 & 3 \end{bmatrix}$$

$$\Rightarrow A^{-1} = \frac{1}{4} \cdot \begin{bmatrix} 3 & -2 & 1 \\ 2 & 0 & -2 \\ -3 & 2 & 3 \end{bmatrix}.$$ **Ans.**

Example 22:

Shot that $A = \begin{bmatrix} 1 & 2 & 2 \\ 2 & 1 & 2 \\ 2 & 2 & 1 \end{bmatrix}$ *satisfies the matrix equation* $A^2 - 4A - 5I = O$, *where I is the unit matrix. Deduce the inverse matrix of A.*

Solution:

$$A^2 = \begin{bmatrix} 1 & 2 & 2 \\ 2 & 1 & 2 \\ 2 & 2 & 1 \end{bmatrix} \times \begin{bmatrix} 1 & 2 & 2 \\ 2 & 1 & 2 \\ 2 & 2 & 1 \end{bmatrix}$$

$$\Rightarrow A^2 = \begin{bmatrix} 1+4+4 & 2+2+4 & 2+4+2 \\ 2+2+4 & 4+1+4 & 4+2+2 \\ 2+4+2 & 4+2+2 & 4+4+1 \end{bmatrix} = \begin{bmatrix} 9 & 8 & 8 \\ 8 & 9 & 8 \\ 8 & 8 & 9 \end{bmatrix}$$

∴ $\mathbf{A^2 - 4A - 5I}$

$$= \begin{bmatrix} 9 & 8 & 8 \\ 8 & 9 & 8 \\ 8 & 8 & 9 \end{bmatrix} - 4\begin{bmatrix} 1 & 2 & 2 \\ 2 & 1 & 2 \\ 2 & 2 & 1 \end{bmatrix} - 5\begin{bmatrix} 1 & 0 & 0 \\ 0 & 1 & 0 \\ 0 & 0 & 1 \end{bmatrix}$$

$$= \begin{bmatrix} 9-4-5 & 8-8-0 & 8-8-0 \\ 8-8-0 & 9-4-5 & 8-8-0 \\ 8-8-0 & 8-8-0 & 9-4-5 \end{bmatrix} = \begin{bmatrix} 0 & 0 & 0 \\ 0 & 0 & 0 \\ 0 & 0 & 0 \end{bmatrix}$$

= **O**, where **O** is the 3 × 3 null matrix. **Hence proved.**

i.e., $\mathbf{A^2 - 4A - 5I = O \Rightarrow 5I = A^2 - 4A}$

Multiplying both sides $\mathbf{A^{-1}}$ we get

$\mathbf{5A^{-1} = A - 4I}$, ∵ $\mathbf{AA^{-1} = I}$

$$= \begin{bmatrix} 1 & 2 & 2 \\ 2 & 1 & 2 \\ 2 & 2 & 1 \end{bmatrix} - 4 \begin{bmatrix} 1 & 0 & 0 \\ 0 & 1 & 0 \\ 0 & 0 & 1 \end{bmatrix}$$

$$= \begin{bmatrix} 1-4 & 2-0 & 2-0 \\ 2-0 & 1-4 & 2-0 \\ 2-0 & 2-0 & 1-4 \end{bmatrix} = \begin{bmatrix} -3 & 2 & 2 \\ 2 & -3 & 2 \\ 2 & 2 & -3 \end{bmatrix}$$

$$\Rightarrow A^{-1} = \frac{1}{5} \begin{bmatrix} -5 & 2 & 2 \\ 2 & -3 & 2 \\ 2 & 2 & -3 \end{bmatrix}.$$ **Ans.**

Example 23:

If the matrix $\mathbf{A} = \begin{vmatrix} 1 & 2 \\ 4 & 3 \end{vmatrix}$, *use Cayley Hamilton's Theorem to calculate* $\mathbf{A}^{-1}$.

Solution:

Here $|\mathbf{A} - \lambda \mathbf{I}| = \begin{vmatrix} 1-\lambda & 2 \\ 4 & 3-\lambda \end{vmatrix}$.

$= (1 - \lambda)(3 - \lambda) - 2.4 = 3 - 4\lambda + \lambda^2 - 8 = \lambda^2 - 4\lambda - 5$

$\therefore$ The characteristic equation of the matrix $\mathbf{A}$ is

$\lambda^2 - 4\lambda - 5 = 0$

$\therefore$ From Cayley Hamilton's Theorem we conclude that

$\mathbf{A}^2 - 4\mathbf{A} - 5\mathbf{I} = \mathbf{O}$, where $\mathbf{O}$ is 2×2 null matrix.

$\Rightarrow 5\mathbf{I} = \mathbf{A}^2 - 4\mathbf{A}$

Multiplying both sides by $\mathbf{A}^{-1}$, we get

$5\mathbf{A}^{-1} = \mathbf{A} - 4\mathbf{I}$, $\because$ $\mathbf{A}\,\mathbf{A}^{-1} = \mathbf{I}$

$$= \begin{bmatrix} 1 & 2 \\ 4 & 3 \end{bmatrix} - 4 \begin{bmatrix} 1 & 0 \\ 0 & 1 \end{bmatrix} = \begin{bmatrix} 1-4 & 2-0 \\ 4-0 & 3-4 \end{bmatrix}$$

$$\Rightarrow 5A^{-1} = \begin{bmatrix} -3 & 2 \\ 4 & -1 \end{bmatrix} \Rightarrow A^{-1} = \frac{1}{5} \begin{bmatrix} -3 & 2 \\ 4 & -1 \end{bmatrix}.$$ **Ans.**

Example 23:

Verify Cayley Hamilton's Theorem for the matrix $A = \begin{bmatrix} 0 & 0 & 0 \\ 3 & 1 & 0 \\ -2 & 1 & 4 \end{bmatrix}$.

Hence or otherwise evaluate A^{-1}.

Solution:

Here $|\mathbf{A} - \lambda\mathbf{I}| = \begin{vmatrix} -\lambda & 0 & 0 \\ 3 & 1-\lambda & 0 \\ -2 & 1 & 4-\lambda \end{vmatrix}$

$= -\lambda\,(1-\lambda)\,(4-\lambda) = -\lambda\,(4 - 5\lambda + \lambda^2) = -(\lambda^3 - 5\lambda^2 + 4\lambda)$

$\therefore$ The characteristic equation of the matrix **A** is

$$\lambda^3 - 5\lambda^2 + 4\lambda = 0 \qquad \text{...(i)}$$

Now $\mathbf{A}^2 = \begin{bmatrix} 0 & 0 & 0 \\ 3 & 1 & 0 \\ -2 & 1 & 4 \end{bmatrix}\begin{bmatrix} 0 & 0 & 0 \\ 3 & 1 & 0 \\ -2 & 1 & 4 \end{bmatrix} = \begin{bmatrix} 0 & 0 & 0 \\ 3 & 1 & 0 \\ -5 & 5 & 16 \end{bmatrix}$

And $\mathbf{A}^3 = \mathbf{A}^2.\mathbf{A} = \begin{bmatrix} 0 & 0 & 0 \\ 3 & 1 & 0 \\ -5 & 5 & 16 \end{bmatrix}\begin{bmatrix} 0 & 0 & 0 \\ 3 & 1 & 0 \\ -2 & 1 & 4 \end{bmatrix}$

$= \begin{bmatrix} 0 & 0 & 0 \\ 3 & 1 & 0 \\ -17 & 21 & 64 \end{bmatrix}$

$\therefore \mathbf{A}^3 - 5\mathbf{A}^2 + 4\mathbf{A}$

$= \begin{bmatrix} 0 & 0 & 0 \\ 3 & 1 & 0 \\ -17 & 21 & 64 \end{bmatrix} - 5\begin{bmatrix} 0 & 0 & 0 \\ 3 & 1 & 0 \\ -5 & 5 & 16 \end{bmatrix} + 4\begin{bmatrix} 0 & 0 & 0 \\ 3 & 1 & 0 \\ -2 & 1 & 4 \end{bmatrix}$

$= \begin{bmatrix} 0 & 0 & 0 \\ 3 & 1 & 0 \\ -17 & 21 & 64 \end{bmatrix} + \begin{bmatrix} 0 & 0 & 0 \\ -15 & -5 & 0 \\ 25 & -25 & 80 \end{bmatrix} + \begin{bmatrix} 0 & 0 & 0 \\ 12 & 4 & 0 \\ -8 & 4 & 16 \end{bmatrix}$

$= \begin{bmatrix} 0+0+0 & 0+0+0 & 0+0+0 \\ 3-15+12 & 1-5+4 & 0+0+0 \\ -17+25-8 & 21-25+4 & 64-80+16 \end{bmatrix} = \begin{bmatrix} 0 & 0 & 0 \\ 0 & 0 & 0 \\ 0 & 0 & 0 \end{bmatrix}$

$= \mathbf{O}$, where $\mathbf{O}$ is the null matrix.

Hence A satisfies the characteristic equation given by (i).

Hence Cayley Hamilton's Theorem is satisfied by the matrix A

i.e., $\quad \mathbf{A}^3 - 5\mathbf{A}^2 + 4\mathbf{A} = \mathbf{O} \qquad \text{...(ii)}$

Multiplying each term of (ii) by $\mathbf{A}^{-1}$ we get

$$\mathbf{A}^2 - 5\mathbf{A} + 4\mathbf{A} = \mathbf{O} \qquad \because \mathbf{A}\mathbf{A}^{-1} = \mathbf{I}$$

$$\Rightarrow \quad 4\mathbf{I} = -\mathbf{A}^2 + 5\mathbf{A}$$

$\Rightarrow 4A^{-1} = -A + 5I$, multiplying each term by A^{-1} and $AA^{-1} = I$

$$= -\begin{bmatrix} 0 & 0 & 0 \\ 3 & 1 & 0 \\ -2 & 1 & 4 \end{bmatrix} + 5\begin{bmatrix} 1 & 0 & 0 \\ 0 & 1 & 0 \\ 0 & 0 & 1 \end{bmatrix}$$

$$\Rightarrow 4A^{-1} = \begin{bmatrix} 0 & 0 & 0 \\ -3 & -1 & 0 \\ 2 & -1 & -4 \end{bmatrix} + \begin{bmatrix} 0 & 0 & 0 \\ 0 & 5 & 0 \\ 0 & 0 & 5 \end{bmatrix} = \begin{bmatrix} 5 & 0 & 0 \\ -3 & 4 & 0 \\ 2 & -1 & 1 \end{bmatrix}$$

$$\Rightarrow A^{-1} = \frac{1}{4}\begin{bmatrix} 5 & 0 & 0 \\ -3 & 4 & 0 \\ 2 & -1 & 1 \end{bmatrix} = \begin{bmatrix} \frac{5}{4} & 0 & 0 \\ -\frac{3}{4} & 1 & 0 \\ \frac{1}{2} & -\frac{1}{4} & \frac{1}{4} \end{bmatrix}.$$

Ans.

Example 24:

If $A = \begin{vmatrix} 8 & -6 & 2 \\ -6 & 7 & -8 \\ 2 & -4 & 3 \end{vmatrix}$ *find the characteristic roots of A.*

Solution:

Hence $|A - \lambda I| = \begin{vmatrix} 8-\lambda & -6 & 2 \\ -6 & 7-\lambda & -8 \\ 2 & -4 & 3-\lambda \end{vmatrix}$

$= (8 - \lambda) \{(7 - \lambda)(3 - \lambda) - 32\} + 6 \{-6 (3 - \lambda) + 16\} + 2 \{24 - 2 (7 - \lambda)\}$

$= (8 - \lambda) \{-11 - 10\lambda + \lambda^2\} + 6 \{6\lambda - 2\} + 2 \{10 + 2\lambda\}$

$= -88 - 69\lambda + 18\lambda^2 - \lambda^3 + 36\lambda - 12 + 20 + 4\lambda = -\lambda^3 + 18\lambda^2 - 29\lambda - 80$

$\therefore$ The characteristic equation of the matrix **A** is

$$\lambda^3 - 18\lambda^2 + 29\lambda + 80 = 0 \qquad \text{...(i)}$$

Its roots are the required characteristic roots of **A** (students can calculate it if they have read solution of cubic equations, so left as an exercise for the students).

Now $A^2 = \begin{bmatrix} 8 & -6 & 2 \\ -6 & 7 & -8 \\ 2 & -4 & 3 \end{bmatrix}\begin{bmatrix} 8 & -6 & 2 \\ -6 & 7 & -8 \\ 2 & -4 & 3 \end{bmatrix} = \begin{bmatrix} 104 & -98 & 70 \\ -106 & 117 & -92 \\ 46 & -56 & 45 \end{bmatrix}$

And $A^3 = A^2.A = \begin{bmatrix} 104 & -98 & 70 \\ -106 & 117 & -92 \\ 46 & -52 & 45 \end{bmatrix}\begin{bmatrix} 8 & -6 & 2 \\ -6 & 7 & -8 \\ 2 & -4 & 3 \end{bmatrix}$

$= \begin{bmatrix} 1560 & -1590 & 1202 \\ -1734 & 1823 & -1424 \\ 770 & -820 & 643 \end{bmatrix}$, on evaluating

$\therefore \mathbf{A^3 - 18\,A^2 + 29\,A + 80\,I}$

$$= \begin{bmatrix} 1560 & -1590 & 1202 \\ -1734 & 1823 & -1424 \\ 770 & -820 & 643 \end{bmatrix} - 18\begin{bmatrix} 104 & -98 & 70 \\ -106 & 117 & -92 \\ 46 & -52 & 45 \end{bmatrix} + 29\begin{bmatrix} 8 & -6 & 2 \\ -6 & 7 & -8 \\ 2 & -4 & 3 \end{bmatrix} + 80\begin{bmatrix} 1 & 0 & 0 \\ 0 & 1 & 0 \\ 0 & 0 & 1 \end{bmatrix}$$

$$= \begin{bmatrix} 1560 & -1590 & 1202 \\ -1734 & 1823 & -1424 \\ 770 & -820 & 643 \end{bmatrix} + \begin{bmatrix} -1872 & 1764 & -1260 \\ 1908 & -2106 & 1656 \\ -828 & 936 & -810 \end{bmatrix} + \begin{bmatrix} 232 & -174 & 58 \\ -174 & 203 & -232 \\ 58 & -116 & 87 \end{bmatrix} + \begin{bmatrix} 80 & 0 & 0 \\ 0 & 80 & 0 \\ 0 & 0 & 80 \end{bmatrix}$$

$= \begin{bmatrix} 0 & 0 & 0 \\ 0 & 0 & 0 \\ 0 & 0 & 0 \end{bmatrix} = \mathbf{O}$, where $\mathbf{O}$ is the null matrix.

Hence A satisfies the characteristic equation given by (i).

Hence Cayley Hamilton's Theorem is satisfied by the given matrix A *i.e.,* from (i) we get $\mathbf{A^3 - 18A^2 + 29A + 80I = O}$

$\Rightarrow \quad \mathbf{80I = -A^3 + 18A^2 - 29A}$

$\Rightarrow \mathbf{80A^{-1} = -A^2 + 18A - 29I}$, multiplying both sides by A^{-1}

$$= -\begin{bmatrix} 104 & -98 & 70 \\ -106 & 117 & -92 \\ 46 & -52 & 45 \end{bmatrix} + 18\begin{bmatrix} 8 & -6 & 2 \\ -6 & 7 & -8 \\ 2 & -4 & 3 \end{bmatrix} - 29\begin{bmatrix} 1 & 0 & 0 \\ 0 & 1 & 0 \\ 0 & 0 & 1 \end{bmatrix}$$

$$= \begin{bmatrix} -104+144-29 & 98-108+0 & -70+36+0 \\ 106-108\_0 & -117+126-29 & 92-144+0 \\ -46+36+0 & 52-72+0 & -45+54-29 \end{bmatrix}$$

$$\Rightarrow 80\ \mathbf{A^{-1}} = \begin{bmatrix} 11 & -10 & -34 \\ -2 & -20 & -52 \\ -10 & -20 & -20 \end{bmatrix} \Rightarrow \mathbf{A^{-1}} = (1/80) \begin{bmatrix} 11 & -10 & -34 \\ -2 & -20 & -52 \\ -10 & -20 & -20 \end{bmatrix}.$$

Ans.

Example 25:

Verify that $A = \begin{bmatrix} 1 & 2 & 0 \\ 2 & -1 & 0 \\ 0 & 0 & -1 \end{bmatrix}$ *satisfies its own characteristic equation*

(b) Is it true of every square matrix?

(c) State the that applies here.

Solution:

Here is in we can prove that the characteristic equation of the matrix A is

$$\lambda^3 + \lambda^2 - 5\lambda - 5 = 0 \qquad ...(i)$$

Now $\mathbf{A^2} = \begin{bmatrix} 1 & 2 & 0 \\ 2 & -1 & 0 \\ 0 & 0 & -1 \end{bmatrix} \begin{bmatrix} 1 & 2 & 0 \\ 2 & -1 & 0 \\ 0 & 0 & -1 \end{bmatrix} = \begin{bmatrix} 5 & 0 & 0 \\ 0 & 5 & 0 \\ 0 & 0 & 1 \end{bmatrix}$

And $\mathbf{A^3} = \mathbf{A^2.A} = \begin{bmatrix} 5 & 0 & 0 \\ 0 & 5 & 0 \\ 0 & 0 & 1 \end{bmatrix} \begin{bmatrix} 1 & 2 & 0 \\ 2 & -1 & 0 \\ 0 & 0 & -1 \end{bmatrix} = \begin{bmatrix} 5 & 10 & 0 \\ 10 & -5 & 0 \\ 0 & 0 & -1 \end{bmatrix}$

$\therefore \mathbf{A^3 + A^2 - 5A - 5I}$

$$= \begin{bmatrix} 5 & 10 & 0 \\ 10 & -5 & 0 \\ 0 & 0 & -1 \end{bmatrix} + \begin{bmatrix} 5 & 0 & 0 \\ 0 & 5 & 0 \\ 0 & 0 & 1 \end{bmatrix} - 5 \begin{bmatrix} 1 & 2 & 0 \\ 2 & -1 & 0 \\ 0 & 0 & -1 \end{bmatrix}$$

$$- 5 \begin{bmatrix} 1 & 0 & 0 \\ 0 & 1 & 0 \\ 0 & 0 & 1 \end{bmatrix}$$

$$= \begin{bmatrix} 5+5-5-5 & 10+0-10+0 & 0+0+0+0 \\ 10+0-10+0 & -5+5+5-5 & 0+0+0+0 \\ 0+0+0+0 & 0+0+0+0 & -1+1+5-5 \end{bmatrix} = \begin{bmatrix} 0 & 0 & 0 \\ 0 & 0 & 0 \\ 0 & 0 & 0 \end{bmatrix}$$

= **O**, where **O** is the null matrix.

Hence the matrix A satisfies its characteristic equation given by (i).

(b) Every matrix satisfies its characteristic equation.

(c) Cayley Hamilton's Theorem.

Example 26:

Find the characteristic equation of the matrix.

$$A = \begin{bmatrix} 1 & 0 & 2 \\ 0 & 1 & 2 \\ 1 & 2 & 0 \end{bmatrix}$$ *and hence compute its cube.*

Solution:

Here we have

$$|\mathbf{A} - \lambda\mathbf{I}| = \begin{vmatrix} 1-\lambda & 0 & 2 \\ 0 & 1-\lambda & 2 \\ 1 & 2 & 0-\lambda \end{vmatrix}$$

$$= (1 - \lambda) \{- \lambda (1 - \lambda) - 4\} + 1 \{- 2 (1 - \lambda)\}$$

$$= - \lambda (1 - \lambda)^2 - 6 (1 - \lambda) = - \lambda^3 + 2\lambda^2 - 5\lambda - 6$$

∴ The characteristic equation of the matrix **A** is

$$\lambda^3 - 2\lambda^2 - 5\lambda + 6 = O \qquad \text{...(i)}$$

∴ By Cayley-Hamilton theorem we have

$$\mathbf{A}^3 - 2\mathbf{A}^2 - 5\mathbf{A} + 6\mathbf{I} = \mathbf{O} \qquad \text{...(ii)}$$

Now $\mathbf{A}^2 = \begin{bmatrix} 1 & 0 & 2 \\ 0 & 1 & 2 \\ 1 & 2 & 0 \end{bmatrix} \times \begin{bmatrix} 1 & 0 & 2 \\ 0 & 1 & 2 \\ 1 & 2 & 0 \end{bmatrix}$

$\Rightarrow \mathbf{A}^2 = \begin{bmatrix} 3 & 4 & 2 \\ 2 & 5 & 2 \\ 1 & 2 & 6 \end{bmatrix}$

∴ Form (ii) we have

$\mathbf{A}^3 = 2\mathbf{A}^2 + 5\mathbf{A} - 6\mathbf{I}$

$$= 2\begin{bmatrix}3 & 4 & 2\\2 & 5 & 2\\1 & 2 & 6\end{bmatrix} + 5\begin{bmatrix}1 & 0 & 2\\0 & 1 & 2\\1 & 2 & 0\end{bmatrix} - 6\begin{bmatrix}1 & 0 & 0\\0 & 1 & 0\\0 & 0 & 1\end{bmatrix}$$

$$= \begin{bmatrix}6 & 8 & 4\\4 & 10 & 4\\2 & 4 & 12\end{bmatrix} + \begin{bmatrix}5 & 0 & 10\\0 & 5 & 10\\5 & 10 & 0\end{bmatrix} + \begin{bmatrix}-6 & 0 & 0\\0 & -6 & 0\\0 & 0 & -6\end{bmatrix}$$

$$= \begin{bmatrix}5 & 8 & 14\\4 & 9 & 14\\7 & 14 & 6\end{bmatrix}.$$ **Ans.**

Example 27:

Verify Cayley's Hamilton's Theorem for the following A and compute $2A^8 - 3A^5 + A^4 + A^2 - 4I$.

$$A = \begin{bmatrix}1 & 0 & 2\\0 & -1 & 1\\0 & 1 & 0\end{bmatrix}.$$

Solution:

Here $|\mathbf{A} - \lambda\mathbf{I}|$

$$= \begin{vmatrix}1-\lambda & 0 & 2\\0 & -1-\lambda & 1\\0 & 1 & 0-\lambda\end{vmatrix} 0 = (1-\lambda)\begin{vmatrix}-1-\lambda & 1\\1 & -\lambda\end{vmatrix}$$

$$= (1-\lambda)[\lambda + \lambda^2 - 1] = -\lambda^3 + 2\lambda - I$$

$\therefore$ The characteristic equation of the matrix **A** is

$$\lambda^2 - 2\lambda + 1 = 0 \qquad \text{...(i)}$$

Now $\mathbf{A}^2 = \mathbf{A.A}$

$$= \begin{bmatrix}1 & 0 & 2\\0 & -1 & 1\\0 & 1 & 0\end{bmatrix}\begin{bmatrix}1 & 0 & 2\\0 & -1 & 1\\0 & 1 & 0\end{bmatrix} = \begin{bmatrix}1 & 2 & 2\\0 & 2 & -1\\0 & -1 & 1\end{bmatrix}$$

$\therefore \mathbf{A}^3 = \mathbf{A}^2.\mathbf{A}$

$$= \begin{bmatrix}1 & 2 & 2\\0 & 2 & -1\\0 & -1 & 1\end{bmatrix}\begin{bmatrix}1 & 0 & 2\\0 & -1 & 1\\0 & 1 & 0\end{bmatrix} = \begin{bmatrix}1 & 0 & 4\\0 & -3 & 2\\0 & 2 & -1\end{bmatrix}$$

$\therefore \mathbf{A}^3 - 2\mathbf{A} + \mathbf{I}$

$$= \begin{bmatrix} 1 & 0 & 4 \\ 0 & -3 & 2 \\ 0 & 2 & -1 \end{bmatrix} - 2\begin{bmatrix} 1 & 0 & 2 \\ 0 & -1 & 1 \\ 0 & 1 & 0 \end{bmatrix} + \begin{bmatrix} 1 & 0 & 0 \\ 0 & 1 & 0 \\ 0 & 0 & 1 \end{bmatrix}$$

$$= \begin{bmatrix} 1-2+1 & 0+0+0 & 4-4+0 \\ 0+0+0 & -3+2+1 & 2-2+0 \\ 0+0+0 & 2-2+0 & -1+0+1 \end{bmatrix} = \begin{bmatrix} 0 & 0 & 0 \\ 0 & 0 & 0 \\ 0 & 0 & 0 \end{bmatrix}$$

= **O**, where **O** is the null matrix.

Hence, A satisfiesthe characteristic equation of A given by (i) and so Cayley Hamilton Theorem for the matrix A is verfied.

Thus we have $A^3 - 2A + I = O$ ...(ii)

Now $2A^8 + 3A^5 + A^4 + A^2 - 4I$

$= 2A^5 (A^3 - 2A + I) + 4A^6 - 5A^5 + A^4 + A^2 - 4I$

$= 2A^5 (O) + 4A^3 (A^3 - 2A + I) - 5A^5 + 9A^4 - 4A^3 + A^2 - 4I$, from (ii)

$= 4A^3 (O) - 5A^2 (A^3 - 2A + I) + 9A^4 - 14A^3 + 6A^2 - 4I$

$= -5A^2 (O) + 9A (A^3 - 2A + I) - 14A^3 + 24A^2 - 9A - 4I$

$= 9A (O) - 14 (A^3 - 2A + I) + 24A^2 - 37A + 10I$

$$= -\mathbf{14\,(O)} + \mathbf{24}\begin{bmatrix} 1 & 2 & 3 \\ 0 & 2 & -1 \\ 0 & -1 & 1 \end{bmatrix} - 37\begin{bmatrix} 1 & 0 & 2 \\ 0 & -1 & 1 \\ 0 & 1 & 0 \end{bmatrix} + 10\begin{bmatrix} 1 & 0 & 0 \\ 0 & 1 & 0 \\ 0 & 0 & 1 \end{bmatrix}$$

$$= \begin{bmatrix} 24-37+10 & 48+0+0 & 48-74+0 \\ 0+0+0 & 48+37+10 & -24-37+0 \\ 0+0+0 & -24-37+0 & 24+0+10 \end{bmatrix}$$

$$= \begin{bmatrix} -13 & 48 & -26 \\ 0 & 95 & -61 \\ 0 & -61 & 34 \end{bmatrix}.$$ **Ans.**

Example 28:

Evaluate the matrix $2A^4 - 7A^3 - 4A^2$, where

$$A = \begin{bmatrix} 1 & 0 & 2 \\ 0 & 1 & 2 \\ 1 & 2 & 0 \end{bmatrix}.$$

Solution:

As in it can be caluculated that the matrix **A** satisfies

$$\mathbf{A}^3 - 2\mathbf{A}^2 - 5\mathbf{A} + 6\mathbf{I} = \mathbf{O} \quad \text{...(i)}$$

Now $\mathbf{2A^4 - 7A^3 - 4A^2 = (A^3 - 2A^2 - 5A^2 - 5A + 6I)\ (2A - 3I)}$

$$- \mathbf{27A + 18I}$$

$= -\ \mathbf{27A + 18I}$, from (i). Now calculate.

Example 29:

If $A = \begin{bmatrix} 1 & 0 & 0 \\ 1 & 0 & 1 \\ 0 & 1 & 0 \end{bmatrix}$, *show that for every integer* $n \geq 4$, $A^n = A^{n-2} + A^3 - A$. *Hence evaluate* A^{20}.

Solution:

Here we have

$$|\mathbf{A} - \lambda\mathbf{I}| = \begin{vmatrix} 1-\lambda & 0 & 0 \\ 1 & 0-\lambda & 0 \\ 0 & 1 & 0-\lambda \end{vmatrix}$$

$= (1 - \lambda)\ \{(-\lambda)\ (-\lambda) - 1\} = (1 - \lambda)\ (\lambda^2 - 1)$

$= -\ \lambda^3 + \lambda^2 + \lambda - 1$

$\therefore$ The characteristic equation of **A** is $\lambda^3 - \lambda^2 - \lambda + 1 = 0$

And so by Cayley-Hamilton's theorem we have

$\mathbf{A^3 - A^2 - A + I = O}$

$\Rightarrow \mathbf{A\ (A^2 - I) = A^2 - I}$ ...(i) **(Note)**

Premultiplying both sides of (i) by $\mathbf{A^{r-3}}$, we have

$$\mathbf{A^{r-2}\ (A^2 - I) = A^{r-3}\ (A^2 - I)} \quad \text{...(ii)}$$

Putting r = n, n – 1, n – 2, ..., 4 in (ii) we get

$\mathbf{A^{n-2}\ (A^2 - I) = A^{n-3}\ (A^2 - I)}$

$\mathbf{A^{n-2}\ (A^2 - I) = A^{n-4}\ (A^2 - I)}$

... ...

... ...

$\mathbf{A^2\ (A^2 - I) = A\ (A^2 - I)}$

Multiplying these n – 3 indenties we have

$\mathbf{A^{n-2}\ (A^2 - I) = A\ (A^2 - I)}$ **(Note)**

$\Rightarrow \mathbf{A^n - A^{n-2} = A^3 - A \Rightarrow A^n = A^{n-2} + A^3 - A}$**, for all n** $\geq$ **4**

Now we have $\mathbf{A^n - A^{n-2} = A\ (A^2 - I)}$ ...(iii)

Putting n = 20, 18, 16, ... 4 in (ii) we get **(Note)**

$$\mathbf{A^{20} - A^{18} = A\ (A^2 - I)}$$

$$\mathbf{A^{18} - A^{16} = A\ (A^2 - I)}$$

... ...

... ...

$$\mathbf{A^{4} - A^{2} = A\ (A^2 - I)}$$

Adding these nine identities, we get

$$\mathbf{A^{20} - A^{2} = 9\ A\ (A^2 - I)} \quad \textbf{(Note)}$$

$$\mathbf{= 9\ A^3 - 9\ A = 9\ (A^2 + A - I) - 9A,\ from\ (i)}$$

$$\mathbf{\Rightarrow A^{20} = 10\ A^2 - 9I} \quad ...(iv)$$

$$\text{Now } \mathbf{A^2} = \begin{bmatrix} 1 & 0 & 0 \\ 1 & 0 & 1 \\ 0 & 1 & 0 \end{bmatrix} \times \begin{bmatrix} 1 & 0 & 0 \\ 1 & 0 & 1 \\ 0 & 1 & 0 \end{bmatrix} = \begin{bmatrix} 1 & 0 & 0 \\ 1 & 1 & 0 \\ 1 & 0 & 1 \end{bmatrix}$$

$$\therefore \text{From } A^{20} = 10 \begin{bmatrix} 1 & 0 & 0 \\ 1 & 1 & 0 \\ 1 & 0 & 1 \end{bmatrix} - 9 \begin{bmatrix} 1 & 0 & 0 \\ 0 & 1 & 0 \\ 0 & 0 & 1 \end{bmatrix}$$

$$= \begin{bmatrix} 10 & 0 & 0 \\ 10 & 10 & 0 \\ 10 & 0 & 10 \end{bmatrix} + \begin{bmatrix} -9 & 0 & 0 \\ 0 & -9 & 0 \\ 0 & 0 & -9 \end{bmatrix}$$

$$= \begin{bmatrix} 1 & 0 & 0 \\ 10 & 1 & 0 \\ 10 & 0 & 1 \end{bmatrix}. \quad \textbf{Ans.}$$

Example 30:

Compute A^{-2} *i above.*

Solution:

In above we have already prove that

$$\mathbf{A^3 - A^2 - A + I - O} \quad ...(i)$$

Also in we have proved that

$$\mathbf{A^{-1}} = -\frac{1}{a^n}\ \{\mathbf{A^{n-1} + a_1\ A^{n-2} + ... + a_{n-1}\ I}\}, \quad ...(ii)$$

provided $A^n + a_1, A^{n-1} + a_2 A^{n-2} + ... + a_{n-1} A + a_n I = O$ ...(iii)

Comparing (i) and (iii) we have

$a_n = -1; a_{n-1} = 1, a_{n-2} = -1$ etc. $\Rightarrow a_1 = -1, a_2 = 1$ etc.

$\therefore$ From (ii) we have

$$A^{-1} = -\frac{1}{(-1)}\{A^{n-1} + (-1)A^{n-2} + ... + (1)I\}, \text{ where } n = 3$$

$$= \{A^2 - A + I\}$$

$$= \begin{bmatrix} 1 & 0 & 0 \\ 1 & 1 & 0 \\ 1 & 0 & 1 \end{bmatrix} - \begin{bmatrix} 1 & 0 & 0 \\ 1 & 0 & 1 \\ 0 & 1 & 0 \end{bmatrix} + \begin{bmatrix} 1 & 0 & 0 \\ 0 & 1 & 0 \\ 0 & 0 & 1 \end{bmatrix},$$

$$\Rightarrow A^{-1} = \begin{bmatrix} 1 & 0 & 0 \\ 0 & 2 & -1 \\ 1 & -1 & 2 \end{bmatrix} \quad ...(iv)$$

Also from (i) pre-multiplying each term by A^{-2} we have

$$A - I - A^{-1} = O \quad \textbf{(Note)}$$

$$\Rightarrow A^{-2} = -A + I + A^{-1}$$

$$= -\begin{bmatrix} 1 & 0 & 0 \\ 1 & 0 & 1 \\ 0 & 1 & 0 \end{bmatrix} + \begin{bmatrix} 1 & 0 & 0 \\ 0 & 1 & 0 \\ 0 & 0 & 1 \end{bmatrix} + \begin{bmatrix} 1 & 0 & 0 \\ 0 & 2 & -1 \\ 1 & -1 & 2 \end{bmatrix}$$

$$= \begin{bmatrix} 1 & 0 & 0 \\ -1 & 3 & -2 \\ 1 & -2 & 3 \end{bmatrix}. \quad \textbf{Ans.}$$

EXERCISES

1. Verify that the matrix $A = \begin{bmatrix} 1 & 1 & 2 \\ 3 & 1 & 1 \\ 2 & 3 & 1 \end{bmatrix}$, satisfies the Cayley Hamilton Theorem.
2. When do you say that two matrices **A** and **B** are similar? Prove that the similar matrices have the same characteristic roots.
3. Find the characteristic polynomial of the matrix **A** and hence compute $2A^6 - 3A^5 + A^4 + A^2 - 4I$, where **A** is the matrix $\begin{bmatrix} 1 & 0 & 2 \\ 0 & -1 & 1 \\ 0 & 1 & 1 \end{bmatrix}$.

   Also determine one of the characteristic roots and corresponding characteristic vectors.

4. Determine the characteristic roots of the matrix

$$\begin{bmatrix} 1 & 0 & 0 & -1 \\ -1 & 0 & 0 & 0 \\ 0 & -1 & 0 & 0 \\ 0 & 0 & -1 & 0 \end{bmatrix}.$$

5. Let A and **B** be two square matrices over the field of real numbers, and let **B** be non-singular, obtain the characteristic roots of

$$\text{A, if } \mathbf{B}^{-1}\,\mathbf{AB} = \begin{bmatrix} -3 & 0 & 0 \\ 0 & 1 & 0 \\ 0 & 0 & 7 \end{bmatrix}.$$

6. Use Cayley Hamilton Theorem to find A^{-1} if

$$\mathbf{A} = \begin{bmatrix} 2 & 1 & 2 \\ 0 & 2 & 3 \\ 0 & 0 & 5 \end{bmatrix}.$$

7. Find the characteristic roots and characteristic vectors of the following :

$$\begin{bmatrix} 1 & 0 & -2 \\ 0 & 0 & 2 \\ -2 & 0 & 4 \end{bmatrix}.$$

8. Find the characteristic roots of $\mathbf{A} = \begin{bmatrix} a_1 & a_2 & a_3 \\ 0 & b_2 & 0 \\ 0 & 0 & c_3 \end{bmatrix}$. **Ans.** a_1, b_2, c_3

9. Find the eigenvalues of the matrix $\begin{bmatrix} 0 & 1 & 2 \\ 1 & 0 & -1 \\ 2 & -1 & 0 \end{bmatrix}$.

Ans. $2, -1 \pm \sqrt{3}$

10. Find the latent roots of the matrix

$$A = \begin{bmatrix} 0 & \sin\alpha & \cos\alpha\sin\beta \\ -\sin\alpha & 0 & \cos\alpha\cos\beta \\ -\cos\alpha\sin\beta & -\cos\alpha\cos\beta & 0 \end{bmatrix}.$$ **Ans.** $0, i, -i$

11. Show that the characteristic roots of the matrix

$$\begin{bmatrix} 1 & 2 & 3 \\ 0 & -4 & 2 \\ 0 & 0 & 7 \end{bmatrix}$$ are $1, -4$ and 7.

12. Show that the matrices

$$\begin{bmatrix} 0 & c & b \\ c & 0 & a \\ b & a & 0 \end{bmatrix}, \begin{bmatrix} 0 & a & c \\ a & 0 & b \\ c & b & 0 \end{bmatrix}, \begin{bmatrix} 0 & b & a \\ b & 0 & c \\ a & c & 0 \end{bmatrix}$$

have the same characteristic equation.

13. Obtain the characteristic equation and roots of the matrix

$$\mathbf{A} = \begin{bmatrix} 2 & -1 & 1 \\ -1 & 2 & -1 \\ 1 & -1 & 2 \end{bmatrix}.$$

14. Show that the matrix $\mathbf{A} = \begin{bmatrix} 1 & 2 \\ 1 & 1 \end{bmatrix}$ satisfies Cayley Hamilton Theorem.

15. Let $\mathbf{A} = \begin{bmatrix} 2 & 2 & 0 \\ 2 & 1 & 1 \\ -7 & 2 & -3 \end{bmatrix}$. Find the characteristic equation of **A** and verify that the matrix A satisfies this equation. Also find the characteristic roots and the corresponding vectors of **A**.

16. Using Cayley-Hamilton Theorem or otherwise determine the inverse of the matrix

$$\mathbf{A} = \begin{bmatrix} 2 & 2 & 1 \\ -2 & 1 & 2 \\ 1 & -2 & 2 \end{bmatrix}.$$

17. Verify Cayley-Hamilton Theorem in the case of the matrix

$$A = \begin{bmatrix} 0 & 1 & 2 \\ 2 & -3 & 0 \\ 1 & 1 & 1 \end{bmatrix}. \text{ Hence find } A^{-1}.$$

18. *Find the eigen-vectors of the following matrices:*

$$(a) \begin{bmatrix} -3 & 2 & 2 \\ -6 & 5 & 2 \\ -7 & 4 & 4 \end{bmatrix}; \quad (b) \begin{bmatrix} 1 & 3 & 0 \\ 3 & -2 & -1 \\ 0 & -1 & 4 \end{bmatrix}.$$

19. Find the characteristic vectors of the matrix

$$\mathbf{A} = \begin{bmatrix} 8 & -6 & 2 \\ -6 & 7 & -4 \\ 2 & -4 & 3 \end{bmatrix}$$

20. Find the eigen-values and eigenvectors for the matrix

$$\mathbf{A} = \begin{bmatrix} 3 & -5 & -4 \\ -5 & -6 & -5 \\ -4 & -5 & 3 \end{bmatrix}.$$

21. Find the invariant vector of the matrix

$$\mathbf{A} = \begin{bmatrix} 2 & 2 & 1 \\ 1 & 3 & 1 \\ 1 & 1 & 1 \end{bmatrix}.$$

22. Show that the matrix $\begin{bmatrix} 1 & 2 & 1 \\ -1 & 0 & 3 \\ 2 & -1 & 1 \end{bmatrix}$ satisfies Cayley-Hamilton Theorem.

23. Evaluate the matrix $\mathbf{A}^5 - 27\mathbf{A}^3 + 65\mathbf{A}^2$, where

$$\mathbf{A} = \begin{bmatrix} 0 & 0 & 1 \\ 3 & 1 & 0 \\ -2 & 1 & 4 \end{bmatrix}.$$

Ans. $\begin{bmatrix} 40 & 0 & 48 \\ 129 & -3 & 0 \\ 86 & -43 & -132 \end{bmatrix}$

24. If the matrix $\mathbf{A} = \begin{bmatrix} 2 & -1 & 1 \\ -1 & 2 & -1 \\ 1 & -1 & 2 \end{bmatrix}$, find the characteristic roots of **A**.

Verify Cayley Hamilton Theorem and hence compute A^{-1}.